6th International Conference on Aperiodic Crystals 2009

(APERIODIC 09)

AA002606

Journal of Physics: Conference Series Volume 226

Liverpool, United Kingdom
13-18 September 2009

ISBN: 978-1-61738-483-7
ISSN: 1742-6588

6th International Conference on Aperiodic Crystals 2009

(APERIODIC 09)

Journal of Physics: Conference Series Volume 226

Liverpool, United Kingdom
13-18 September 2009

ISBN: 978-1-61738-483-7
ISSN: 1742-6588

IOP Institute of Physics

Printed from e-media with permission by:

Curran Associates, Inc.
57 Morehouse Lane
Red Hook, NY 12571

Some format issues inherent in the e-media version may also appear in this print version.

Printed by Curran Associates, Inc. (2010)

Institute of Physics
Dirac House, Temple Back
Bristol BS1 6BE UK

Phone: 44 1 17 929 7481
Fax: 44 1 17 920 0979

techtracking@iop.org

Additional copies of this publication are available from:

Curran Associates, Inc.
57 Morehouse Lane
Red Hook, NY 12571 USA
Phone: 845-758-0400
Fax: 845-758-2634
Email: curran@proceedings.com
Web: www.proceedings.com

TABLE OF CONTENTS

QUASICRYSTALS

MODULATED STRUCTURES

MATHEMATICAL & THEORETICAL ASPECTS OF APERIODIC ORDER

APPROXIMATES AND COMPLEX PHASES

PREFACE

LIVERPOOL
13–18 SEPT

APERIODIC '09

Aperiodic'09, the sixth International Conference on Aperiodic Crystals, took place in Liverpool 13–18 September 2009. It was the first major conference in this interdisciplinary research field held in the UK. The conference, which was organised under the auspices of the Commission on Aperiodic Crystals of the International Union of Crystallography (IUCr), followed on from Aperiodic'94 (Les Diablerets, Switzerland), Aperiodic'97 (Alpe d'Huez, France), Aperiodic'2000 (Nijmegen, The Netherlands), Aperiodic'03 (Belo Horizonte, Brazil) and Aperiodic'06 (Zao, Japan). The next conference in the series will take place in Australia in 2012. The Aperiodic conference series is itself the successor to a series of Conferences on Modulated Structures, Polytypes and Quasicrystals (MOSPOQ), which were held in Marseilles (France) in 1984, Wroclaw (Poland) in 1986, Varanasi (India) in 1988 and Balatonszeplak (Hungary) in 1991.

The remit of the conference covers two broad areas of research on aperiodic crystals, incommensurately modulated and composite crystals on the one hand, and quasicrystals on the other hand, sharing the property that they are aperiodically ordered solids. In addition, the conference also featured recent research on complex metal alloys, which are in fact periodically ordered solids. However, the term complex refers to their large unit cells, which may contain thousands of atoms, and as a consequence complex metal alloys share some of the properties of quasicrystalline solids.

Aperiodic'09 attracted about 110 participants from across the world, including 20 UK-based scientists (the second largest group after Japan who sent 21 delegates). A particular feature of the conference series is its interdisciplinary character, and once again the range of disciplines of participants included mathematics, physics, crystallography and materials science.

The programme started with three tutorial lectures on Sunday afternoon, presenting introductory overviews on quasicrystals (Eiji Abe), complex metal alloys (Alessandra Beni) and incommensurately modulated structures (Gervais Chapuis). While these were mainly aimed at younger researchers in the field, the lectures were very well attended and appreciated by the participants.

The main programme ran from Monday morning until Friday lunchtime. It comprised 13 invited and 40 contributed plenary talks, and more than 40 posters, which were presented at two afternoon/evening poster sessions. The topics covered in the programme range from mathematical foundations, mathematical models, new materials, sample preparation, structure determination, physical properties and surface properties to industrial applications. Every presenter was invited to submit an article for this proceedings volume, and the 36 peer-reviewed papers in this volume present a cross-section of the range of presentations at the conference. They have been arranged into four categories, (i) quasicrystals, (ii) modulated structures, (iii) mathematical and theoretical aspects of aperiodic order, and (iv) approximants and complex phases.

Prizes for best student presentations were awarded to Heinrich Orsini-Rosenberg (ETH Zurich) for his poster *Tailor-made sevenfold approximants: ab-initio investigations on formation and stability* and to Holger Euchner (Universität Stuttgart) for his contributed talk on *Lattice dynamics in complex metallic alloys – vibrational properties of* $Zn_{11}Mg_2$. In addition to a cash prize, Heinrich Orsini-Rosenberg received an icosahedral teapot, which was manufactured and donated by David Warrington, and Holger Euchner received a book prize.

The meeting started with a welcome reception in the University's recently refurbished Victoria Gallery and Museum. A public lecture *Simple sets of shapes that tile the plane but cannot ever repeat* by Professor Sir Roger Penrose FRS attracted a wide audience and gave a fascinating insight into the discovery of the Penrose tiling, which is still the paradigm of aperiodic order in two dimensions and used in modelling quasicrystalline materials, some ten years before quasicrystals were first seen in experiment. The conference excursion and dinner centred on the historic waterfront area of this 800-year old city that was designated European Capital of Culture in 2008. The enjoyment of the ferry trip on the Mersey and subsequent walk to the Albert dock museums was greatly enhanced by the fine weather, which lasted throughout the entire conference week. The guests of honour at the conference dinner were Professor Sir Roger Penrose FRS and Professor Alan L Mackay FRS, bringing together two distinguished UK scientists who made seminal contributions to the subject. Besides the scientific programme, the conference also offered a presentation by the author Ann Lingard, whose latest novel *The Embalmer's Book of Recipes* features a mathematician working on quasicrystals at the University of Liverpool as its main character.

A related EPSRC-supported workshop on Mathematical Aspects of Aperiodic Order, organised by Alex Clark, John Hunton and Uwe Grimm, was held in Leicester in the week preceding the conference, and attracted about 40 participants.

It is not possible to organise a conference without the help and support of a large number of people, and we would like to take this opportunity to acknowledge the support of everyone who was involved in organising and running this conference. We would like to thank the IUCr Aperiodic Commission, and in particular its chair Professor Marc de Boissieu, who not only entrusted us with hosting the event, but acted as the International Advisory Committee and provided valuable advice and guidance. We are indebted to the members of the International Programme Committee, who helped to put together an exciting programme by suggesting invited speakers and selecting presentations for contributed plenary talks. We are immensely grateful to all members of the National Organising Committee, who provided essential support during the organising stage and during the conference itself, and we would particularly like to thank the conference secretary, Mrs Angie Reid, for her excellent administrative support before, during and after the conference, and for her help in the production of this volume. We are very much indebted to out co-editors, Dr Olga Degtyareva (Edinburgh) and Dr Hem Raj Sharma (Liverpool), for their contributions to the editing of this volume. Finally, and most of all, we would like to thank all participants of the conference for coming to Liverpool, for contributing by presenting their recent work and taking part in discussions, and for creating such an enjoyable and fruitful collegiate atmosphere throughout the week.

Last but not least we would like to express our gratitude for the generous financial support we received from the Open University, the University of Liverpool and the Institute of Physics (Mathematical and Theoretical Physics Group and Thin Films and Surfaces Group). The International Union of Crystallography provided financial sponsorship which was allocated to support the participation of 15 young scientists. The European Network of Excellence on Complex Metallic Alloys provided travel and subsistence support to members of participating organisations.

Rónán McGrath
Uwe Grimm

All papers published in this volume of *Journal of Physics: Conference Series* have been peer reviewed through processes administered by the proceedings Editors. Reviews were conducted by expert referees to the professional and scientific standards expected of a proceedings journal published by IOP Publishing.

Real space structure refinement of the basic Ni-rich decagonal Al-Ni-Co phase.

P Kuczera[1], B Kozakowski[1], J Wolny[1], W Steurer[2]

[1]Faculty of Physics and Applied Computer Science, AGH- University of Science and Technology, Krakow, Poland

[2]Laboratory of Crystallography, ETH, Zurich, Switzerland

qczera@gmail.com

Abstract. Structure refinement of the basic Ni-rich Al-Ni-Co decagonal phase was performed in physical space. Statistical approach was used to derive the structure factor. Rhombic Penrose tiling with an edge length of 10.28 Å was chosen as a reference lattice. The best-fit structure model with 106 parameters and 2767 reflections gives $R = 12.3\%$ and $wR = 6.2\%$. The observations-to-parameters ratio is high - 26.1 Density and chemical composition of the resulting structure are very close to experimental values.

1. Introduction

There are 8 different decagonal phases reported in the Al-Ni-Co system [1 and references therein]. The only phase, for which a qualitative structure refinement was performed, is the so called *basic Ni-rich phase*. In the majority of published papers a higher-dimensional approach i.e. the atomic surface modelling method is used for the optimization [2,3]. We present the results of a real space structure refinement of the *basic Ni-rich* decagonal phase based on a diffraction data set without referring to the higher-dimensional properties of decagonal quasicrystals.

Structure factor which was used for the modelling process was calculated on the basis of a statistical method described in [4,5]. We used rhombic Penrose tiling as a reference lattice. Initial atomic positions were set in the vertices of rhombuses divided four times with obedience to the inflation rules. Additionally atoms were divided into 24 groups. For each group we optimized probabilities of vertices occupation, TM concentration, a shift from initial position and two components of Debye-Waller coefficient. Two components are necessary because of the anisotropy of the structure. One component is associated with *xy* plane, i.e. quasiperiodic plane, the other component is associated with *z* direction, i.e. periodic direction. All together 106 parameters were refined. The optimization was performed on a set of 2767 symmetrically independent reflections (merged in 10/m Laue group) taken by a 4 four-circle diffractometer Huber at beamline D3, HASYLAB. The chemical composition of the sample was checked to be $Al_{70.6}Ni_{22.7}Co_{6.7}$ [2].

The diffraction data set is larger in comparison to our previous paper [6] (only 451 symmetrically independent reflections), which makes the refinement much more reliable.

2. Initial structure model

Structure of the *basic Ni-rich* decagonal Al-Ni-Co phase is known to consist of a periodic set of quasiperiodic planes stacked along the z direction. There are two planes within one period of the z axis. These planes are related by a screw axis 10_5. The shortest interatomic distances in the structure of the *basic Ni-rich* decagonal Al-Ni-Co can be determined using a Patterson map [2]. These distances scale in exactly the same way as distances between vertices of Penrose tiling, providing that proper z coordinate is chosen for these vertices. Thus we assumed that the projection of the structure along z axis results in Penrose tiling.

Our structure units are thick and thin rhombuses of Penrose tiling divided three times with obedience to the inflation rules. All positions from the 0-th, 1-th, 2-nd, and 3-rd division are occupied in the initial model. We, however, also put atoms in several positions of the fourth division to fulfil the density restriction. 70 atoms decorating our structure units were divided into 24 groups. This division was made according to adherence rules for rhombuses [4,5], overlapping rules for kite-clusters [7], and an assumption that the decoration is symmetric with respect to the longer diagonal of thick rhombus and shorter of the thin one. The adherence rules were used for all decorating atoms since they are essential for the correctness of the structure factor calculation. Also the assumption of a symmetric decoration was strictly obeyed. The overlapping rules for clusters were used solely for atoms from the 0-th, 1-th, 2-nd, and 3-rd division, which form a skeleton of the structure. The division of atoms into groups is illustrated in the figure 1.

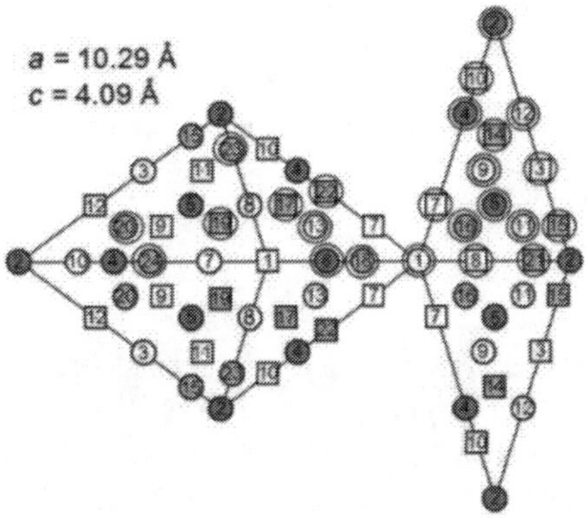

Figure 1. The division of atoms into 24 groups

The edge length of our structure units (a) is 10.29 Å and the period along the z axis (c) 4.09 Å. Circles and squares distinguish the layers at $z = 0.25c$ and $z = -0.25c$ respectively. Dark grey symbol colour denotes TM atoms light grey Al/TM atoms and white Al atoms. Some positions are marked by additional circle. The results of structure refinement are given for these positions. In the initial model all occupation probabilities are equal to 1. Atoms from 0-th, 1-th, 2-nd, and 3-rd division are either Al or TM (white or dark grey symbols, groups 1-13). The initial concentration of TM atom at the positions of 4-th division is 0.25 (light grey symbols, groups 14-24). Atoms in the group 22 are slightly moved out of the edge, inside the thick rhombus. This is why there is no atom from the group 22 in the thin rhombus. A detailed explanation of this assumption is given in the paragraph 5.

3. Structure factor

The structure factor was calculated in the real space on the basis of statistical approach. The concept of average unit cell for Penrose tiling and a detailed derivation of the structure factor for any decoration is extensively studied in [5]. Here we shall only show the resulting formula:

$$F(\mathbf{k}) = D_{\mathrm{Ph}} \sum_{\alpha} \left[F_{\mathrm{L}}^{\alpha} \sum_{l=1}^{n_{\mathrm{L}}} f_l D_l p_l \exp\left(i\mathbf{k}\cdot\mathbf{r}_l^{\alpha}\right) + F_{\mathrm{S}}^{\alpha} \sum_{s=1}^{n_{\mathrm{S}}} f_s D_s p_s \exp\left(i\mathbf{k}\cdot\mathbf{r}_s^{\alpha}\right) \right]$$

$F^{\alpha}_{\mathrm{L(S)}}$ is the Fourier transform of a triangular probability distribution associated with thin (S) or thick (L) rhombus at orientation α (there are 10 different possible orientations of each rhombus[7]), $\mathbf{r}^{\alpha}_{l(s)}$ is the position of an atom in a given structure unit at orientation α, $p_{l(s)}$ is the weight of a given atom in the structure unit, $n_{\mathrm{L(S)}}$ is the number of atoms decorating thick or thin rhombus, **k** is a scattering vector, $f_{l(s)}$ is the average atomic form factor. A formula for $f_{l(s)}$ reads:

$$f_{l(s)} = p_{\mathrm{calc}}^{l(s)} (p_{\mathrm{TM}}^{l(s)} f_{\mathrm{TM}} + (1 - p_{\mathrm{TM}}^{l(s)}) f_{\mathrm{Al}})$$

p_{calc} is the occupation probability of a given vertex by an atom, p_{TM} is concentration of TM atoms at a given position, f_{TM} is the atomic form factor for TM atoms (weighted Co and Ni), f_{Al} atomic form factor for Al atoms. $D_{l(s)}$ is the Debye-Waller factor:

$$D_{l(s)} = \exp\left(-\tfrac{1}{16\pi^2}\left(k_x^2 + k_y^2\right) b_{xy}^{l(s)} - \tfrac{1}{16\pi^2} k_z^2 b_z^{l(s)}\right)$$

b_{xy} is the average displacement parameter in the quasiperiodic plane, b_z is the displacement parameter in the z-direction. A relation between these parameters and and the square mean shift of the atom from the equilibrium position $\left\langle u^2_{xy(z)} \right\rangle$ reads:

$$b_{xy(z)} = 8\pi^2 \left\langle u^2_{xy(z)} \right\rangle$$

D_{Ph} is a phason Debye-Waller factor. Details on calculating perpendicular components of the scattering vector can be found in [5]:

$$D_{\mathrm{Ph}} = \exp\left(-\tfrac{1}{16\pi^2}\left(k_{\perp x}^2 + k_{\perp y}^2\right) b_{\mathrm{Ph}}\right)$$

where b_{Ph} is the phason displacement parameter.

4. Optimized function

We used a conjugated gradient minimization algorithm. The optimized function reads:

$$\Delta = wR^2 + PF_1 + PF_2 + PF_3 + PF_4$$

where:

$$wR^2 = \frac{\sum_i^{N_{\mathrm{peaks}}} \left(\frac{F_{\mathrm{obs}}^i - F_{\mathrm{calc}}^i}{\sigma^i} \right)^2}{\sum_i^{N_{\mathrm{peaks}}} \left(\frac{F_{\mathrm{obs}}^i}{\sigma^i} \right)^2}$$

$$PF_1 = C_1 \sum_j^{N_{\mathrm{atom}}} \left(\delta_{\mathrm{TM}}^j + \delta_{\mathrm{obs}}^j \right)$$

$$\delta_{TM\,(obs)}^j = \begin{cases} 0; & p_{\mathrm{TM(calc)}}^j \in <0,1> \\ \left(p_j \cdot p_{\mathrm{TM(calc)}}^j\right)^2; & p_{\mathrm{TM(calc)}}^j < 0 \\ \left(p_j \cdot (1 - p_{\mathrm{TM(calc)}}^j)\right)^2; & p_{\mathrm{TM(calc)}}^j > 1 \end{cases}$$

$$PF_2 = C_2 \sum_j^{N_{atom}} \left(\delta^j_{XY} + \delta^j_Z\right)$$

$$\delta^j_{XY(Z)} = \begin{cases} 0; & b^j_{xy(z)} > 0 \\ \left(p_j \cdot b^j_{xy(z)}\right)^2; & b^j_{xy(z)} < 0 \end{cases}$$

$$PF_3 = C_3\left(d_r - d_t\right)^2$$

$$PF_4 = C_4\left(c_r - c_t\right)^2$$

F_{obs} and F_{calc} are the amplitudes of observed and calculated structure factor respectively, σ is the standard deviation of a given diffraction peak, N_{peaks} is the number of diffraction peaks in the data set and N_{atoms} is the number of atoms decorating our structure units, d_r and d_t are the calculated and theoretical concentration of TM atoms respectively, finally c_r and c_t are the calculated and theoretical point density. PF_1, PF_2, PF_3, PF_4 keep the concentration of TM atoms, displacement coefficients, chemical composition and point density respectively in the physical boundaries. Numbers C_1, C_2, C_3, C_4, are the weights of a given component in the optimized function.

5. Refinement results

The results of optimization are given in the table 1.

Table 1. Refinement results

Gr.	x[Å]	y[Å]	z[Å]	Δx[Å]	Δy [Å]	b_{xy}[Å^2]	b_z[Å^2]	p_{calc}	p_{TM}
1	0.000	0.000	1.023	*0.000	*0.000	2.6(2)	3.8(1)	*1.00	*0.00
2	3.180	9.786	-1.023	*0.000	*0.000	0.58	0.49	*1.00	*1.00
3	5.145	3.735	-1.023	0.082	*-0.253	1.2(1)	2.8(2)	0.79	*0.00
4	1.965	6.051	1.023	0.026	*0.079	0.19	0.41	*1.00	0.75
5	3.180	2.315	-1.023	-0.062	0.130	0.43	0.37	*1.00	*1.00
6	-3.931	0.000	-1.023	-0.08(4)	*0.000	2.0(1)	0.3(1)	*1.00	*1.00
7	0.751	2.315	-1.023	0.033	*0.100	0.4(1)	3.7(2)	0.84	*0.00
8	2.428	0.000	-1.023	-0.130	*0.000	0.18	5.0(2)	*1.00	*0.00
9	2.717	3.735	1.023	0.14(1)	0.002	0.66	4.6(2)	*1.00	*0.00
10	2.428	7.481	-1.023	0.003	*0.008	0.5(1)	5.5(3)	0.87	*0.00
11	4.394	1.430	1.023	-0.144	-0.16(1)	0.8(1)	4.4(1)	0.85	*0.00
12	4.394	6.051	1.023	-0.006	*0.019	0.01	4.9(3)	*1.00	*0.00
13	-4.394	1.430	1.023	0.154	-0.179	0.0(1)	2.2(2)	*1.00	*0.00
14	3.180	5.166	-1.023	-0.11(1)	-0.19(1)	2.4(2)	2.0(3)	*1.00	0.21
15	5.886	1.430	-1.023	0.025	*-0.077	2.3(2)	0.7(1)	0.4(1)	0.2(1)
16	1.965	1.430	1.023	-0.191	-0.00(1)	0.2(1)	0.2(1)	*1.00	0.23
17	-5.608	2.315	-1.023	0.58(1)	0.08(2)	1.4(2)	2.4(1)	0.50	0.26
18	-2.428	0.000	1.023	0.160	*0.000	0.6(1)	0.1(1)	*1.00	0.49
19	-8.325	1.554	-1.023	0.175	-0.016	0.51	0.3(1)	*1.00	0.71
20	-12.255	1.430	1.023	-0.12(1)	0.23(2)	1.2(1)	0.8(2)	0.50	0.42
21	4.857	0.000	-1.023	0.09(3)	*0.000	1.8(3)	1.0(2)	0.4(1)	0.0(1)
22	-3.931	2.861	-1.023	-0.52(3)	-0.27(2)	1.0(2)	1.5(2)	0.23	0.2(1)
23	-7.862	4.620	1.023	-0.72(2)	0.15(2)	0.9(2)	1.4(3)	0.39	0.20
24	-11.216	0.000	1.023	-0.25(2)	*0.000	2.3(3)	1.1(2)	0.27	0.00

In columns x, y, z the positions of ideal Penrose tiling are given. In Δx, Δy columns components of the shift vectors from these ideal positions are listed. In the table 1 shift vectors are given for atoms representing a group (marked with additional black circle in the figure 1). Shifts for other atoms in a given group can be obtained by a proper rotation of the listed shift vectors. Columns b_{xy} and b_z contain the values of Debye-Waller coefficients, whereas in p_{calc} and p_{TM} columns the occupation probability and TM concentration are given respectively.

Values marked with (*) have not been optimized. For shift vectors there are several reasons for which a parameter is not optimized. Atoms in the first two groups decorate the vertices of our structure units, which cannot be shifted. In some other groups the y direction of a shift vector is not optimized. Atoms decorating edges or symmetry axes of the structure units can only move along them. Therefore, if an atom decorates an edge of a structure unit, then the y component of the shift vector is related to the x component (which was optimized) by a *tangent* function. If, on the other hand, an atom decorates a symmetry axis of a structure unit, then the y component of the shift vector is fixed to zero. Above restrictions for shift vectors are required for correctness of structure factor calculation. Group number 22 is an exception. It showed large Debye-Waller coefficients already in the first steps of the refinement. Therefore these atoms were artificially slightly moved out of the edge, inside the thick rhombus. Then we allowed optimization of the position. This proved to be effective, because the refined values of Debye-Waller coefficients are reasonable, and the refined positions are still inside the thick rhombus, which guarantees the correctness of the structure factor calculation.

For some groups also the TM concentration or occupation probabilities have not been refined. If in the first steps of optimization the occupation probability or the TM concentration for a given group was smaller than 0.05 or greater than 0.95, then in the next steps these parameters were set to 0 and 1 respectively and not optimized.

If for an optimized parameter in table 1 a standard deviation is not given in the bracket, then this means, that it is lower than 0.001 for shift vectors or 0.01 for probabilities and Debye-Waller coefficients.

A part of the refined structure is shown in the figure 2.

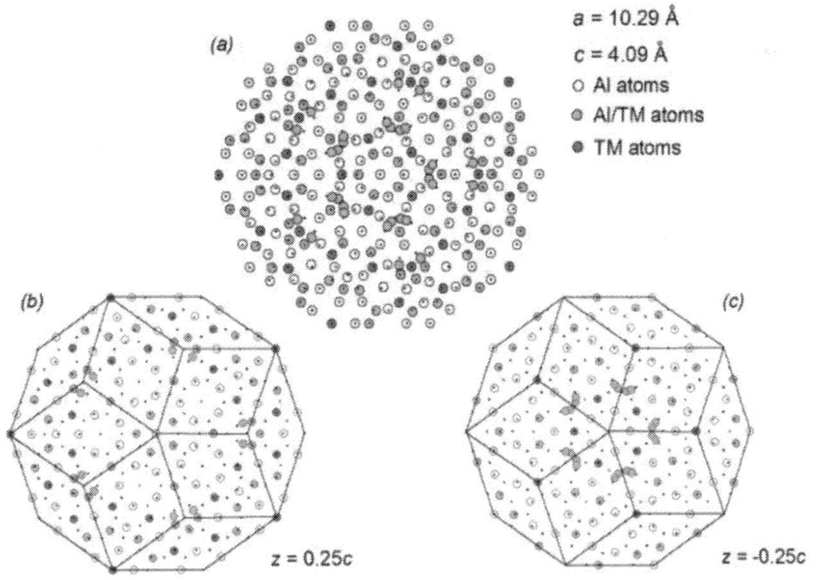

Figure 2. Refined structure

Dark grey circles denote TM atoms, light grey Al/TM atoms and white Al atoms. Black dots indicate ideal Penrose tiling positions. Figure 2a shows a projection of the structure along the z axis. Figures 2b and 2c show the layers at $z = 0.25c$ and $z = -0.25c$ respectively. The edge length (black lines in the figures 2b and 2c) was refined to be 10.29 Å and the period c along the z axis 4.09 Å.

Shifts from ideal Penrose tiling positions are rather small and not greater than 8% of the edge length of rhombuses (and in most cases not greater than 4%). Generally shifts for atoms from first three inflation divisions (groups 1 – 13), which form a 'skeleton' of the structure are much smaller than for atoms from fourth division (groups 14 – 24).

Occupation probabilities for these skeleton atoms are 1 in most cases and always greater than 0.75. For other groups partial occupation is common. Similarly TM concentration is also well defined for skeleton atoms. Only for one group its value is not equal to 0 or 1. For other groups mixed occupation is possible. A sum of occupancy probabilities for pairs of atoms indicated by pairs of light grey circles, which touch each other or overlap in figures 2b and 2c, is lower than one. These are flip positions. In each tile atoms occupy either first or second position. If there are any distances shorter than 2.43 Å, then a sum of occupation probabilities of atom pairs revealing to short interatomic distance is lower than 1

In the resulting structure there is a certain degree of disorder within the structure units - thin and thick rhombuses. This could be called a local or short-range disorder. On the other hand, these rhombuses form the Penrose tiling, thus, in our approach, the structure is also perfectly long-range ordered.

Debye-Waller coefficients (b_{xy} and b_z) for most groups are smaller than 3Å^2, which is an acceptable value for inorganic materials. Greater values may indicate a certain degree of static disorder in the structure. Some b_z i.e. the component associated with z direction exceed the value of 3Å^2. This could mean that the atomic layers of the structure are puckered rather than flat. In [1] even an 8 Å superstructure is assumed. This means that the period along z axis is doubled. However, there are no half - integer reflections in the diffraction pattern, thus in experiment only an average 4 Å structure is observed. In this approach larger b_z values result from overlapping layers of the superstructure having opposite z displacements. In our paper only the average 4 Å structure is investigated.

A phason displacement parameter b_{Ph}, extinction parameter g and scale factor between the calculated and measured diffraction pattern were the global parameters optimized. A total number of parameters refined was 106 and the observations-to-parameters ratio is high and equal 26.1. We obtained R = 12.3% and the wR = 6.2%. Global parameters, R and wR values calculated for different numbers of strongest reflections are listed in the table 2. The stability of our solution was also checked. In the last steps of the refinement C_3 and C_4 for PF_3 and PF_4 were set to 0. Despite that, the refined structure has the exact chemical composition $Al_{70.7}TM_{29.3}$ (chemical composition of the sample was measured to be $Al_{70.6}TM_{29.4}$ [1]). The point density of our structure is 0.0656 Å^{-3} and overall density 3.94 g/cm^3. Experimental values are 0.0661 Å^{-3} and 3.94 g/cm^3 respectively. The F_{obs}/F_{calc} plot for the final structure in the logarithmic scale is shown in the figure 3.

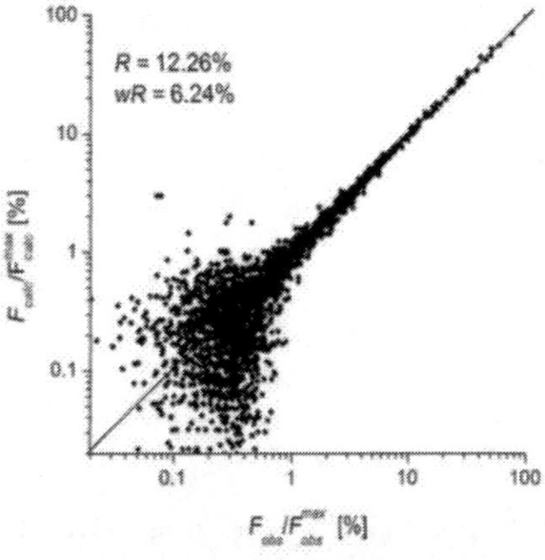

Figure 3. F_{calc}/F_{obs} plot in logarithmic scale

Table 2 Refinement parameters

b_{Ph} [Å^2]	0.40(4)
g	2.(9)·10^{-5}
R [%] all 2767 reflections	12.3
wR [%] all 2767 reflections	6.2
R [%] $I>3\sigma_I$ 1194 reflections	4.3
wR [%] $I>3\sigma_I$ 1194 reflections	5.7
R [%] $I>10^{-4}\,I_{max}$ 837 reflections	4.1
wR [%] $I>10^{-4}\,I_{max}$ 837 reflections	5.4

In comparison to our previous paper [6] the diffraction data set is much larger while the number of refined parameters is similar. Therefore the observations-to-parameters ratio is higher, and whole refinement is more reliable. The division of atoms was slightly changed, and the number of groups was increased from 19 to 24. Debye-Waller coefficients are lower, *R* and *wR* values are better even for 837 strongest reflections (table 2), than for 451 reflections in the previous paper. This proves a better structure model choice.

6. Conclusions

In this paper we have described the results of the structure analysis of a decagonal *basic Ni-rich* Al-Ni-Co quasicrystal. The refinement was conducted with only real space parameters. We used a decorated rhombic Penrose tiling as an initial structure for the refinement process. With the statistical approach we could derive the structure factor unrelated to the perpendicular space. This is a very convenient approach as it allows us to treat atomic positions as phase coefficients of the Fourier transform. Furthermore, the structure factor we used was derived in a strict; i.e. without any approximations, mathematical way.

We obtained $R = 12.3\%$ and the $wR = 6.2\%$ for 2767 reflections. 106 parameters were refined. The resulting structure has the exact chemical composition $Al_{70.7}TM_{29.3}$. The refined structure has the point density 0.0656 $Å^{-3}$ and overall density 3.94 g/cm^3, which is very close to experimental values.

Acknowledgements

This work has been partially supported by Ministry of Science and Higher Education of Poland and also by CMA-NoE activities in AGH-UST.

References

[1] Katrych S, Steurer W 2004 *Z. Kristallogr.* **219** 606
[2] Cervellino A, Haibach T, Steurer W 2002 *Acta Cryst.* **B**58 8
[3] Takakura H,Yamamoto A,Tsai A P 2001 *Acta Cryst.* **A**57 576
[4] Kozakowski B, Wolny J 2006 *Phil. Mag.* **86** 549
[5] Kozakowski B, Wolny J arXiv:cond-mat/0503464v1 [cond-mat.other]
[6] Wolny J, Kozakowski B, Kuczera P, Takakura H 2008 *Z. Kristallogr.* **223** 847
[7] Duda M, Kozakowski B, Wolny J 2007 *J. of Non-Crystalline Solids* **353** 2500

Journal of Physics: Conference Series **226** (2010) 012002 doi:10.1088/1742-6596/226/1/012002

Ab initio structure solution of decagonal quasicrystals by iterative dual space methods: performance tests on synthetic and experimental data

F Fleischer[1], **T Weber and W Steurer**

Laboratory of Crystallography, Department of Materials, ETH Zurich, 8093 Zurich, Switzerland

E-mail: frank.fleischer@mat.ethz.ch

Abstract. Iterative *ab initio* structure solution algorithms were tested on their performance power in phasing diffraction data. Statistical investigations on the reproducibility of the recovered phases allow an estimate about the reliability of the results. The working principle is demonstrated on synthetic structures and on experimental diffraction data of decagonal AlCuCo, AlFeNi and AlCoNi quasicrystals.

1. Introduction

The extension of Fourier recycling techniques to arbitrary dimensions [1,2] provides an easy access to quasiperiodic structures. The structure solution of quasicrystals in higher dimensions leads to the reconstruction of geometrical objects – the occupation domains (ODs) – in position, shape and electron density (ED). The reconstruction of non-point-like EDs for quasicrystals is therefore more demanding than structure solutions of ordinary periodic structures. Algorithms, like low-density elimination (LDE) and charge flipping (CF) overcome this problem and were successfully applied for solving quasiperiodic structures [e.g. 1,3,4], however, their working principle is not yet fully understood.

In this paper we present a comprehensive study on the basic CF and LDE algorithms applied on quasicrystals. Systematic performance tests of iterative phase retrieval algorithms – reported in detail for a broad range of simulated diffraction data [5] – are extended to experimental quasicrystal data.

2. The charge flipping and low-density elimination algorithms

The CF and LDE algorithms are Fourier recycling techniques, where the iteration process alternates between reciprocal and physical space. In the first iteration step the experimental form factor amplitudes are combined with random phases. Then, the corresponding ED is calculated by Fourier transformation and modified as follow: a threshold value δ is set to define the charges that will be manipulated. CF *flips* the sign of all charges smaller than δ, i.e. positive charges become negative and *vice versa*. LDE sets all negative charges to zero. Then, a new set of structure factors is calculated by

[1] To whom any correspondence should be addressed.

Fourier transformation of the perturbed ED. In a final step the resulting phase of each structure factor is combined with the measured intensity and serves as an input for the next iteration cycle. The iteration process is continued until convergence is reached.

All calculations were done with the program package SUPERFLIP [6], which allows performing CF and LDE calculations in arbitrary dimensions. Each data set is first extended to a complete data set corresponding to its Laue symmetry. In the following, data sets are handled in space group *P*1. The result of a typical CF and LDE run is a set of phased structure factors with the constraint $|F_{\text{refined}}(\mathbf{H})|^2 = |F_{\text{obs}}(\mathbf{H})|^2$. All calculations were performed on an ED grid with a resolution of about 0.25 Å along each direction. The number of iteration steps was restricted to 1000, which was found to be adequate to reach convergence. Since the origin of the resulting ED is not defined, the resulting EDs were shifted to a reference ED and inverted if required. The following investigations use the basic algorithms without any performance enhancements to study the *ab initio* abilities of CF and LDE.

3. Test structures

The first part of our investigations is based on synthetic data what allows a direct comparison of resulting phases with the correct ones. In the second part experimental diffraction data from decagonal quasicrystals are used.

3.1. Synthetic diffraction data

The artificial decagonal quasiperiodic structure corresponds to the structure reported in [7] with the modification that all ODs are fully occupied with Al only. An overall atomic displacement parameter $U_{\text{iso}} = 0.02$ Å^2 and a phasonic Debye-Waller factor $U_{\text{phason}} = 0.02$ Å^2 was used for the calculation of the structure factors. All structure factors of the reciprocal asymmetric unit were calculated within $|Q^{\text{par}}| < 1.4$ Å^{-1} and $|Q^{\text{perp}}| < 1.2$ Å^{-1}. The reflections were scaled such that the strongest reflection except $F(\mathbf{0})$ corresponds to $I = 10000$ counts. We added a normally distributed random number with variance $\sigma^2 = I_{\text{calc}}$ to the intensities to simulate more realistic experimental data. To test the influence of incomplete data, test data sets were restricted to the 600 and 150 strongest reflections. We performed 50 CF and LDE runs for each model structure.

3.2. Experimental diffraction data

Bragg data sets of the decagonal phases AlCoNi, AlCuCo and AlNiFe were used as a basis for the experimental test structures. The single crystal X-ray diffraction data of AlCuCo were taken from [8]. The single crystal X-ray diffraction data of AlNiFe were collected with an Oxford Diffraction diffractometer ("Onyx"-CCD detector, graphite monochromator, Mo K_α radiation) from a single crystal with composition $Al_{73}Ni_{22}Fe_5$, which was annealed at 920°C and quenched afterwards. The data set for AlCoNi was measured at the Swiss-Norwegian beam line at the European Synchrotron Radiation Facility ESRF, Grenoble, France on a crystal with composition $Al_{71}Co_{13}Ni_{16}$, which was quenched from 730°C. This phase was identified as the type-I structure. For our studies, only main reflections, i.e. Bragg reflections corresponding to the basic structure, were used.

All diffraction data were indexed using a 5D decagonal lattice with $a_1^* \ldots a_5^*$ listed in Table 1. We performed 100 CF and LDE runs on all experimental data sets.

4. Quality factors

In previous tests on synthetic structures [5], we defined quality parameters to compare the resulting phases of a CF and LDE run with correct phases. The phase difference between the retrieved and the known correct phase of a reflection averaged over *n* runs is calculated by

$$\Delta\varphi(\mathbf{H}) = \left|\varphi_{true}(\mathbf{H}) - \arg\left(\frac{1}{n}\sum_{i=1}^{n} F_{i\,refined}(\mathbf{H})\right)\right| \quad (1)$$

Table 1. Overview of the quality parameters of the chosen experimental data sets of the decagonal quasicrystals AlCoNi, AlCuCo, AlNiFe

	AlCoNi	AlCuCo	AlNiFe
Laue group	10/mmm		
$a_1^* \ldots a_4^*$	0.1635(3) Å^{-1}	0.2656(2) Å^{-1}	0.2651(3) Å^{-1}
a_5^*	0.2444(2) Å^{-1}	0.24107(3) Å^{-1}	0.2437(4) Å^{-1}
Independent reflections	1355	259	898
R_{int}	0.061	0.062	0.033
Q_{max}^{par}	2.00 Å^{-1}	1.98 Å^{-1}	1.50 Å^{-1}
Q_{max}^{perp}	0.51 Å^{-1}	2.14 Å^{-1}	1.50 Å^{-1}

In practice it is essential to have a quality parameter that does not require the knowledge of the correct phases. We defined a quality parameter $\kappa(\mathbf{H})$, which quantifies the reproducibility of the phases of each reflection. Since a random number generator is used for initialization, each run has different start conditions and therefore different phases are obtained, even if the structure solution was successful. The definition of the reproducibility parameter is based on the constrain $|F_{\text{refined}}(\mathbf{H})|^2 = |F_{\text{obs}}(\mathbf{H})|^2$ and is defined as the ratio between the average and the experimental structure factor amplitudes.

$$\kappa(\mathbf{H}) = \frac{\left|\frac{1}{n}\sum_{i=1}^{n} F_{i\,refined}(\mathbf{H})\right|^2}{\left|F_{i\,refined}(\mathbf{H})\right|^2} \quad (2).$$

The value $\kappa(\mathbf{H})$ may range between 0 and 1. The phase of a reflection was found with a high consistency if $\kappa(\mathbf{H})$ is large, otherwise the phase shows a strong variation in different runs. An overall reproducibility factor for a complete data set is defined by weighting the reliability factor of each reflection with its intensity:

$$\mathrm{K} = \frac{\sum_j \left|F_{j\,obs}(\mathbf{H})\right|^2 \cdot \kappa_j(\mathbf{H})}{\sum_j \left|F_{j\,obs}(\mathbf{H})\right|^2} . \quad (3)$$

K may range between 0 and 1. An overall K of about 1 corresponds to a data set with a high phase reproducibility of most reflections, whereby a small value indicates that the phase retrieval is almost at random.

The averaging of several runs can drastically increase the quality of the phased data set. Averaging leads to an improvement of phases, but intensities of reflections with a low $\kappa(\mathbf{H})$ become smaller compared to experimental intensities. Therefore, re-scaling of the resulting structure factor amplitudes to the observed ones leads to an improvement of the reconstructed ED [5]. From eq 2 follows:

$$\overline{F}_{scaled}(\mathbf{H}) = \frac{1}{n \cdot \sqrt{\kappa(\mathbf{H})}} \sum_{i=1}^{n} F_{i\,refined}(\mathbf{H}) . \quad (4)$$

Averaging was done using 50 runs for the synthetic data and 100 runs for the experimental data.

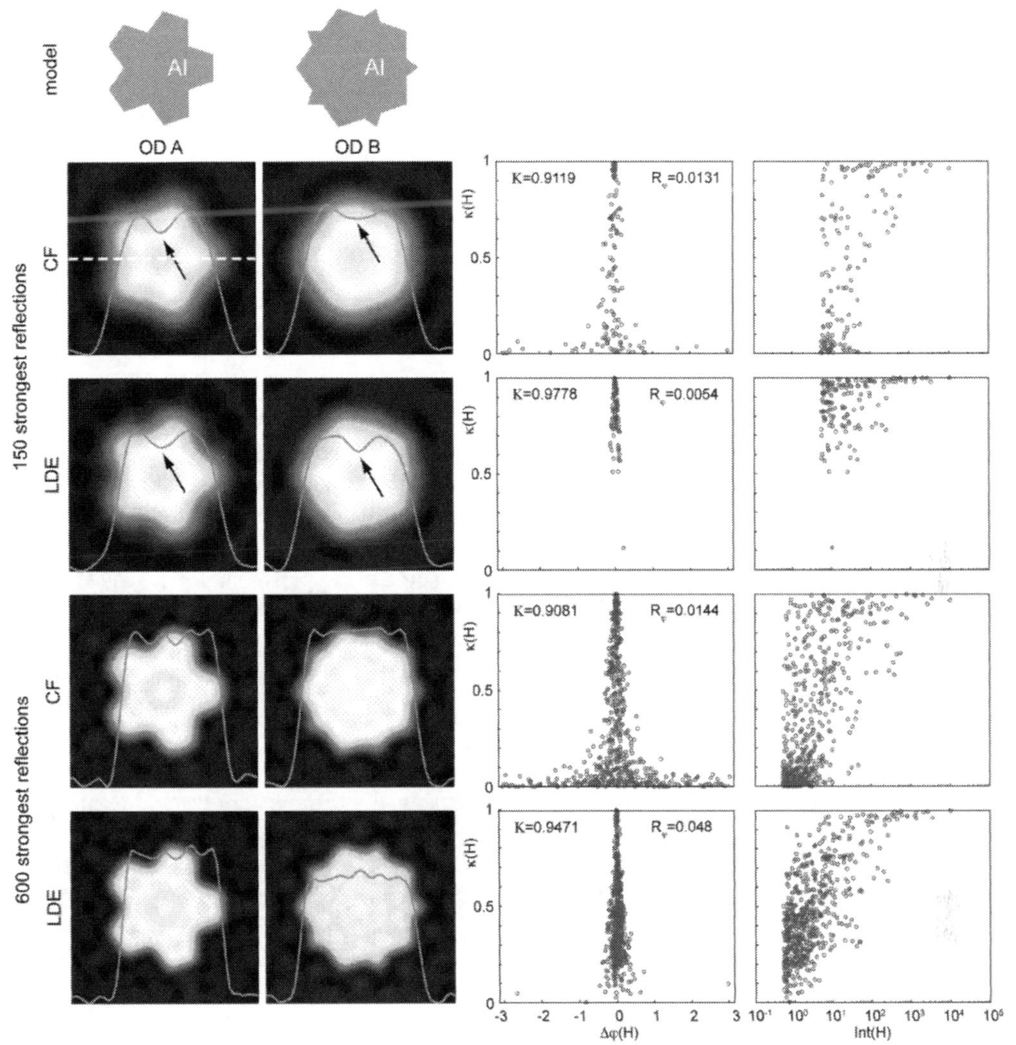

Figure 1 Maps of reconstructed occupation domains for a model structure averaged from the results of 50 CF and LDE runs and rescaled to the experimental amplitude. EDs were calculated from data sets restricted to the 150 and 600 strongest symmetry independent reflections. Since the inversion related ODs A/D and B/C were reconstructed without significant differences in shape and ED, only the non-symmetry related ODs A and B are shown. Line scans (blue lines) were taken across the ODs (white dashed line in one inset) to show details of the ED distribution. Strong truncation effects are observed in the centres of the ODs (arrows) especially for small data sets. Typical distributions of the reliability factor κ(H) as a function of the phase difference Δφ(H) and of the intensity of the reflections are shown in the right columns.

5. Results and discussion

5.1. Synthetic diffraction data

All CF and LDE runs converged within 1000 iteration steps and the structures could be solved in all cases. The reconstructed ED maps of the average solutions (Fig. 1) allow a direct comparison between LDE and CF results. Significant truncation errors at the center of the ODs can be observed in particular in the case of the data set containing only 150 reflections, what could lead to a mis-interpretation of the results. This effect was also demonstrated on experimental data by [4].

We calculated the reproducibility factors $\kappa(\mathbf{H})$ and the phase difference $\Delta\varphi(\mathbf{H})$ for each reflection, based on 50 averaged runs for each synthetic diffraction data set and found a typical distribution (Fig. 1). Indicated by the strong correlation between high $\kappa(\mathbf{H})$-values and phase differences of $\Delta\varphi(\mathbf{H}) \approx 0$, it can be concluded that reproducibility of phases is a measure for the quality of the results. The overall quality factor R_φ, which is defined as a conventional intensity-weighted *R*-factor of the phase difference $\Delta\varphi(H)$, is $R_\varphi \approx 0.014$ for CF and 0.004 for LDE. The overall reproducibility factor K was obtained as $K \approx 0.90$ for CF and $K \approx 0.97$ for LDE, i.e. phases are found with higher consistency when using LDE. Especially weak reflections were found with higher consistency and smaller phase errors.

5.2. Experimental data

The transfer of the procedure to experimental data and the reconstruction of high quality EDs by averaging is shown for decagonal AlCoNi, AlCuCo and AlNiFe quasicrystals. The higher-dimensional structure solution is represented by the reconstruction of the ODs. Though sizes and positions of ODs can be estimated from single runs, an accurate shape and occupation can be estimated only after averaging and re-scaling on the observed structure factor amplitudes (Fig. 2). Apart from some differences in details CF and LDE could consistently solve all experimental test structures. The space group of all reconstructed structures was determined from the reconstructed EDs as $P10_5/mmc$, which is in agreement with previously reported results [8,9].

After averaging all successful runs, LDE shows well-defined ODs with partially occupied regions (e.g. triangles, parallelograms, see arrows in Fig. 2). This richness in detail is not found in the results obtained with CF, what is due to inconsistent phase retrievals, especially in the case of weak reflections. The significance of the detailed observations, however, is not known. Though we believe that most of the details seen in the averaged LDE maps in Fig. 2 represent true structural information, our limited experience with LDE/CF does currently not allow a reliable separation of artifacts from true structural features. On the one hand truncations errors are always present and corresponding artifacts are expected to be present. On the other hand, our performed tests on synthetic structures suggests that CF and, to a larger extent, LDE deliver reliable EDs. The final proof of presence and absence of observed features can only be done with least-squares refinements. Since quasicrystal refinements are rare, time-consuming and often not in the quality as in the case of periodic structures, a comprehensive comparison of structure models obtained by LDE/CF and by least-squares refinements is currently hardly possible. It can, however, be stated that the structure solution of decagonal AlCuCo shows no significant differences to the refined structure reported in [8].

The observed $I(\mathbf{H})$-$\kappa(\mathbf{H})$ distribution are similar to the ones observed from the simulated model structures. High $\kappa(\mathbf{H})$-values are obtained for strong reflections, while weak reflections tend to show smaller $\kappa(\mathbf{H})$-values. It is reasonable to conclude that phases of reflections with high $\kappa(\mathbf{H})$-values are correct, since performance tests on centrosymmetric model structures have shown that a high reproducibility of phases usually correlate with a high quality of the retrieved phases. A much higher overall reliability factor was observed for structures solved with LDE ($K > 0.97$) than for CF ($K \approx 0.79$).

6. Conclusion

Tests on artificial data show a high reliability for the phase retrieval of reflections with consistently reproduced phases indicating that $\kappa(\mathbf{H})$ may be used as a quality factor for the retrieved phases. Iterative phase retrieval algorithms like LDE and CF are useful to derive reasonable structure models from experimental quasicrystal data and an averaging procedure allows the reconstruction of high quality EDs. This was shown for decagonal AlNiFe, AlCoCu and AlCoNi quasicrystals on experimental data with different quality. The quality of such solved structures as initial starting models for least-square techniques has to be evaluated in further tests.

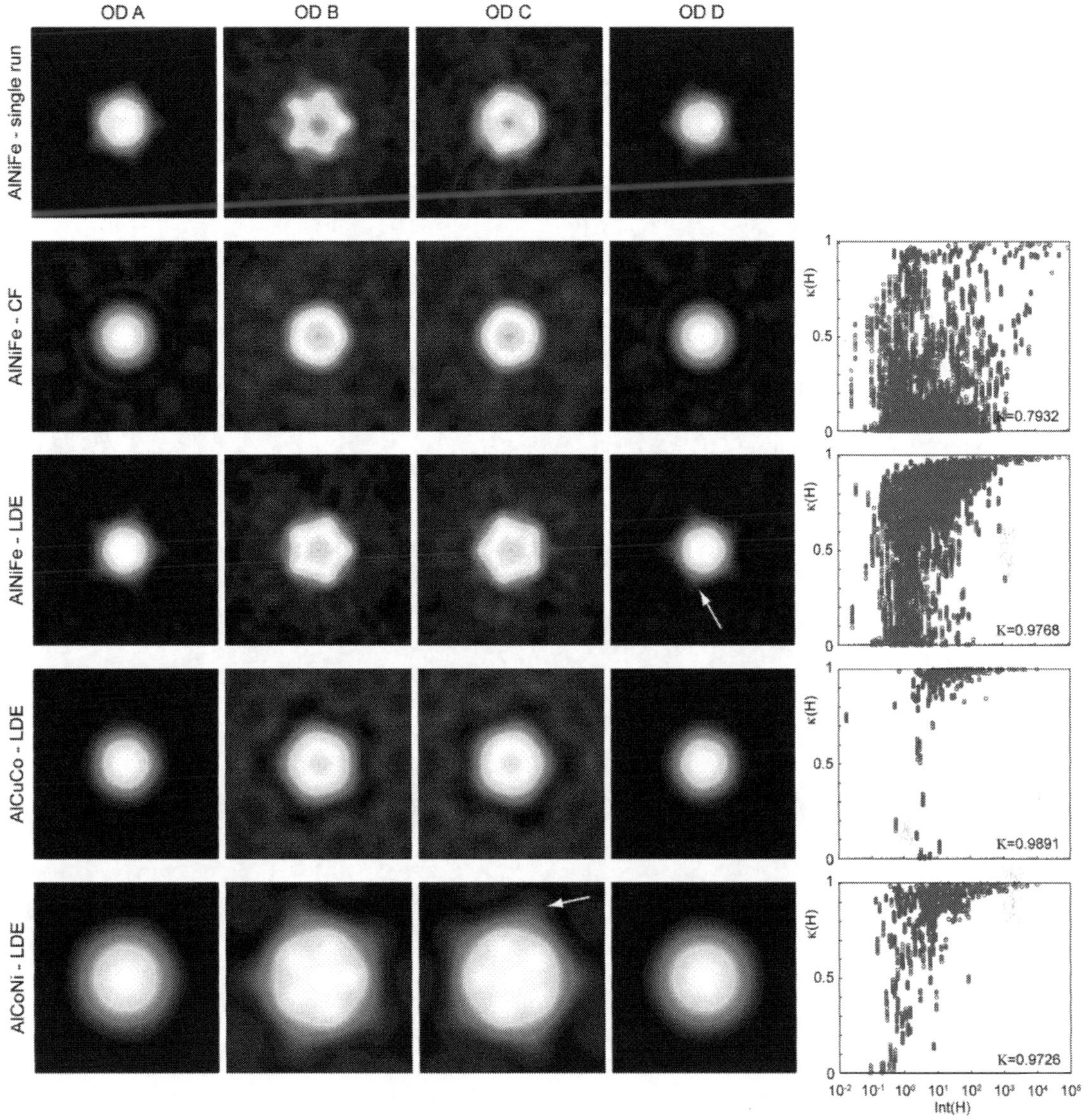

Figure 2 Maps of reconstructed occupation domains for experimental data sets of AlNiFe, AlCuCo and AlCoNi. The ODs reconstructed from a single LDE run and the averaged structure factors of 100 successful CF and LDE runs for AlNiFe are shown as typical examples. Furthermore, the EDs calculated from 100 LDE runs of the decagonal phases AlCuCo and AlCoNi are shown. The density for the ODs B and C are scaled by a factor of two compared to the ODs A and D, as they are occupied by Al, which has a scattering factor of about two times smaller than the transition metals. Arrows mark some details of the ODs for the averaged results of 100 runs. The calculations were performed without any symmetry averaging in $P1$. ODs A and B are located at $+(i/5, i/5, i/5, i/5, 1/4)$ (i = 1, 2) and OD C and D are found at $-(i/5, i/5, i/5, i/5, 3/4)$ (i = 1, 2). The $\kappa(\mathbf{H})$-intensity distribution and the overall K are given in the diagrams. The ODs of AlCoNi are τ-inflated ($\tau = (1+\sqrt{5})/2 = 1.61803\ldots$) compare to AlNiFe and AlCuCo, because a different basis was used.

7. Acknowledgement
We gratefully acknowledge the financial support from the Swiss National Foundation under project SNF 200020-121568.

References
[1] Takakura H, Shiono M, Sato T J, Yamamoto A and Tsai A P 2001 *Phys. Rev. Lett.* **86** 236-239
[2] Palatinus L 2004 *Acta Cryst.* A**60** 604-610
[3] Katrych S, Weber T, Kobas M, Massüger L, Palatinus L, Chapuis G and Steurer W 2007 *J. Alloys Compd.* **428** 164-172
[4] Weber T, Pedersen B, Gille P, Frey F and Steurer W 2008 *Z. Kristallogr.* **223** 863-867
[5] Fleischer F, Weber T, Palatinus L, Deloudi S and Steurer W, 2010 *J. Appl. Cryst.* **43** 89-100
[6] Palatinus L and Chapuis G 2007 *J. Appl. Cryst.* **40** 786-90
[7] Takakura H, Yamamoto A and Tsai A P 2001 *Acta Cryst.* A**57** 576-85
[8] Steurer W and Kuo K H 1990 *Acta Cryst.* B**46** 403-12
[9] Steurer W 2004 *Z. Kristallogr.* **219** 391-446

To the model of a decagonal Al-Cu-Co

Zorka Papadopolos
Institut für Theoretische Physik, Universität Tübingen, Germany
E-mail: ptipz01@tphys.physik.uni-tuebingen.de

Abstract.
It is an approach to the model of a decagonal Al-Cu-Co quasicrystal, reproduced from the data of a scanning transmission electron microscopy (STEM) image. The atomic positions, seen on STEM images should be related to an ideal 2-dimensional pentagonal tiling, obtained by the projection from the 4-dimensional A_4 root-lattice.

1. Introduction

A decagonal Al-Cu-Co had been, among other authors, modelled by Burkov [1]. In his model, the stoichiometry is $Al_{62}Cu_{19}Co_{19}$, and it is periodic in the 10f/5f direction, with the period of circa 4 Å. Burkov claims that his model is based on the electron microscopy [2] and x-ray diffraction data [3, 4, 5], although Kuo et al. [3, 4, 5] consider $Al_{65}Cu_{20}Co_{15}$ with periodicity of circa 8 Å.

A variant of the Burkov model was latter founded over the pentagonal tiling $\mathcal{T}^{*(A_4)}$ of the plane by two golden triangles [8]. The tiling is projected from the 4-dimensional root-lattice A_4, embedded in the 5-dimensional lattice $\mathbb{Z}_5$, see [6] and the Refs quoted therein. From now on, we call the tiling for short, the 'triangular tiling'.

The recent STEM (scanning transmission electron microscopy) data on $Al_{64}Cu_{22}Co_{14}$ [7] are on a decagonal phase with the period of circa 4 Å, as by Burkov. In this paper the tiling $\mathcal{T}^{*(A_4)}$ will be put over the STEM image from Ref [7], and the result will be compared to the way how it was previously done in the Burkov model.

2. The triangular tiling over the STEM image of $Al_{64}Cu_{22}Co_{14}$

The pentagonal tiling $\mathcal{T}^{*(A_4)}$ [8] has as the prototiles two golden triangles. Both golden triangles are isosceles, triangles having two sides equal. The isosceles triangle, that we label by the letter 'D', has the base side (edge) of length ②, and two edges of an equal length τ②, $\tau = (\sqrt{5}+1)/2$. The standard length ② is along a 2f direction. Another isosceles triangle, labelled by the letter 'd' has the base edge of length τ②, and two edges of an equal length ②.

In what follows we will address to the STEM image done along the c-axis. perpendicular to the quasiperiodic planes. This image is presented in Figs 3 and 4 of Ref [7]. In Ref [7] $Al_{64}Cu_{22}Co_{14}$ is identified as a decagonal quasicrystal, periodic along the c-axis, and the length of the period is estimated to be circa 4.00 Å. The strong shining spots are identified on Fig 4 in Ref [7] as the transition metals, and the less shining, or almost not seen spots are interpreted as Al positions. In this paper we will reduce our attention to the strongly shining spots, and among these, to the spots forming local configurations of the pentagonal symmetry.

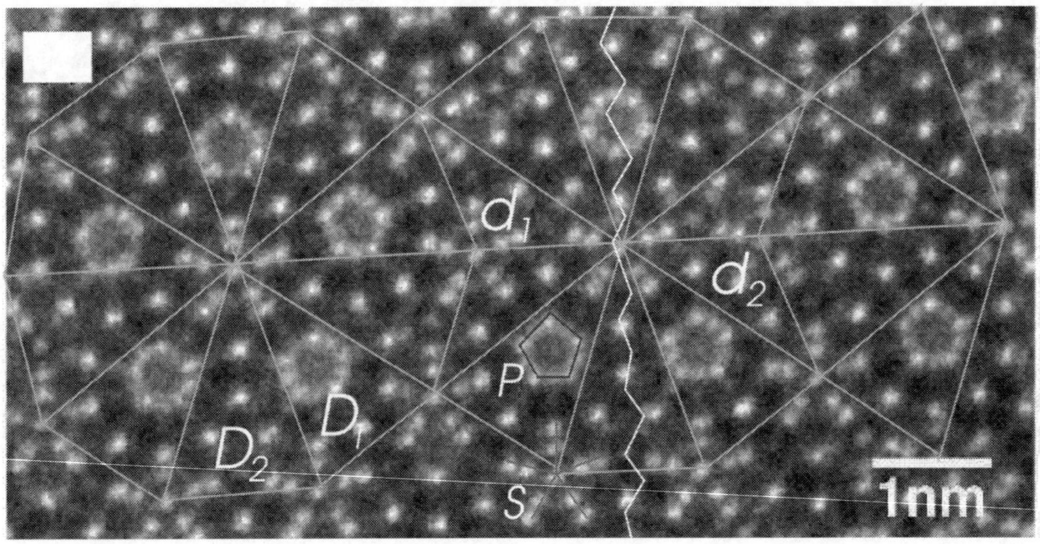

Figure 1. Over the STEM image of $Al_{64}Cu_{22}Co_{14}$ from Ref [7] the tiling $\mathcal{T}^{*(A_4)}$ by two golden triangles is drown. The four decorated prototiles D_1, D_2, d_1 and d_2 are marked. The variants '1' or '2' of decorations are particularly oriented with respect to the pentagram (five pointed star) S, marked in red. The length of the pentagram-arm is circa 0.4 nm. Also the pentagon P is marked in dark blue. The length of the pentagon-edge is circa 0.3 nm.

Let us have a look on STEM image on Fig 3a of Ref [7]. Among the shining spots there appear two smallest local configurations with the pentagonal symmetry, a pentagram (five pointed star) S, and the pentagon P. We mark the pentagram S in red and the pentagon P in dark blue, see Fig 1. On the STEM image a patch of the $\mathcal{T}^{*(A_4)}$ tiling is drown with light blue edges, all along the 2f directions. We see that prototiles, scaled from above defined D and d can have two different decorations by the shining spots (TM positions) depending on their orientation with respect to the pentagram S. For the reason we label these by D_1 and D_2 and by d_1 and d_2, see Fig 1.

By the white broken line is drown a plane orthogonal to the 5f/10f image plane, such that either to the left or to the right hand side of the plane an ideal tiling patch is seen. It is evident from the fact that the single vertex configuration of the tiling that appears on the broken line is not an allowed vertex configuration in the tiling $\mathcal{T}^{*(A_4)}$ [8].

Let us introduce the scale on the tiling. The long edge of the golden triangles congruent to D and d corresponds to the $\tau^3$②, with ② being the standard length along a 2f direction. Note that the prototiles of by τ^2 augmented tiling $\mathcal{T}^{*(A_4)}$ are labeled by the same labels, as the standard tiling $\mathcal{T}^{*(A_4)}$, introduced at the beginning. The long edge of the tiling, compared to the 1nm unit on the STEM image is $\tau^3$②$= 19.51$Å. The value is very close to 19, 6Å, the length of the long edge estimated in the Burkov model [1]. The edge length of the pentagon P, drown in dark blue in Fig 1 is then τ^{-1}②$= 2.85$Å, whereas an 'arm' of the pentagram (five pointed star) S, marked in Fig 1 in red is ⑤$2\tau/(\tau+2) = 3.92$Å, with ⑤ being the standard length along an icosahedral 5f direction. Hence, we conclude that the length of the standard basis vectors in the high dimensional lattice $\mathbb{Z}_5$ is $|e_i| = 6.19$ Å $(i = 1, 2, \ldots, 5)$. Projected into the standard length [5] onto the quasiperiodic plane, [5] :=⑤$2\tau/(\tau + 2) = 3.92$Å, or circa 0.4nm. It is labelled in red on the images.

Let us mark the local configurations on which the vertices of the tiling in Fig 1 are fixed. It turns out that these are the pentagrams S, see Fig 2. Moreover, *all* the S configurations from the STEM image are the vertices of the tiling $\mathcal{T}^{*(A_4)}$.

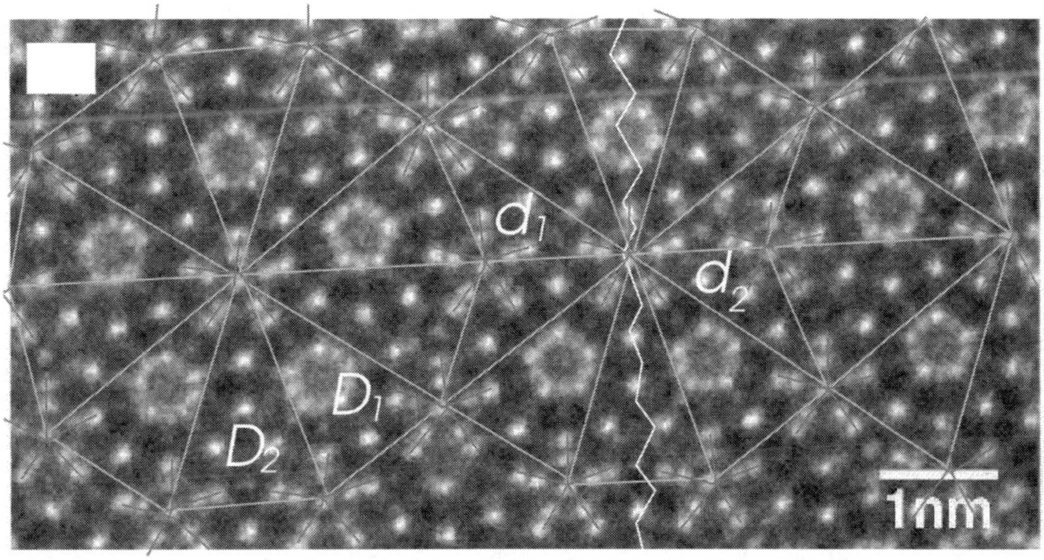

Figure 2. All the S configurations (pentagrams, marked in red) from the STEM image of $Al_{64}Cu_{22}Co_{14}$ are the vertices of the tiling $\mathcal{T}^{*(A_4)}$.

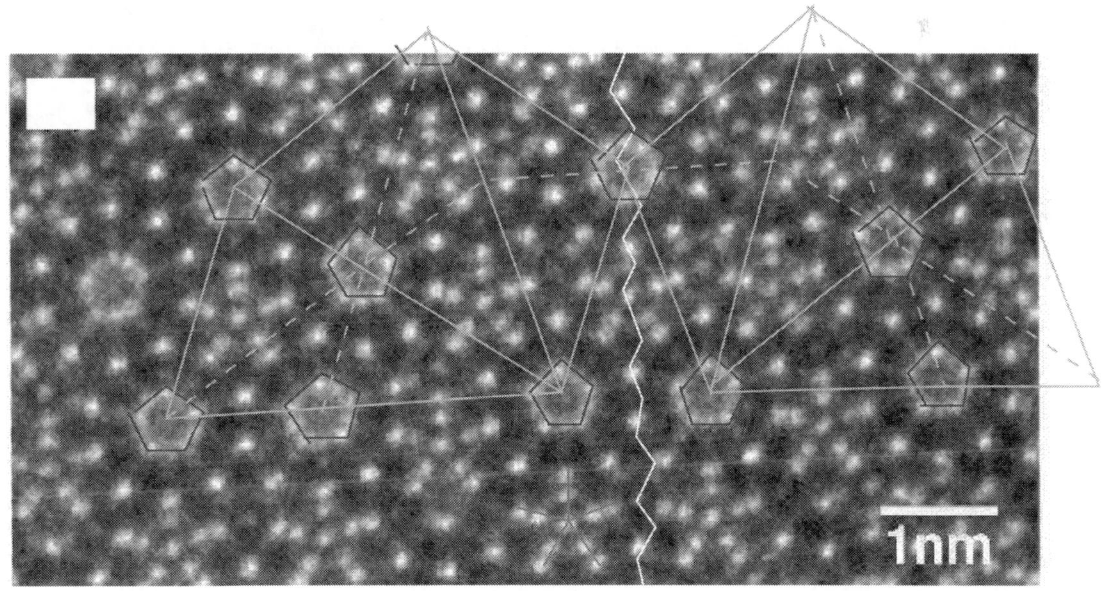

Figure 3. By full lines in light blue is presented a patch of the tiling $\mathcal{T}^{*(A_4)}$ with the vertices on some of the P configurations over the STEM image of $Al_{64}Cu_{22}Co_{14}$. If the dashed light blue lines are added, there appears by τ^{-1} deflated tiling, though with *not all* the vertices on the P configurations.

According to Fig 4 in Ref [7], the pentagon P has at the vertices large shining spots, identified as the TM positions. On a pentagon of the same size, rotated by 36°, the smaller spots are interpreted as Al positions. The union of these two pentagons is a decagon. If one compares Figs 1 and 2 of the present paper to the Fig 3 in Ref [6], a patch of the Burkov model, one finds in Fig 3 in Ref [6] decagons of the same size, as a union of the mentioned two pentagons, but with the TMs placed on *both* pentagons. On a STEM image such a configuration should appear as a strong shining decagon. These decagons are centered on the vertices of the tiling $\mathcal{T}^{*(A_4)}$, see Fig 3 in Ref [6]. Let us try to put a tiling $\mathcal{T}^{*(A_4)}$ on the STEM image of $Al_{64}Cu_{22}Co_{14}$, such that all vertices are on the pentagons P, see Fig 3 in the present paper. It turns out, that the tiling $\mathcal{T}^{*(A_4)}$, fulfilling this condition is augmented by factor τ, compared to the tiling with the vertices on pentagrams S, as in Fig 2. If the first step of deflation is performed on Fig 3, one ends up with the tiling on the same scale as in Figs 1 and 2, but the patch of the tiling does not have all the vertices centered at the pentagons P!

Let us consider the high-resolution electron microscopy image of $Al_{65}Cu_{15}Co_{20}$ presented in Fig 3(a) of Ref [2]. The pentagrams S are not seen on the image, but some decagonal/pentagonal configurations of the same size as the pentagons P of $Al_{64}Cu_{22}Co_{14}$ (see Fig 1) are clearly identified. On the image in Fig 3(a) of Ref [2] the "Penrose 1" tiling (P1) is put on these pentagons/decagons P as the vertices. The edge length of P1 is circa 20 Å. It is evident that the tiling P1 with the edge length scaled by τ^{-1} can *not* be supported by the P configurations at *all* the vertices. In Fig 2(c) of Ref [7] a P1 tiling with the smallest possible edge is plotted over the 5f image of $Al_{64}Cu_{22}Co_{14}$ such that all the vertices of the tiling P1 are on configurations P. It turns out that the edge length is circa 20 Å, as well. Through the inspection can be seen that the quasilattices of the P configurations in both decagonal phases could probably be considered as equivalent. Hence, in both cases the underlying triangular tiling $\mathcal{T}^{*(A_4)}$ by the smallest possible tiles with *all* the vertices on pentagons P has the short edge of 20 Å, as the tiles drown in full lines in Fig 3. For the relation of the $\mathcal{T}^{*(A_4)}$ to the P1 tiling see Ref [9]. The conclusion is that the frequency of the pentagons/decagons P is overestimated in the Burkov model.

3. Outlook

A lattice point x of the reciprocal lattice A_4^R can be written in the basis $\{a_i = e_i - \frac{1}{2}s, i = 1, 2, \ldots, 5, s = e_1 + e_2 + .. + e_5\}$ as $x = \sum_{i=1}^{5} n_i a_i$ (n_i are integers). With respect to the translations of the A_4 lattice this set splits into 5 classes: $r(q^R) := [\sum_{i=1}^{5} n_i]_{mod5}$. The function $r(q^R)$ takes five possible values: 0, 1, 2, 3 and 4. One of these classes labels the lattice points in the lattice A_4, the further 4 classes are the holes of the lattice A_4. Using the single cluster from Fig 4 in Ref [7] and distributing it among the prototiles D_1, D_2, d_1, and d_2 one easily classifies the shining spots of the STEM image into five classes (presented at Aperiodic'09). But, in the overlap of two such clusters there appear some too short interatomic distances. The way out of the problem will be in changing somewhat the atomic positions on a periphery of the suggested cluster.

Another problem with the model will be to fix accurately the Al positions. An STM image of the clean 5f surface, with probably high concentration of Al atoms could be useful.

And finally, for the distribution of the planar images of the atomic positions into different planes orthogonal to the c (5f) axis within the period, a STEM image in the 2f direction is needed.

4. References

[1] Burkov S E 1993 *Phys. Rev. B* **47** 12325
[2] Hiraga K, Sun W, and Lincoln F J 1991 *Jpn. J. Appl. Phys.* **30** L302
[3] He L X, Wu Y K, Kuo K H 1988 *Journal of Material science letters* **7** 1284
[4] Steurer W and Kuo K H 1990 *Acta Crystallogr. Sect. B* **46** 703
[5] Steurer W and Kuo K H 1991 *J. Phys.* **C3** 3397
[6] Kramer P, Quandt A, Schlottmann M, and Schneider T 1995 *Phys. Rev. B* **51** 8815
[7] Taniguchi S and Abe E 2008 *Philos. Mag.* **88** 1949
[8] Baake M, Kramer P, Schlottmann M, and Zeidler D 1990 *Int. J. Mod. Phys. B* **4** 2217
[9] Papadopolos Z, Kasner G, Ledieu J, Cox E J, Richardson N V, Chen Q, Diehl R D, Lograso T A, Ross A R, McGrath R 2002 *Phys. Rev. B* **66** 184207

Impurity phases in icosahedral Ag-In-Yb quasicrystal: influence in surface structure

H R Sharma[1], P J Nugent[1], J A Smerdon[1], M Shimoda[2], S Ohhashi[3], V Fournée[4], J Ledieu[4] and A. P. Tsai[3]

[1]Surface Science Research Centre and Department of Physics, the University of Liverpool, Liverpool L69 3BX, UK

[2]National Institute for Materials Science, 1-2-1 Sengen, Tsukuba, Ibaraki, 305-0047, Japan

[3]Institute of Multidisciplinary Research for Advanced Materials, Tohoku University, Sendai, 980-8577, Japan

[4]Institut Jean Lamour (UMR7198 CNRS-Nancy-Université-UPV-Metz), Ecole des Mines, Parc de Saurupt, 54042 Nancy Cedex, France

E-mail: H.R.Sharma@liv.ac.uk

Abstract. We present scanning tunnelling microscopy (STM) studies of the fivefold surface of icosahedral (*i*) Ag-In-Yb quasicrystals grown under different conditions. The sample grown at a slower rate is found to exhibit impurity phases on the surface, whereas a faster growth rate yields a sample with a structurally perfect surface.

1. Introduction

Recent success in growing single grain samples of the icosahedral (*i*) Ag-In-Yb quasicrystal has opened up an opportunity to extend ultra high vacuum (UHV) surface studies beyond Al-based quasicrystals. These quasicrystals are isostructural to the binary *i*-Cd-Yb quasicrystal. The *i*-Cd-Yb quasicrystal is not suitable for surface studies under UHV because Cd evaporates upon heat treatments due to its high vapour pressure. Therefore, Cd is replaced by equal percentages of Ag and In that fulfils the Hume-Rothery criteria of the valence electron to atom ratio, e/a = 2.0 [1, 2].

We recently reported scanning tunnelling microscopy (STM) and low energy electron diffraction (LEED) studies of the fivefold surface of *i*-$Ag_{42}In_{42}Yb_{16}$ that show a step-terrace structure with quasicrystalline long range order [3, 4]. By comparing the terrace structure and step height distribution with the bulk structure model of *i*-Cd-Yb, it was found that the terraces are formed at bulk planes intersecting the centre of the rhombic triacontahedral clusters, the building blocks of the system [5]. However, due to the unavailability of the bulk model structure of *i*-Ag-In-Yb, it has not been possible to determine the exact distribution of Ag and In atoms.

Here, we report extended studies of the *i*-$Ag_{42}In_{42}Yb_{16}$ quasicrystal. We studied surfaces of two samples synthesized with different growth rates [2, 6]. One sample (hereafter referred to as Sample 1) was grown with a rate of 0.2 mm h^{-1} [2], while another (Sample 2) was at 0.8 mm h^{-1} [6]. Using scanning tunnelling microscopy, we identified phases with periodic structures on the surface of the sample grown at slower rate, whereas the other sample (grown at the faster rate) exhibited a comparatively higher degree of perfection.

2. Experiment

Both samples were grown using the Bridgman method. Details of growth conditions are reported in Ref [2, 6]. The orientation of the fivefold axis was determined using Laue backscattering. Laue patterns confirmed the fivefold diffraction from various areas of the sample [2]. The surfaces of both samples were prepared with the same process. Samples were cut perpendicular to the fivefold axis and polished down to 0.25 μm diamond paste. Subsequently, the surface was prepared in UHV chamber (base pressure 2 $\times$ 10^{-10} mbar by Ar^{+} sputtering (1-3 keV, 30-60 minutes) and annealing (at 715 K for 2-3 hours). The temperature was measured using an optical pyrometer with emissivity set at 0.35. An Omicron STM was employed to image the surface at room temperature.

3. Results and discussion

We first explain results from Sample 1 (the slower-grown sample). The surface was mirror-like after polishing. After annealing at about 400 ^{o}C, the sample develops different domains which were visible to the naked eye. Using a digital optical microscope, we observed at least six domains of various sizes within the sample area of 0.5 cm^2 (Figure 1). A homogenous contrast was observed in the optical microscope after the sample was gently polished. However, the domains routinely reappeared after annealing. This can be explained if atomic vacancies in the bulk diffuse to the surface during annealing and condense to form voids [7] affecting the reflectivity of the surface. The observation of different domains suggests a non-uniform distribution of voids in different domains. We investigated different fivefold samples cut from the same ingot. All sample yielded domains upon annealing.

We employed an electron probe micro-analyzer to determine the composition of four large domains. Two points were analyzed at each domain. The composition of the domains was found to be $Ag_{42.2}In_{42.9}Yb_{14.9}$, $Ag_{41.7}In_{43.1}Yb_{15.2}$, $Ag_{41.7}In_{43.3}Yb_{14.9}$ and $Ag_{41.7}In_{43.4}Yb_{14.9}$. These values are close to the composition of the icosahedral phase. However, the grain boundary showed different compositions. The composition at the boundary of two randomly selected domains was $Ag_{40}In_{38.2}Yb_{21.8}$.

Although the domains had the same composition as the icosahedral phase, only two domains were found to yield quasicrystalline LEED patters and STM images with the step-terrace structure expected from the bulk. The results presented in our previous publication were from one of these domains [3]. Another quasicrystalline domain is rotated slightly with respect to the primary quasicrystalline domain, resulting in a vicinal surface. Remaining domains were structurally poor to the degree that LEED patterns were very weak or absent. We identified local periodic structures on limited areas in these domains.

The vicinal domain was identified using LEED. The specular (reflected) beam from this domain was shifted from the specular position of the primary quasicrystalline domain. As expected for a vicinal surface, STM from this domain shows a high density of steps. Since the sample was polished uniformly using a rotating polishing pad, we rule out the possibility that the polishing treatment resulted in the vicinal surface. This is therefore likely to have originated during the sample growth or during post annealing treatments.

We identified various periodic structures in other domains. Figure 2(a,b) shows two different types of periodic structures imaged on the same domain. Periodic rows can be identified on the as-measured STM images (Figure 2 (a, b)). In order to extract further information, we analyzed Fourier transforms (FT), autocorrelation functions (AF) and Fourier pass filters (FF) of the STM images. The Fourier transforms (Figure 2 (e, f)) and the autocorrelation function (Figure 2 (c)) show sharp maxima revealing a periodic structure with excellent long range order within the experimental length scale. The autocorrelation function of the image reveals a rectangular unit cell of 2.14 nm $\times$ 3.08 nm size (Figure 2(c)). A slight distortion from a perfect rectangle is attributed to thermal drift present during the STM measurements. Using a Fourier pass-filter,

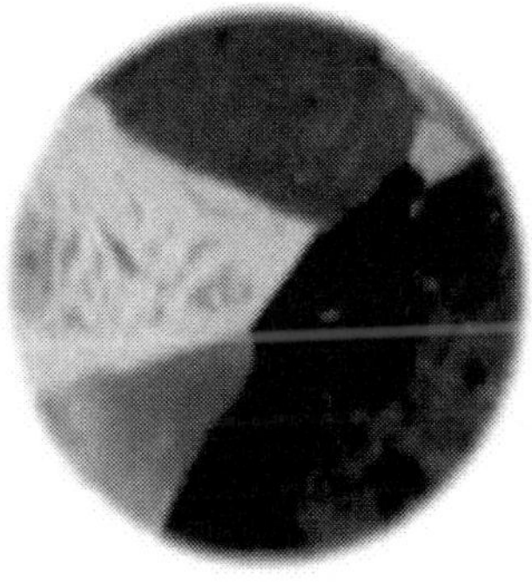

Figure 1. Ag-In-Yb sample demonstrating grain boundaries after annealing at 400 ^{o}C. Area of the sample is about 0.50 cm^2. The edge of the picture was removed to make a circular appearance.

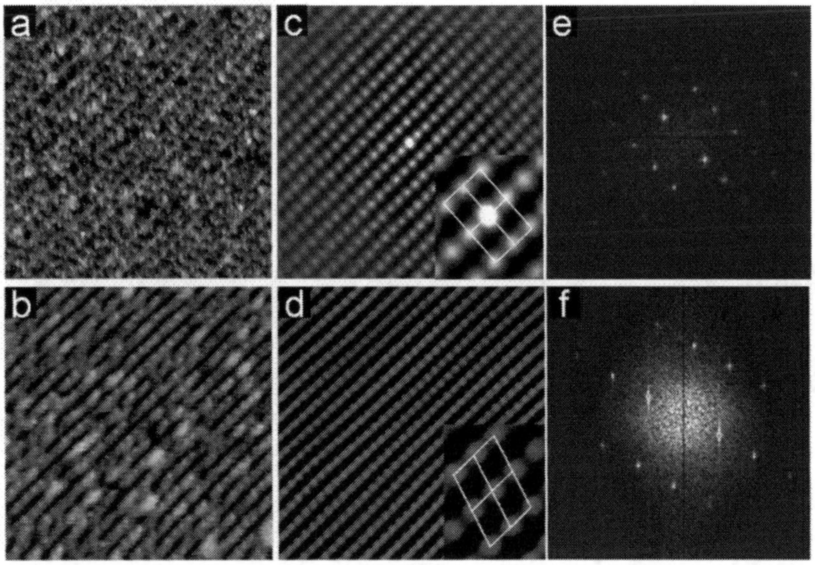

Figure 2. (a, b) STM images taken from a domain showing two different periodic structures (a: 50 nm × 50 nm, b: 40 nm × 40 nm). (c) Autocorrelation function of image 'a'. Inset: magnified image showing unit cell. (d) Image 'b' after Fourier pass-filtering. Inset: magnified image showing unit cell. (e-f) Fourier transform of images 'a' and 'b' respectively.

an oblique unit mesh (1.50 nm × 2.10 nm) has been measured on the STM image presented in Figure 2b. In addition to these periodic structures, we identified a periodic structure with an oblique unit cell (1.50 nm × 4.1 nm) on the terraces which were separated by steps of 1.7 nm. The step-height is related to the periodicity along the surface normal. We also observed weak LEED patterns from periodic domains. However, due to the low quality of the LEED patterns, it was not possible to extract the surface unit cell parameters accurately.

Occasionally, periodic microstructures are found to coexist with quasicrystalline structures. Example of such structures is given in Figure 3. These images were obtained from the primary quasicrystalline domain. Flat terraces are separated by steps of about 0.5 nm height. The Fourier

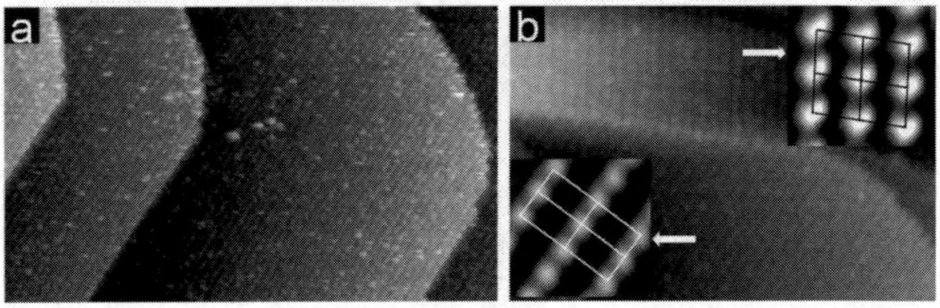

Figure 3. (a-b) STM images collected from a limited area of the primary quasicrystalline domain (160 nm × 90 nm). (b) Coexisting structures of two different periodicities (97 nm × 65 nm). The different structures on lower and upper parts are highlighted by Fourier pass-filtering (insets). The lower part of image 'b' is similar to that observed on the terraces of image 'a'.

pass-filtering analysis reveals rectangular unit cells of 2.47 nm × 1.3 nm size on the terraces. This structure coexists with another periodic structure of oblique unit mesh with lattice parameters equal to 2.7 nm × 2.3 nm. A clear interface of these two structures is observed (Figure 3 (b)). The coexistence of different microstructures was observed also in Al-Pd-Mn grown by the self-flux method [8].

The observed periodicities do not seem to have a straightforward relationship with bulk terminations of the known phases of Ag-In-Yb. In the phase diagram of Ag-In-Yb, the 1/1 and 2/1 approximants are found in the vicinity of the icosahedral phase [2, 6]. They have cubic structures with lattice constant a = 1.57 nm and 2.53 nm, respectively. Other known phases are $AgIn_2$ (tetragonal: a = 0.688 nm, c = 0.562 nm), In_4Ag_9 (cubic: a = 0.992 nm), $InAg_3$ (hexagonal: a = 0.295 nm, c = 0.786 nm), Ag_3In (hexagonal: a = 0.299 nm, c = 0.480 nm) and $YbAg_{5.4}In_{6.6}$ (tetragonal: a = 0.973 nm, c = 0.571 nm) [9].

The observed step height of 0.50 nm is consistent with the periodicity along the [110] direction of the 1/1 approximant. The interplanar distance of a simple cubic lattice of lattice constant 'a' along the [110] direction is $a/\sqrt{2}$. The step height of 0.50 nm is close to the half of $a/\sqrt{2}$ = 0.53 nm for the 1/1 approximant. However, analysis of the bulk model structure of the 1/1 approximant [10] rules out the possibility that the observed step-terrace structure is related to bulk terminations of the 1/1 approximant structure along the high symmetry directions. The bulk truncations at high density planes of the 1/1 approximant along the [110] direction would yield a step-height distribution different to that observed. Our recent STM results on the 1/1 approximant surface perpendicular to the [001] axis [11], which is expected to be the most stable surface, are not also consistent with observed these STM images.

Similarly, no correlation is found between the observed periodic structures and the periodicities of bulk terminations of known phases along high symmetry directions. Therefore, we speculate that the observed structures may correspond to the surface projection of the known phases along non-high symmetry directions or to the unknown phases of Ag-In-Yb. However, we cannot rule out the possibility of surface reconstruction. With the available STM images and the lack of diffraction patterns, it was not possible to confirm this point.

In contrast to Sample 1, Sample 2 is found to be structurally perfect. No domain boundaries were observed in the sample of about 1 cm^2 area even after annealing at 440 ^oC for several hours. STM images and quasicrystalline LEED patterns could be obtained from all areas of the sample. STM images and LEED patterns were similar to those observed from the primary quasicrystalline domain on Sample 1. We studied the twofold and threefold surfaces of the same system cut from the same ingot and found no multi domains after annealing at the same temperature. The size of the twofold and threefold samples was similar to that of the fivefold sample. We also note that samples cut from different parts across the same ingot were investigated in different laboratories. All samples were found to have the same structural quality.

4. Summary

We have used scanning tunnelling microscopy to investigate the surfaces of *i*-Ag-In-Yb quasicrystals synthesized under different conditions. The sample grown at a slower rate yielded various differing structural phases on the surface. The surface of the sample grown at a faster rate is found to be structurally perfect. A systematic study should be performed to establish the possible relationship that exists between the growth conditions and the terminating surface layers. With high quality samples nowadays available, it has been possible to prepare state-of-the art surfaces. These latter are being currently analysed using various surface science techniques including ultraviolet photoemission spectroscopy and medium energy ion scattering. The progress will be reported elsewhere.

Acknowledgement

This work was supported by Engineering and Physical Sciences Research Council (Grant no EP/D071828/1 and EP/DO5253X/1). The European Network of Excellence on "Complex Metallic Alloys", Contract No. NMP3-CT-2005-500145, and the Agence Nationale de la Recherche, Reference No. ANR-07-Blan-0270, are acknowledged for their financial support.

References

[1] Guo J Q and Tsai A P 2000 *Phil. Mag. Lett.* **82** 349
[2] Ohhashi S, Hasegawa J, Takeuchi S and Tsai A P 2007 *Phil. Mag.* **87** 3089
[3] Sharma H R, Shimoda M, Sagisaka K, Takakura H, Smerdon J A, Nugent P J, McGrath R, Fujita D, Ohhashi S and Tsai A P 2009 *Phys. Rev. B* **80** 121401 (R)
[4] Sharma H R, Shimoda M, Ohhashi S and Tsai A P 2007 *Phil. Mag.* **87** 2989
[5] Takakura H, Gomez C P, Yamamoto A, Boissieu M D and Tsai A P 2007 *Nature Mater.* **6** 58
[6] Cui C and Tsai A P 2010 *Journal of Crystal Growth* **312** 131
[7] Ebert Ph, Yurechko M, Kluge F, Cai T, Grushko B, Thiel P A and Urban K 2003 *Phys. Rev. B* **67** 024208
[8] Fournée V, Ross A R, Lograsso T A, Anderegg J W, Dong C, Kramer M, Fisher I R, Canfield P C and Thiel P A 2002 *Phys. Rev. B* **66** 165423
[9] Crystallographic database of the AgInYb, PDF2, National Institute for Materials Science, Tsukuba, Japan
[10] Takakura H, Gomez C P, Yamamoto A, Boissieu M D and Tsai A P *to be published*
[11] Cui C, Shimoda M, Tsai A P *et al, to be published*

Journal of Physics: Conference Series **226** (2010) 012005 doi:10.1088/1742-6596/226/1/012005

Structural model of quasiperiodic Pb monolayer deposited on fivefold i-Al-Pd-Mn surface

M. Krajčí[1,2], J. Hafner[2], J. Ledieu[3], and V. Fournée[3]

[1] Institute of Physics, Slovak Academy of Sciences, Dúbravská cesta 9, SK-84511 Bratislava, Slovak Republic

[2] Fakultät für Physik and Center for Computational Materials Science, Universität Wien, Sensengasse 8/12, A-1090 Wien, Austria

[3] Institut Jean Lamour (UMR7198 CNRS-Nancy-Université-UPVM), Ecole des Mines, Parc de Saurupt, 54042 Nancy Cedex, France

E-mail: `marian.krajci@savba.sk`

Abstract.
On the basis of analysis of experimental STM images and *ab-initio* calculations we propose a structural model of the quasiperiodic Pb monolayer grown on the fivefold i-Al-Pd-Mn surface at a coverage close to the saturation. The skeleton of the Pb monolayer can be seen as a network of the "star-fish" (SF) clusters. The atomic structure of the monolayer is based on a decorated P1 tiling. The model can reproduce also the experimentally observed quasiperiodic τ-scaled P1 ordering (τ is the golden mean). The bright spots seen in STM images appear at Pb atoms in the centers of those SF clusters where the substrate has Al atoms at these positions.

1. Introduction

Adsorbates on quasicrystalline surfaces can potentially create novel forms of matter with interesting physico-chemical properties. Much effort has been devoted to grow single-element adlayers on quasicrystalline surfaces [1–7]. Pb is one of few elements that adopts a pseudomorphic structure when deposited on quasicrystalline substrates [5, 7]. In previous work [5] a quasiperiodic Pb monolayer on the fivefold i-Al-Pd-Mn surface has been successfully prepared and characterized by the scanning tunneling microscopy (STM) and other experimental techniques. The initial adsorption of Pb on the five-fold Al-Pd-Mn surface has been investigated both experimentally and by ab-initio calculations [7]. It was found that Pb monolayer nucleates in pentagonal islands dubbed as starfish clusters. The atomic structure of these clusters has been resolved and the calculations confirmed their structural stability. In this contribution we go a step further and propose a structural model of the whole monolayer.

The atomic arrangement in an adlayer is a result of a complex interplay between the structure of the support, the adatom/substrate interactions, and the lateral interaction between adatoms in the adlayer. Understanding the structure of the substrate is a prerequisite for understanding the structure of the adlayer. STM studies of the clean five-fold Al-Pd-Mn surface in images with atomic resolution have identified two characteristic motifs, the pentagonal "dark stars" (DS) [8, 9] and the "white flower" (WF) [10, 11]. WF motif has been assigned to truncated pseudo-Mackay clusters which are, together with Bergman clusters, the constituting elements of the bulk icosahedral structure. This assignment has been confirmed by detailed simulations of the STM contrast based on *ab-initio* density-functional calculations [12]. These simulations also showed that DS corresponds to a surface vacancy originating from the irregular atomic arrangement around the low-coordinated Mn atoms in the center of the pseudo-Mackay clusters.

The quasiperiodic arrangement of the atoms at the surface can be described by a planar tiling. The choice of this tiling is, however, not unique. One possible choice relating to the cluster structure of the bulk quasicrystal is a P1 tiling [13] consisting of regular pentagons, pentagonal stars, boats and thin golden rhombi. The positions of centers of the Bergman clusters projected to the surface plane coincide with the vertices of the P1 tiling, the centers of the pentagonal tiles correspond to the positions of the pseudo-Mackay clusters. Other choice of the quasiperiodic tiling is a decagonal DHBS tiling which naturally describes ordering of atoms in the top surface plane. This choice will be discussed more in detail in Sec. 3.

2. Experimental STM studies

The growth of a Pb monolayer on the fivefold surface of the Al-Pd-Mn quasicrystal has been investigated in detail in previous studies [5,7]. At the early stages of the deposition, Pb adatoms diffuse across terraces and form nano-sized islands. Because of their protruding "arms" and their overall pentagonal shape, these clusters were dubbed "starfish" (SF). These clusters exhibit a similar size and orientation across successive terraces independently of the deposition flux used. These observations suggest an heterogeneous nucleation, hence a preferred nucleation site. The local atomic arrangements around single Pb atoms (i.e. at the first stage of the SF nucleation) and around completed SF motifs have been investigated on several atomically resolved STM images. It appears that these pentagonal islands are located on equatorially truncated pseudo-Mackay cluster surrounded by five truncated Bergman clusters, i.e. on top of WF motifs. This conclusion is supported by additional measurements of the SF-to-SF distances across terraces [7].

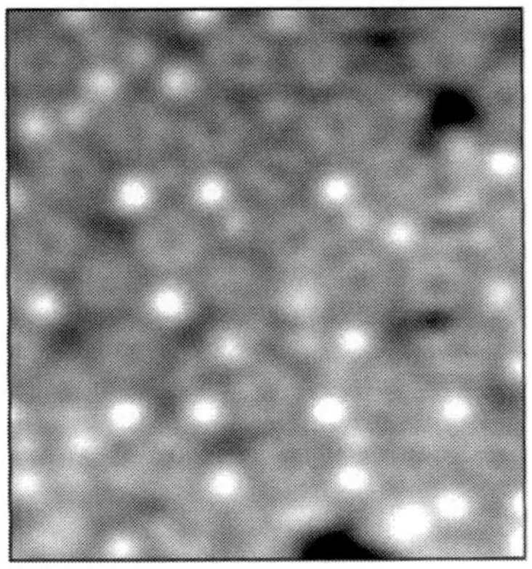

Figure 1. 10 nm x 10 nm High resolution STM image of a Pb monolayer deposited on the fivefold surface of the i-Al-Pd-Mn quasicrystal at 653 K.

Upon further deposition in the sub-monolayer coverage regime, the largest distinguishable pentagonal features can be understood as a set of SF clusters. Similarly, the skeleton of the complete Pb monolayer can be seen as a self-assembled network of the SF clusters. As the size of the SF cluster can be characterized by a pentagon with the edge of the P1 tiling (7.76 Å), it is then possible to assume that the atomic structure of the monolayer is a decorated P1 tiling. However, the quasiperiodic structure of the Pb monolayer as observed by STM is characterized by a τP1 tiling, which is the P1 tiling with the edge τ inflated compared to the structure of the substrate, τ=1.6018 .... is the golden mean. Hence, the atomic decorations of the τP1 and P1 tilings are still to be determined. An inspection of the STM images also reveals a rather large degree of structural disorder in the monolayer (see Fig.1). Although a part of the disorder could have a thermal origin, a significant contribution should come from an imperfect match between the local structures of the substrate and the adlayer.

3. Modeling of the structure of Pb monolayer based on *ab-initio* DFT calculations

The theoretical modeling of the structure of the Pb monolayer is based on density-functional theory (DFT). We have used the Vienna Ab-initio simulation Package VASP [14,15] to perform ab-initio electronic structure calculations, structural optimizations and simulation of the STM images. More details of the method can be found in our previous papers [7,16,17]. A model of the surface has been derived from the Katz-Gratias-Boudard model [18,19] of bulk i-Al-Pd-Mn. The atomic structure of the fivefold surface is derived from the structure of an icosahedral approximant by cleaving at a plane perpendicular to one of the fivefold axes. Details of construction of models of the bulk and the surface can be found in our previous papers [12,17,19].

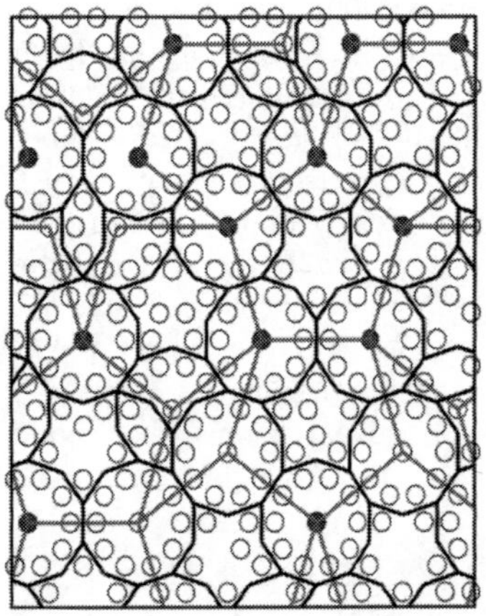

Figure 2. The atomic structure of the top atomic plane of the five-fold Al-Pd-Mn surface. The quasiperiodic arrangement of atoms in this plane can be most naturally described by a decagonal DHBS tiling (black). The centers of the decagonal tiles occupied either by Mn (closed circles) or Al (open circles) atoms coincide with the the vertices of the τP1 tiling (red, gray).

Fig. 2 is the atomic structure of the top atomic plane of the five-fold Al-Pd-Mn surface. The model of the surface has been constructed from the 5/3-approximant to the bulk quasicrystal. The quasiperiodic arrangement of atoms in the top plane can be most naturally described by a planar DHBS tiling composed of decagons (D), squashed hexagons (H), boat-shaped tiles (B) and pentagonal stars (S) [17]. The top plane is Al-rich and contains a few Mn, but no Pd atoms. Such chemical composition of the five-fold Al-Pd-Mn surface has been confirmed also experimentally [20]. The D tiles can be occupied either by Mn or by Al atoms. The decagonal tiles have a direct relation to the WF features seen in the STM images of the clean surfaces. In the topmost plane the vertices of the D tiles coincide with the centers of small pentagons of Al atoms (with an edge-length of 2.96 Å) forming "leaves" of the white flowers. The centers of WF's represented by the D tiles are the nucleation centers for the Pb starfish clusters. The pentagonal-star (S) tiles in Fig. 2 have two orientations. They are decorated by five Al atoms located in the star arms, plus 0 to 2 other Al atoms in the interior of the tile. The irregular decoration of the S tiles is a consequence of the low and irregular coordination of Mn atoms in the centers of the pseudo-Mackay clusters, located below or above the surface. The S tiles with empty interiors are seen in STM images of the clean surface as DS's. Fig. 2 shows also the τP1 tiling and its relation to the DHBS. The centers of the D tiles coincide with the the vertices of the τP1 tiling. The τP1 tiling with the edge of 12.55 Å describes the quasiperiodic ordering of the Pb monolayer.

Our previous studies [7] of initial Pb adsorption have been performed on a slab model of the fivefold surface derived from the 2/1 approximant to the quasicrystalline Al-Pd-Mn structure.

However, the 2/1 approximant is too small for modeling a quasiperiodic ordering at the scale of the τP1 tiling. In the present study we use a model of the surface derived from the 3/2-approximant. The model has a shape of a slab with size of 32.86 Å × 38.63 Å and thickness of 4.2 Å. Although this small thickness (5 atomic planes only) is not sufficient to achieve converged binding energies for the adsorbants for simulation of STM images (i.e. reproducing the charge density distribution several Å above the surface) it was found to be thick enough. Positions of all atoms in the slab, except the atoms in two bottom atomic planes were relaxed together with the positions of the adsorbed atoms. Neighboring images of the slab are separated by a 14 Å thick vacuum layer. The computational cell has an orthorhombic shape, it contains 350 atoms of the substrate plus up to 106 atoms of the adsorbed monolayer.

The geometry of one SF cluster can be characterized by a pentagon of the P1 tiling. The P1 tiling consists of two types of pentagons differing in orientation. We distinguish them as pentagons pointing "up" and "down". We shall denote them as uP and dP tiles, respectively. The orientation of the SF cluster corresponds the uP pentagon. The uP tiles are concentric with the D tiles of the DHBS tiling. Some of the dP pentagons correspond to the surface vacancies. The surface vacancies act as trap sites binding ad-atoms very strongly. We assume that these sites are occupied in very early stages of the deposition. Simulation of STM images have shown that a single Pb atom binded inside the surface vacancy does not give any special contrast. Above the central Pb atom filling the vacancy the dP tiles can accommodate 5 Pb atoms forming a regular pentagon. This pentagon can adopt one of two possible orientations discussed below. The decoration of other tiles is to a large extent enforced by the decoration of the pentagonal tiles.

We have considered many possible configurations of adsorbed Pb atoms. The stability of the adsorbed Pb layer was probed by performing a conjugate-gradient relaxation of the structure of the adlayer/substrate complex under the action of the Hellmann-Feynman forces. Because of the limited extent of this paper we present a result for one configuration only representing a structure of a densely packed Pb monolayer with the coverage of 0.083 atoms/Å^2 ($\Theta \approx 0.63$). We note that the experimentally measured coverage of 1 ML at the full saturation is ≈0.09 atoms/Å^2.

A structural model of the Pb monolayer adsorbed on the surface at the coverage of 0.083 atoms/Å^2 is presented in Fig. 3. One can observe that Pb atoms occupy vertices and mid-positions of the edges of the P1 tiling. The mid-edge positions of acute angles of the tiles cannot be occupied simultaneously. Although a Pb atom could occupy a bridge position between these mid-edge sites (more accurately - a vertex position of the DHBS tiling), in most cases it prefers one of the mid-edge sites in a close contact with other Pb atoms decorating interiors of the pentagons. Obviously, the lateral Pb-Pb interaction is stronger than the Pb-substrate interaction. This is the reason why the P1 tiling is more appropriate for describing the quasiperiodic structure of the monolayer than the DHBS tiling.

The internal decoration of the uP tiles corresponds to the SF clusters. At high coverage Pb atoms are adsorbed also at the hollow sites in the centers of the SF clusters. To obtain high packing it was necessary to decorate the dP tiles with Pb pentagons with the opposite orientation compared to those inside the uP pentagons. This orientation allows us to insert additional Pb atoms in mid-edge positions between two neighboring uP and dP pentagons. At a lower coverage an opposite orientation of the Pb pentagon inside the dP tile is also possible. In dependence on the coverage the pentagonal decoration of the dP tile can thus adopt either orientations.

Fig. 4 shows that in the STM image bright spots appear in the centers of some uP pentagons. The explanation of the appearance of the bright spots in STM images, see also Fig. 1, can be found in the structure of the substrate. At the clean surface in the centers of the D tiles (Fig. 2) or concentric uP tiles either a Mn or Al atom can occur. The bright spots appear in those uP tiles

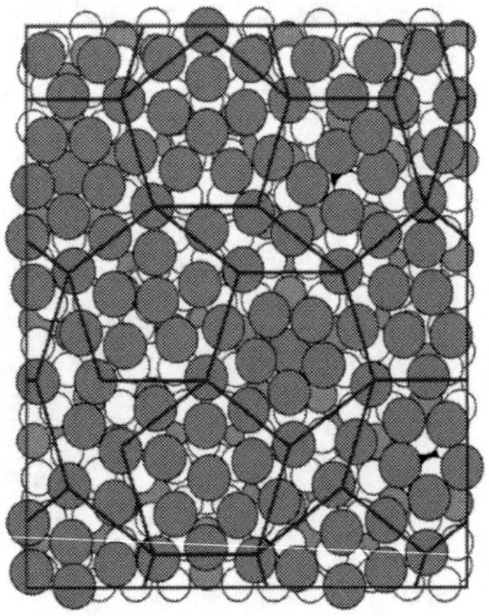

Figure 3. The structure of adsorbed Pb monolayer at the coverage of 0.083 atoms/Å^2. The quasiperiodic ordering is represented by the P1 (black lines).

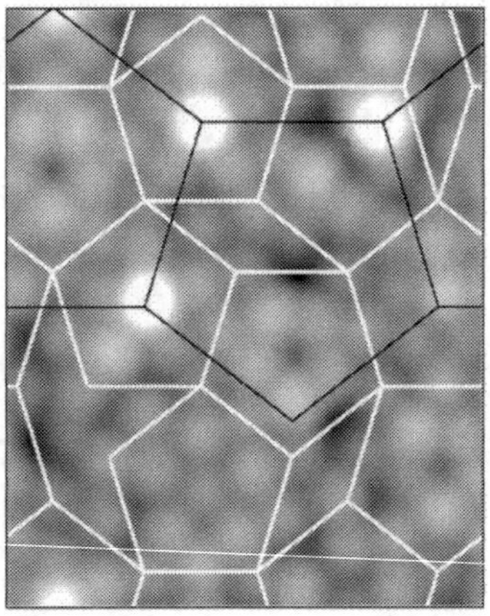

Figure 4. A simulated STM image of the Pb monolayer from the previous figure. The bright spots can form vertices of the τP1 tiling.

that have Al atoms at these positions. The uP tiles centered in the substrate by Mn sites give only weak contrast. The central Al atom is less strongly bonded to the neighboring Al atoms in the substrate than the Mn atom. After the structural relaxation of the adlayer/substrate complex while the central Mn atoms remain approximately in the surface plane the central Al atoms are lifted $\approx$0.6 Å above the surface plane pushing correspondingly up also the Pb atoms above. The distance between two neighboring bright spots is just the edge of the τP1 tiling. The bright spots can thus form vertices of the τP1 tiling as it has been experimentally observed, see Fig. 3 in [5]. The bright spots seen in STM images can also have an origin in defects. The regular decoration in some tiles can be substantially distorted. In Fig. 3 the regular pentagonal decoration of the dP tile at upper right corner of the figure was rearranged and one Pb atom was pushed up. A bright spot originating in an irregular arrangement of atoms in the monolayer can appear at any position. However, the observed defect formation can have also another interpretation. What is a decoration defect from the viewpoint of one tiling can be a regular site in another tiling. The formation of the defect observed in Fig. 4 can be interpreted as a reordering of the quasiperiodic arrangement of atoms expressed by the P1 tiling to another quasiperiodic arrangement at a larger scale described by the τP1 tiling. A similar reordering we observed also in alkali metals quasiperiodic monolayers [17] where the P1 quasiperiodic ordering was transformed to the DHBS ordering.

4. Conclusion

We have studied the structure of the quasiperiodic Pb monolayers grown on the fivefold i-Al-Pd-Mn surface. We have constructed structural models of the monolayer at a coverage close to the saturation. The skeleton of the Pb monolayer can be seen as a network of the SF clusters. The atomic structure of the monolayer is a decorated P1 tiling. The structural models reproduce also the quasiperiodic τP1 ordering observed also in the experimental STM images. The bright spots in the STM images appear in the centers of the uP (or D) tiles where the substrate has

Al atoms at these positions. The bright spots can originate also from irregular arrangements of atoms in the monolayer.

Acknowledgments

We acknowledge the European Network of Excellence on Complex Metallic Alloys (CMA) contract NMP3-CT-2005-500145 and the Agence Nationale de la Recherche, reference ANR-07-Blan-0270, are acknowledged for their financial support. The work has been supported by the Austrian Ministry for Education, Science and Art through the Center for Computational Materials Science. M. K. thanks also for support from the Grant Agency for Science of Slovakia (No. 2/5096/25) and from the Slovak Research and Development Agency (Grant No. APVV-0413-06, CEX-Nanosmart).

References

[1] Franke K J, Sharma H R, Theis W, Gille P, Ebert Ph, Rieder K H, 2002 *Phys. Rev. Lett.* **89** 156104
[2] Cai T, Ledieu J, McGrath R, Fournée V, Lograsso T A, Ross, Thiel P A 2003 *Surf. Sci.* **526** 115
[3] Ledieu J, Hoeft J T, Reid D E, Smerdon J A, Diehl R D, Lograsso T A, Ross A R, and McGrath R 2004 *Phys. Rev. Lett.* **92**, 135507
[4] Smerdon J A, Parle JK, Wearing L H, Lograsso T A, Ross A R, McGrath R 2008 *Phys. Rev.* B **78** 075407
[5] Ledieu J, Leung L, Wearing L H, McGrath R, Lograsso T A, Wu D, Fournée V 2008 *Phys. Rev.* B **77** 073409
[6] Shukla A K, Dhaka R S, D'Souza S W, Singh Sanjay, Wu D, Lograsso T A, Krajčí M, Hafner J, Horn K, Barman S R, 2009 *Phys. Rev.* B **79** 134206
[7] Ledieu J, Krajčí M, Hafner J, Leung L, Wearing L H, McGrath R, Lograsso T A, Wu D, Fournée V 2009 *Phys. Rev.* B **79** 165430
[8] Schaub T M, Bürgler D E, Güntherodt H -J, and Suck J -B 1994 *Phys. Rev. Lett.* **73**, 1255
[9] Ledieu J and R. McGrath R 2003 *J. Phys.: Condens. Matter* **15**, S3113
[10] Papadopolos Z, Kasner G, Ledieu J, Cox E J, Richardson N V, Chen Q, Diehl R D, Lograsso T A, Ross A R, and McGrath R 2002 *Phys. Rev. B* **66**, 184207
[11] Kasner G and Papadopolos Z 2006 *Phil. Mag.* **86**, 813
[12] Krajčí M, Hafner J, Ledieu J, McGrath R 2006 *Phys. Rev.* B **73** 024202
[13] Papadopolos Z, Pleasants P, Kasner G, Fournée V, Jenks C J, Ledieu J, and McGrath R 2004 *Phys. Rev.* B **69**, 224201
[14] Kresse G and J. Furthmüller J 1996 *Comp. Mat. Sci.* **6**, 15; *Phys. Rev.* B **54**, 11 169.
[15] Kresse G and Joubert D 1999 *Phys. Rev.* B **59**, 1758
[16] Krajčí M and Hafner J 2007 *Phys. Rev.* B **75** 224205
[17] Krajčí M and Hafner J 2008 *Phys. Rev.* B **77** 134202
[18] Boudard M, de Boissieu M, Janot C, Heger G, Beeli C, Nissen H -U, Vincent H, Ibberson R, Audier M, and Dubois J M 1992 *J. Phys.: Cond. Matter* **4**, 10149
[19] Krajčí M, Windisch M, Hafner J, Kresse G, and Mihalkovič M, *Phys. Rev.* B **51**, 17355
[20] Unal B, Jenks C J, Thiel P A 2008 *Phys. Rev.* B **77** 195419

Journal of Physics: Conference Series **226** (2010) 012006 doi:10.1088/1742-6596/226/1/012006

C_{60} adsorption on an aperiodically modulated Cu surface

J.A. Smerdon[1], J.K. Parle[1], L.H. Wearing[1], L. Leung[1], T.A. Lograsso[2], A.R. Ross[2], R. McGrath[1]

[1]Surface Science Research Centre and Department of Physics, The University of Liverpool, Liverpool L69 3BX, UK, [2]Ames Laboratory, Iowa State University, Ames, IA 50011, USA

E-mail: `mcgrath@liv.ac.uk`

Abstract. Copper deposited on the five-fold surface of icosahedral Al-Pd-Mn forms domains of a structure whose surface has a one-dimensional aperiodic modulation. It is shown that C_{60} deposited on this aperiodic film has highly reduced mobility as compared to C_{60} deposited on periodic Cu surfaces. This finding is explained in terms of the recently proposed structural model of this system.

1. Introduction

The interaction of adsorbing species with solid surfaces is of fundamental importance in epitaxial growth and nanostructure formation. The competition between adsorbate-adsorbate and adsorbate-substrate interactions leads to a wide spectrum of behaviour. When lattice misfit is minimal, epitaxial growth may occur; where misfit is appreciable the resulting interfacial strain may be accommodated pseudomorphically, by misfit dislocations, or by rotational epitaxy along high-symmetry directions of the substrate. Such effects have been extensively studied in simple model systems such as rare gas adsorbates on graphite [1] and alkali atoms on metal surfaces [2].

Since its discovery [3], C_{60} has also been employed as a model system in the study of ordering of two-dimensional molecular solids on surfaces. The interaction potential in solid C_{60} is composed of a long-range attractive van der Waals component and a repulsive short-range component, the counterbalance of which stabilises the molecular solid. On a surface, these interactions also compete with the adsorbate-substrate interactions and a wide range of behaviour is found, as the strength of this adsorbate-substrate interaction varies with the chemistry and geometry of the surface. On single crystal metal surfaces, behaviour ranges from epitaxial growth [4] to rotational epitaxy [5], to interfacial reconstruction with adlayer buckling and rotation [6]. On (111) semiconductor surfaces, epitaxial growth is also observed [7], while on anisotropic surfaces such as Si(100)(2×1) the C_{60} overlayer has been found to be uniaxially incommensurate along the the dimer row directions [8].

It is of great interest to extend such studies to the case of the interaction of C_{60} with surfaces which have non-periodic, through well-ordered geometries. On such surfaces, the possibility of the accommodation of lattice misfit through rotational epitaxy is removed, and hence the formation of incommensurate structures is unavoidable. The surfaces of quasicrystals fall into this category. Quasicrystals are metallic alloys whose surface structure can be thought of as a physical realization of a Penrose tiling, which is a two-dimensional quasiperiodic pattern [9].

With the constraint of periodicity removed, such materials often display five-fold and ten-fold rotational symmetries which are not observed in periodic solids. The surfaces of quasicrystals have been found to be essentially terminations of the bulk structure, without reconstruction and minimal relaxation [10–14].

Unfortunately, C_{60} does not form ordered overlayers on quasicrystal surfaces. On the five-fold surface of AlPdMn [15, 16] and the ten-fold surface of AlNiCo [17], strong chemical interaction with the substrate binds the molecules strongly, inhibiting diffusion and leading to the formation of disordered overlayers. However, studies are emerging in which C_{60} is not deposited directly on to the clean quasicrystalline surface, but instead upon a quasicrystalline interfacial layer such as Pb/Al-Ni-Co, one purpose of which is to tune the reactivity of the quasicrystalline system [18].

It has previously been found that Cu, Ni, Fe and Co, when deposited on the five- and ten-fold surfaces of Al-based quasicrystals, tend to adopt the rotational epitaxial growth mode [19–27], in which crystalline islands grow in various azimuthal orientations matching the rotational symmetry of the underlying quasicrystal. For Cu and Co, further interesting behaviour is observed in a modulation of the islands to exhibit some aperiodic ordering at their surfaces. The Cu/AlPdMn system has been well-studied using scanning tunneling microscopy (STM) [19], low-energy electron diffraction (LEED) [20], medium-energy ion-scattering spectroscopy (MEIS) [21] and dynamical LEED (LEED-IV) [22,28]. Upon adsorption on the five-fold surface of icosahedral Al-Pd-Mn, no ordering is observable in a Cu film up to 3 ML by STM or LEED. As coverage increases beyond this value, pentagonal angles in domain edges become apparent, as does a row structure (manifested as a z-modulation) on the surface of the film. The rows are spaced according to a one-dimensional Fibonacci sequence with separations $S = 4.6$ Å and $L = 7.4$ Å as measured by STM. The LEED pattern reappears concurrently with observable ordering in STM images, and is composed of a pattern based on a periodic Cu-Cu separation along rows and a continuous streaking due to diffraction from the 1D Fibonacci sequence, in 5 azimuthal orientations arising from the alignment with high-symmetry axes on the quasicrystal surface.

Smerdon *et al.* proposed a structure for the thin film based on MEIS results that identified the crystalline phase making up the rows as fcc Cu with the (100) face parallel to the surface [21]. This picture of the film does not explain why the row structure is observable by STM, nor why there is no ordering below 3 ML coverage. However, analysis of recent LEED-IV results [28] suggests that the phase is in fact body-centred tetragonal (bct) Cu that grows at an angle to the surface normal. This model provides an explanation of both observations mentioned above.

In this work, we explore the differences between C_{60} adsorption on single crystal ideal Cu surfaces from existing studies and C_{60} adsorption on the quasiperiodically arranged form of Cu.

2. Experimental conditions

The adsorption studies were carried out in an Omicron variable temperature STM (VT-STM) UHV chamber. The base pressure of the system was 1×10^{-10} mbar. The $Al_{70}Pd_{21}Mn_9$ quasicrystal sample, produced at Ames laboratory using the Bridgman method, was polished successively with 6 μm, 1 μm and 1/4 μm diamond paste before introduction to vacuum and thereafter was prepared in cycles consisting of 45 minutes sputtering with 3 keV Ar^+ ions followed by at least 4 hours annealing to 950 K, using electron-beam heating, up to a total annealing time of 20 hours.

All evaporation of adsorbates was carried out with the sample at or close to room temperature. The Cu evaporation was performed using a simple filament source consisting of a W wire wrapped around a piece of OFHC Cu. The C_{60} evaporation was performed using a thin W filament wrapped around a Pyrex crucible containing C_{60}, with temperature regulated using a K-type thermocouple tightly strapped to the crucible. During evaporation the W filament glowed yellow and was approximately 3 cm away from the sample. Evaporating temperatures in the range 500 K - 520 K were used and the chamber pressure did not exceed 4×10^{-10} mbar during evaporation.

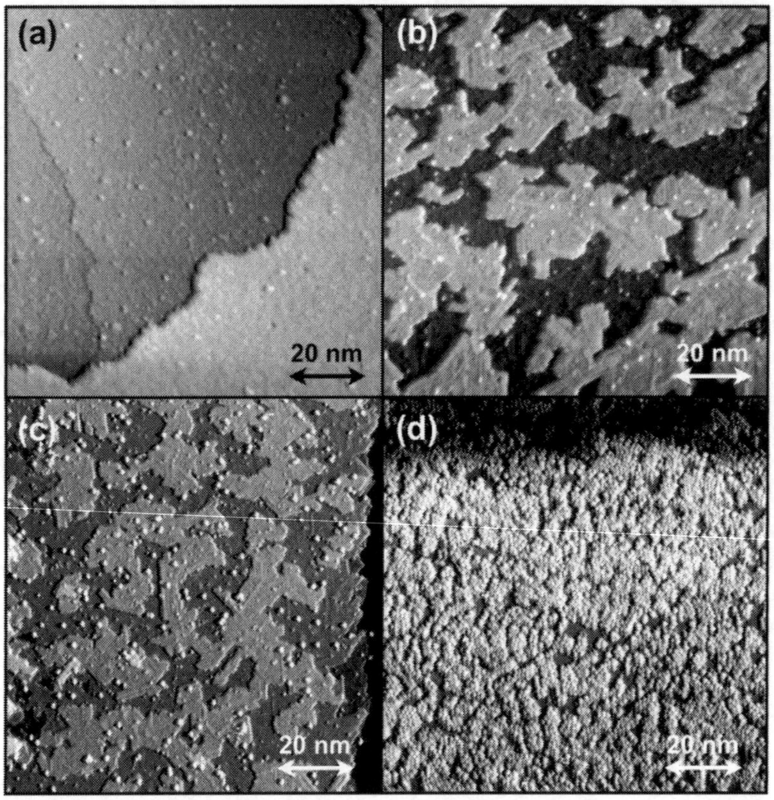

Figure 1. (a); A 100 nm×100 nm image of a clean Al-Pd-Mn surface, (b); A 100 nm×100 nm image of the 4.5 ± 0.1 ML Cu layer deposited on the Al-Pd-Mn surface. The lighter rows have previously been identified as strain-related defects [19], the white spots are contamination, (c); 0.025 ML C_{60} adsorbed atop the Cu layer, showing an absence of island formation, (d); 0.8 ML C_{60} adsorbed atop the Cu layer, showing some evidence of orientation of rows of C_{60}.

The sample was maintained at room temperature throughout the experiment.

3. Results and discussion

Fig. 1 shows the starting surface. Following the deposition of 4.5 ± 0.1 ML Cu, the previously reported [19] one-dimensionally aperiodic structure is observed via STM as shown in figure 1(b). The Cu film on this substrate is highly complex, and has only recently been adequately described [28]. The film structure is suggested to be a collection of orthorhombic domains oriented off-normal to the substrate such that the bct(100) terraces are cut according to the substrate Fibonacci sequence, producing a similar one-dimensional sequence on the film surface parallel to the substrate/film interface. This suggests that there are potentially a considerable amount of inequivalent adsorption sites of high coordination at the surface of this film.

Following formation of the Cu film, C_{60} molecules were deposited, up to a maximum coverage of 0.8 ± 0.1 ML coverage. At lower coverages of 0.02 ± 0.005 ML (as shown in figure 1(c)), C_{60} molecules are observed to be of a constant height above the substrate of 0.6 ± 0.1 Å, independent of tunneling conditions. C_{60} molecules do not coalesce to form islands, leading us to conclude that the molecules are not mobile at room temperature on this surface.

Most molecules also exhibit intra-molecular features in STM, indicating that the molecules are not rotating freely on the film. This is observed in general with C_{60} adsorption on transition metal substrates [6,29–31]; however, there is then some consistency in the rotational alignment of the molecules corresponding to a particular adsorption site. We do not observe such consistency.

At higher coverages of between 0.5-0.8 ML, some C_{60} row structure becomes apparent in some parts of the film, as shown in figure 1(d). Adjacent rows are found to have a difference in height; the average z-difference as shown in fig. 2 is 1.3 Å, indicating that these higher rows cannot be a second layer of C_{60}.

These observations contrast with those for adsorption on periodic Cu substrates. The absence of C_{60} island formation at room temperature and low coverage is unusual contrary to that observed for C_{60} on single-crystal Cu(100), (110) and (111) surfaces [29, 31, 32]. In all these cases, although the molecules are tightly-bound enough to the substrate to adopt a particular orientation (or set of orientations) visible in terms of cage structure, the molecules may still diffuse to decorate step edges and form islands at sub-monolayer coverage.

This indicates that in the present system, diffusion of C_{60} molecules is inhibited. As the Cu film is much rougher than a single crystal surface, it might be expected that roughness is the inhibiting factor. Were the roughness parallel to the Cu rows comparable to the roughness of a flat terraced Cu surface, although diffusion *across* the rows would be inhibited, C_{60} molecules might be expected to diffuse more or less freely *along* them. In this case it is likely that the domain steps would be decorated first, similarly to the single-crystal Cu systems. This is not the case; with a thresholded Sobel edge-finding algorithm, $(26 \pm 5)\%$ of the surface is found to be within one C_{60} diameter of a step edge, and the population of C_{60} molecules located within one C_{60} diameter of a step edge is found to be $(30 \pm 10)\%$, so, within the measurement error, there are the same number of C_{60} molecules near step edges as there would be through random adsorption. In fact, the z-range measured using STM along the rows on tilt-corrected data is on average 30 pm, three times more than that measured in a similar fashion over an area of fcc(110) Cu. We assume that this inherent roughness must therefore be a sufficient barrier to diffusion such that, following impingement, molecules remain at the closest local potential energy minimum. This is a unique result for a transition metal surface, that may be explained by the model of the Cu film proposed by Pussi *et al.* [28]. As explained above, this model comprises domains of bct Cu oriented in five directions consistent with the substrate high-symmetry axes. The domains are offset parallel to the substrate surface, resulting in a cut plane that gives the Fibonacci sequence in the pattern of steps at the film surface. This means that the entire film surface is a pattern of step edges. It is expected that C_{60} molecules would be immobile once located at a step edge.

C_{60} molecules chemisorbed on transition metal substrates generally exhibit either a three-lobed feature or a ring feature parallel to the crystallographic plane of the surface [6, 29, 31]. The three-lobed feature indicates that the molecule is chemisorbed with hexagonal faces parallel to the surface, the ring feature indicates that pentagonal faces are parallel to the surface. This shows that, on these substrates, an axis of high symmetry (either three-fold or five-fold in the examples given, but two-fold axes are also possible) is oriented perpendicular to the surface crystallographic plane, resulting in rotational symmetry of individual molecules around the normal to the surface crystallographic plane. Such rotational symmetry of individual molecules is very rarely observed in this study, as can be seen in the Laplace filtered image on the right side of fig. 2. This evidence also supports the model that the Cu film structure is vicinal, with a primary crystallographic axis at an angle to the quasicrystal five-fold plane [28]s.

The row formations exhibited by C_{60} at higher coverage are reminiscent of those observed for adsorption on the Cu(100) and Ni(110) surfaces [31, 32]. Upon adsorption on Cu(100) and Ni(110) surfaces at room temperature, C_{60} causes reconstruction of the substrate surface, identifiable by the differing heights of C_{60} molecules above the surface observed. For example, a single row of C_{60} molecules may sit higher than a row either side. If the height difference

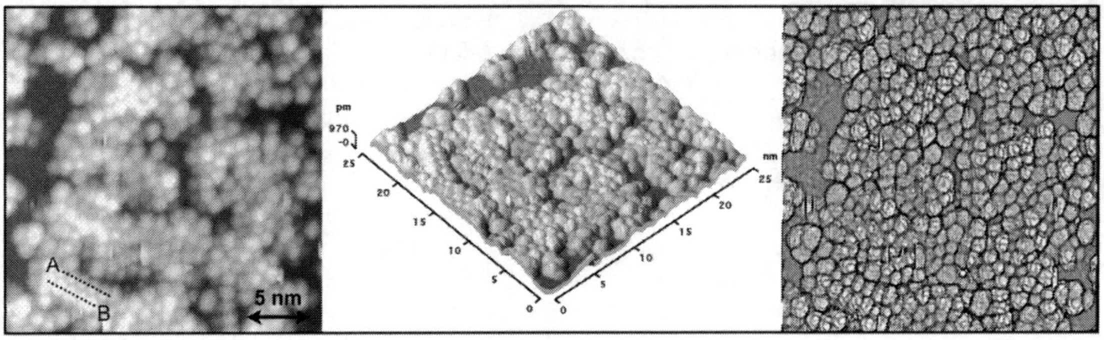

Figure 2. Left: A 25 nm×25 nm image of 0.8 ML C_{60} adsorbed atop a 4.5 ± 0.1 ML Cu film on the five-fold Al-Pd-Mn surface (fast-scan direction is vertical). The z-separation of profiles along A and B is on average 1.3 Å; Centre: The image rendered in 3 dimensions, showing the flatness of the film; Right: A Laplace filtered version of the image, showing intra-molecular features.

is incompatible with a C_{60} interplanar step height, then it must be ascribed to the underlying surface. In the case of C_{60}/Cu(100) [31], for each height of C_{60} within a C_{60} island, i.e., for each unique type of adsorption site in terms of substrate reconstruction, the C_{60} molecule adopts a particular orientation with respect to the substrate. To a degree, we observe such phenomena in the present system; for example, the row of molecules denoted B in fig. 2 sits 1.3 Å higher than its neighbours. The step height of a second layer of C_{60} is 8.2 Å [33] and the step height of the Cu film is 1.9 ± 0.1 Å [19]. The interlayer spacing of the bct structure is 0.68 Å [28], meaning that atomic rows could be removed from the Cu layer giving positions of C_{60} with a z-quantization of 0.68 Å. Hence a reconstruction is a plausible explanation for such observations, especially in the context of other studies of C_{60} on transition metals.

4. Conclusions

C_{60} molecules on a Cu film on the five-fold surface of icosahedral Al-Pd-Mn. C_{60} molecules show markedly different adsorption behaviour as compared to deposition on low-index single crystal Cu surfaces. The molecules have extremely limited mobility at room temperature and do not form islands or decorate step edges, instead remaining where they impinge on the surface. At higher coverages (above 0.5 ML C_{60}), there is some suggestion of reconstruction of the underlying Cu film in the observed differing heights of adjacent rows of consistently orientationally aligned C_{60} molecules. The features of adsorption of the C_{60} molecules can be interpreted in terms of the recent model proposed by Pussi *et al.* [28].

Acknowledgments

The EPSRC is acknowledged for funding this project under Grant No EP/D05253X/1.

References

[1] Novaco A and McTague J 1977 *Phys. Rev. Lett.* **38** 1286
[2] Diehl R D and McGrath R 1997 *J. Phys.: Condens. Matter* **9** 951–968
[3] Kroto H W, Heath J R, O'Brien S C, Curl R F and Smalley R E 1985 *Nature* **318** 162
[4] David T, Gimzewski J K, Purdie D, Reihl B and Schlittler R R 1994 *Phys. Rev. B* **50** 5810–5813
[5] Johansson M K J, Maxwell A J, Gray S M, Brühwiler P A, Mancini D C, Johansson L S O and Mårtensson N 1996 *Phys. Rev. B* **54** 13472–13475
[6] Pai W W, Hsu C L, Lin M C, Lin K C and Tang T B 2004 *Phys. Rev. B* **69**
[7] Xu H, Chen D M and Creager W N 1993 *Phys.Rev.Lett.* **70** 1850–1853

[8] Klyachko D and Chen D M 1995 *Phys. Rev. Lett.* **75** 3693–3696
[9] Shechtman D, Blech I, Gratias D and Cahn J W 1984 *Phys. Rev. Lett.* **53** 1951
[10] Diehl R D, Ledieu J, Ferralis N, Szmodis A W and McGrath R 2003 *J. Phys.: Condens. Matter* **15** R63–R81
[11] Papadopolos Z, Kasner G, Ledieu J, Cox E J, Richardson N V, Chen Q, Diehl R D, Lograsso T A, Ross A R and McGrath R 2002 *Phys. Rev. B* **66** 184207
[12] Papadopolos Z, Pleasants P, Kasner G, Fournée V, Jenks C J, Ledieu J and McGrath R 2004 *Phys. Rev. B* **69** 224201
[13] Sharma H R, Shimoda M and Tsai A P 2007 *Adv. Phys.* **56** 403–464
[14] Krajčí M, Hafner J, Ledieu J and McGrath R 2006 *Phys. Rev. B* **73** 024202
[15] Ledieu J, Muryn C A, Thornton G, Diehl R D, Lograsso T A, Delaney D W and McGrath R 2001 *Surf. Sci.* **472** 89–96
[16] McGrath R, Ledieu J, Cox E J, Haq S, Diehl R D, Jenks C J, Fisher I, Ross A R and Lograsso T A 2002 *J. Alloys Compounds* **342** 432–436
[17] Cox E J, Ledieu J, Dhanak V R, Barrett S D, Jenks C J, Fisher I and McGrath R 2004 *Surf. Sci.* **566-568** 1200–1205
[18] Smerdon J A, Leung L, Parle J K, Jenks C J, McGrath R, Fournée V and Ledieu J 2008 *Surf. Sci.* **602** 2496–2501
[19] Ledieu J, Hoeft J T, Reid D E, Smerdon J A, Diehl R D, Lograsso T A, Ross A R and McGrath R 2004 *Phys. Rev. Lett.* **92** 135507
[20] Ledieu J, Hoeft J T, Reid D E, Smerdon J A, Diehl R D, Ferralis N, Lograsso T A, Ross A R and McGrath R 2005 *Phys. Rev. B* **72** 035420
[21] Smerdon J A, Ledieu J, McGrath R, Noakes T C Q, Bailey P, Drexler M, McConville C F, Lograsso T A and Ross A R 2006 *Phys. Rev. B* **74** 035429
[22] Pussi K, Reid D E, Ferralis N, McGrath R, Lograsso T A, Ross A R and Diehl R D 2008 *Philos. Mag.* **88** 2103 – 2110
[23] Smerdon J A, Ledieu J, Hoeft J T, Reid D E, Wearing L H, Diehl R D, Lograsso T A, Ross A R and McGrath R 2006 *Philos. Mag.* **86** 841–848
[24] Wearing L H, Smerdon J A, Leung L, Dhesi S S, Ledieu J, Bencok P, Fisher I, Jenks C J and McGrath R 2008 *J. Phys.: Condens. Matter* **20** 015005
[25] Wearing L H, Smerdon J A, Leung L, Lograsso T A, Ross A R and McGrath R 2007 *Surf. Sci.* **601** 3450–3455
[26] Weisskopf Y, Erbudak M, Longchamp J N and Michlmayr T 2006 *Surf. Sc.* **600** 2592–2597
[27] Weisskopf Y, Burkardt S, Erbudak M and Longchamp J N 2007 *Surf. Sci.* **601** 544–551
[28] Pussi K, Gierer M and Diehl R D 2009 *J. Phys.: Condens. Matter*
[29] Larsson J A, Elliot S D, Greer J C, Repp J, Meyer G and Allenspach R 2008 *Phys. Rev. B* **77** 115434
[30] Maruyama Y, Ohno K and Kawazoe Y 1995 *Phys. Rev. B* **52** 2070
[31] Abel M, Dmitriev A, Fasel R, Lin N, Barth J V and Kern K 2003 *Phys. Rev. B* **67** 245407
[32] Murray P, Pedersen M and Laegsgaard E 1997 *Phys. Rev. B* **55** 9360
[33] Szuba S, Czajka R, Kasuya A, Wawro A and Rafii-Tabard H 1999 *Appl. Surf. Sci.* **144-145** 648

About the chemistry of the terminating 5f-planes in *i*-AlPdMn : A comparison between models and surface studies

M. Quiquandon[1], **B. Unal**[2,4], **P. Thiel**[2,3,4] **and D. Gratias**[1]

[1] LEM CNRS-ONERA, 29 avenue de la Division Leclerc, F-92322 Châtillon Cedex, France
[2] Department of Materials Science and Engineering, Iowa State University, Ames, Iowa 50011, USA
[3] Department of Chemistry, Iowa State University, Ames, Iowa 50011, USA
[4] Ames Laboratory, Iowa State University, Ames, Iowa 50011, USA

E-mail: `marianne.quiquandon@onera.fr`

Abstract. Structural arguments are proposed that support the hypothesis that the equilibrium 5f-surfaces of *i*-AlPdMn icosahedral alloys are actual cuts of the bulk structure with almost pure Al termination planes [1, 2] corresponding to actual bulk properties with no need for possible Al segregation on the surface. Most of the 5f termination planes in *i*-AlPdMn are made of two close planes containing essentially Al at the upper one and Al and Pd at the second one (see for instance [3]). Here, perfect unrelaxed structural models that differ only in the chemical distribution of the atomic species are devised, close to the description proposed in [4]. Once compared to experimental X-rays data [5], the calculations show that although the chemical decoration has a rather limited influence on the global R-factor, there is a clear tendency toward favoring models with almost pure Al on the atomic surface located on the node n, in excellent agreement with the surface studies.

1. Introduction

The *i*-AlPdMn phase is one of the most studied systems among the quasicrystals. Especially, their fivefold surface has been investigated by various surface sensitive techniques [1, 2, 6, 7, 8]. The interest to their surface has arisen from interesting surface properties such as good oxidation resistance and low friction (see, for instance [9]).

Surface studies of sputter-annealed 5f-surfaces of *i*-AlPdMn possesses a step-terrace morphology [3, 6, 7, 10]. The structural nature of the terraces has been explored by many techniques, among them low energy electron diffraction analysis of intensity-voltage curves (LEED-IV) where Gierer *et al.* [1, 2] noticed that two types of closely spaced, dense planes form the terminating layers, from among the large ensemble of non-identical planes which forms the bulk structure. In other words, the bulk structure can be classified into groups of self-similar planes, and one of these groups was the most likely candidate for the surface structure. Very similar observations were reported in X-ray photoelectron diffraction (XPD) studies [11]. However, at present, there is a debate as to which group of planes is favored [12, 13, 14, 15].

The chemical decoration of the terminating layers has been investigated by LEED-IV technique [1, 2]. The best fit of the data to the Boudard model [5] revealed that the planes containing 93 at.% Al and 7 at.% Mn on average are the top planes in the terminating layers while Pd is restricted to the second plane. Very similar results were obtained by low energy ion scattering (LEIS) [8]. To be specific, both techniques indicated that all Pd resides only in the second plane; the top plane is free of Pd. Regarding the top plane Mn content, LEIS set the upper limit in one study as 6.3 at.%. A later LEIS study indicated that the top planes contain almost pure Al with 1 at.% Mn [8]. LEED-IV data analysis set an upper limit of 10 at.%. It is worth noting that, unlike LEED-IV, LEIS technique does not require a structural model: data analysis is model independent. Recent STM studies for film nucleation, especially Si [16] and Bi [17], suggest that the topmost surface plane contains some Mn: the average Mn concentration in the topmost plane is between 0 and 10 at.%, and our recent bulk structural model analysis concludes that it is most probably less than 4 at.% [3].

Our present goal is the following. We shall first set up a simple but realistic model that allows for large chemical variations while keeping constant the overall stoichiometry and all other diffraction parameters and check the theoretical X-rays diffraction spectra against the experimental ones for bulk i-AlPdMn. We will then analyze the variations of the R-factors as function of specific occupancy factors to check the consistency between the rich Al planes as observed by surface studies and the predictions of the bulk models.

2. The atomic model

The geometrical fact that makes the present work possible for a valuable test for consistency, is that each of the dense 5f-planes is generated by *one unique AS*, either n or n' according to the stacking [10, 18]. The consequence is that the composition of these termination planes is dictated by the nature of the atomic species attached to *one AS only*. Based on the most recent analyzes [3] from the actual structural models, an agreement has emerged that the n ASs generate the termination upper planes.

We therefore designed a generic model of i-AlPdMn, characterized by two independent parameters $0 \leq x \leq 1$ and $0 \leq y \leq 1$ representing the occupation factors of the atomic species Pd and Mn on the ASs n and n', in a way that keeps constant both the overall composition, and the shape and volume of the relevant chemical cells in perpendicular space. The model is used to study the variations of R-factor as a function of x and y obtained from the comparison with the experimental X-rays diffraction data collected by M. Boudard *et al.* [5]. The idea is to progressively modify the chemical decorations between n and n' with no change in the shapes of the cells nor in the overall composition: then, the variations in diffraction intensities are exclusively due to the sole chemical contrast.

The skeleton of the atomic positions of the model in the real space is the usual simple one generated in 6D space by three AS located at respectively $n = (0,0,0,0,0,0)$, $n' = (1,0,0,0,0,0)$ and $bc = 1/2\ (\bar{1},1,1,1,1,\bar{1})$ as shown on figure 1.

The chemistry of the model is based on the cell decomposition of the M and M' clusters (pseudo-Mackay clusters) as discussed in previous model studies [4]. This cell decomposition is made of seven cells shown on figure 1:

- the central cell $\omega = c_1 \cup c_2$ is a small triacontahedron of volume $2\tau - 3$ located in the middle of the n and n' AS's. It generates the M (if located on n) and M' (if located on n') centers.
- the cell noted $\alpha = c_2 \cup c_3 \cup c_4$ that generates the large icosahedra of the M if on n', or of the M' if on n;
- the cell noted $\beta = c_4 \cup c_5 \cup c_6$ that generates the external icosidodecahedra of the M if on n or of the M' if on n';
- the cell noted $\gamma = c_6 \cup c_7$ that generates the partially occupied (7/20) inner dodecahedra of the M if on n', or M' if on n.

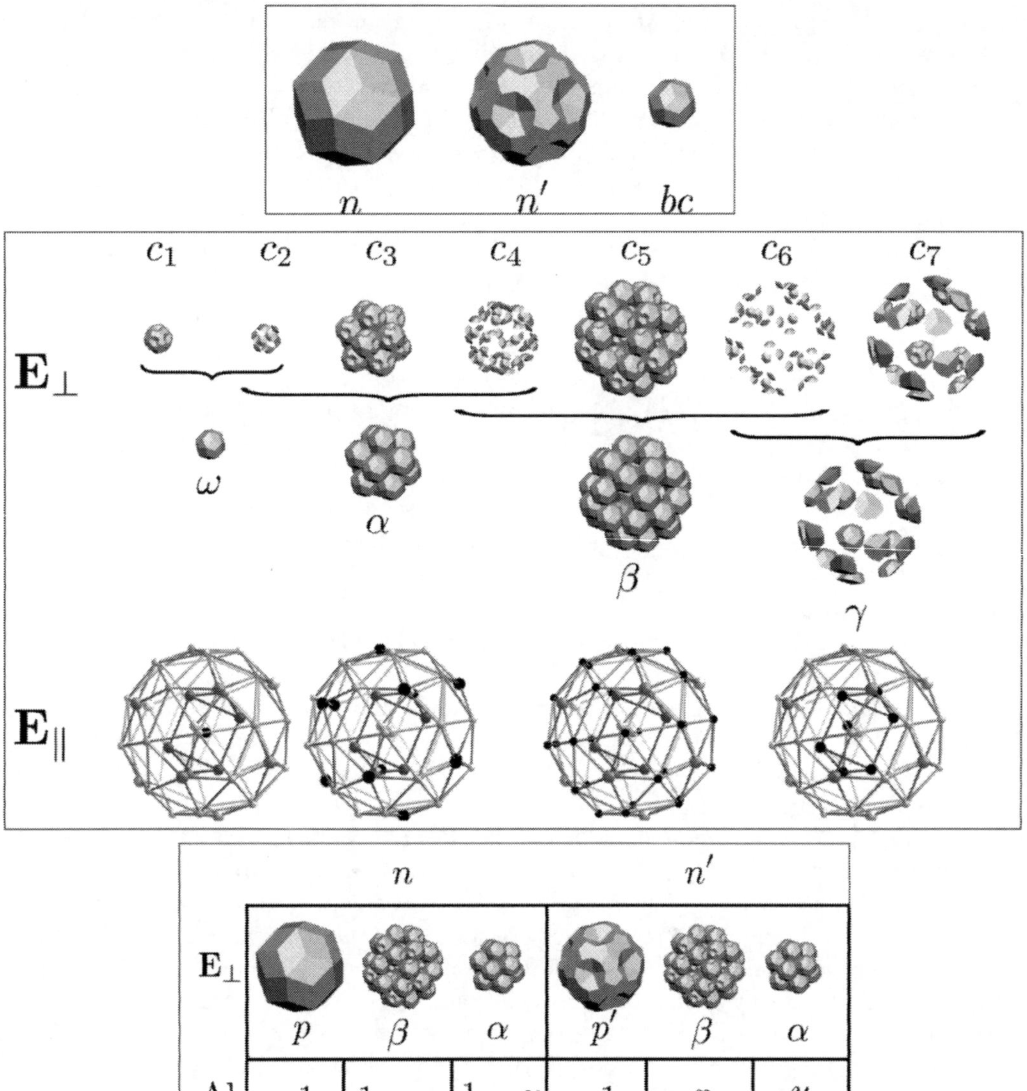

	p	β	α	p'	β	α
Al	1	$1-x$	$1-y$	1	x	y
Pd	0	x	0	0	$1-x$	0
Mn	0	0	y	0	0	$1-y$

Figure 1. Top: the three main atomic surfaces of the models, from left to right, n , n' and bc. Middle: the cell decomposition in $\mathbf{E}_\perp$ of the geometric model for the M and M' clusters in $\mathbf{E}_\parallel$. Bottom: Definition of the model. The model depends on two independent occupancy parameters x and y that vary between 0 and 1 at constant stoichiometry $Al_{68.83}Pd_{22.75}Mn_{8.42}$ (as compared with the experimental value $Al_{70.3}Pd_{21.4}Mn_{8.3}$ in [19]) and density of 5.07. For $x = 0$, $y = 0$, the AS n is pure Al thus generating a structure where the dense 5f planes are occupied by pure Al. At the opposite extreme, values of $x = 1$ and $y = 1$ correspond to a model where the AS n is maximally charged in Pd and Mn atoms and n' is pure Al. Surfaces studies are in favor of the first case ($x = y = 0$).

Table 1. Chemical decoration of M and M' clusters as a function of the x and y parameters. The centers and partial inner dodecahedra of M and M' are always occupied by Al.

Model	(x, y)	$(0,0)$	$(0,1)$	$(1,0)$	$(1,1)$
M cluster					
icosahedron	$12y$ Al + $12(1-y)$ Mn	12 Mn	12 Al	12 Mn	12 Al
icosidodecahedron	$30(1-x)$ Al + $30x$ Pd	30 Al	30 Al	30 Pd	30 Pd
M' cluster					
icosahedron	$12(1-y)$ Al + $12y$ Mn	12Al	12Mn	12Al	12Mn
icosidodecahedron	$30x$ Al + $30(1-x)$ Pd	30Pd	30Pd	30Al	30Al

These four major cells intersect two by two by c_2, c_4 and c_6. The decision has been made here to build the chemical order using the cells ω, α, β and γ [1].

The small AS at bc is always filled with Pd (see for instance [5]). The ASs at n and n' are decomposed using β and α cells. The chemical distribution is done the following way (see figure 1). We first assign Al to both n and n' and then replace x percent of the volume of β_n by Pd and y percent of the volume of α_n by Mn. Correlatively, to insure constant composition, we replace $1-x$ percent of the volume of $\beta_{n'}$ by Pd and $1-y$ percent of the volume of $\alpha_{n'}$ by Mn. The chemical decoration of the model is therefore entirely defined by the pair (x, y) varying in the domain $\{(0,0), (1,0), (1,1), (0,1)\}$. These four corners correspond to M and M' clusters decorated with Al, Pd and Mn atoms as described on table 1. Finally, since ω and γ are embedded in larger cells corresponding to Al species, the M and M' centers and inner partial dodecahedra are invariably occupied by Al atoms. Exchanging x and y with their complements to 1 results in exchanging the chemical distributions on the external icosahedra and the icosidodecahedra of the M and M' clusters. Hence, the $(0,0)$ model corresponds to pure Al on n and all Pd and Mn on n', whereas its complementary model $(1,1)$ corresponds to pure Al on n' and all Pd and Mn on n.

The model is compared to the actual structure by calculating the theoretical diffraction intensities and comparing them to the experimental data collected by Boudard *et al.* [5] using the reliability factors, noted here R-factors, expressed in % and defined by $R_a = \sum(a_{exp} - a_{theo})^2 / \sum a_{exp}^2$ where a_{exp} and a_{theo} are the absolute values of the experimental and theoretical structure factors of the observed reflections. These reliability factors $R(x, y)$ for X-rays data are plotted as 2D surfaces on figure 2.

The first point to be noted is that the R-factors vary between 6.51 % for model $(0,0)$ to 16.56 % for model $(1,1)$. This difference in the R-factors, of 10%, is relatively small given that the difference between the two models is a large change in the chemical decoration, *i.e.* complete exchange in the chemistry of M and M' clusters. This relative weak influence of the chemistry on the overall diffraction data could be the origin of the multiplication of different atomic models of i-AlPdMn concerning the chemical order. The difference in R-factors of 10% between the extreme situations of having Pd and Mn atoms located on n' for model $(0,0)$ and on n for model $(1,1)$ establishes a relative scale for judging the effects of other aspects of chemical order.

The second point to be noted is that the surface $R(x, y)$ shows little change with respect to y but a significant one with respect to x. From models $(0,0)$ and $(1,0)$ where Mn atoms are on the icosahedra of the M clusters, to the $(0,1)$ and $(1,1)$ models where Mn atoms are on the icosahedra of the M' clusters, the variation in R-factors is only $8.89 - 6.51 = 2.38\%$ and

[1] Actually because of very small overlaps between ω, α, β and γ, we use in the computations the two cells c_5 and c_1 in place of respectively β and ω; this insures perfect chemical distribution and generates no pertinent differences in our context.

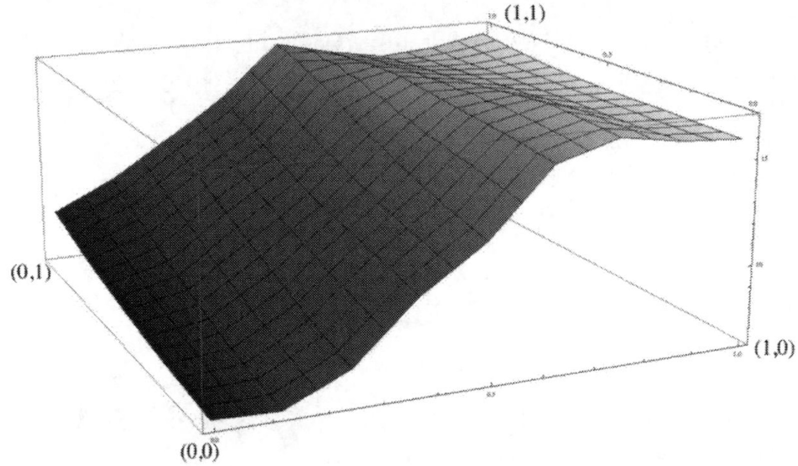

Figure 2. Values $R(x, y)$ of the R-factor as a function of the two variables (x,y) for X-rays diffraction data with 314 beams. Values of the R-factor at the four extreme (x, y) points are: (0,0) 6.51, (0,1) 8.89, (1,0) 15.87 and (1,1) 16.56.

$16.56 - 15.87 = 0.69\%$, respectively. Although the smallest R-factor, 6.51%, is obtained for Mn atoms located on n' (Mn on the icosahedra of the M clusters), transferring some Mn to n (at the expense of Al) has a very small impact on the R-factor. In other words, the X-ray diffraction data are not well-suited to showing the Mn/Al ordering on either M or M'. On the contrary, the R-factor changes strongly from models $(1, 0)$ and $(1, 1)$ to models $(0, 0)$ and $(0, 1)$. The changes in R-factors along x (exchanging Pd / Al between the M and M' icosidodecahedra) are $15.87 - 6.51 = 9.36\%$ and $16.56 - 8.89 = 7.67\%$, respectively. The minimum R-factor of 6.51% is clearly in favor of locating all Pd atoms on n'.

These results are nicely consistent with the surface analysis data that show almost pure Al termination 5f-planes with a small amount of Mn.

In the simplest scheme of AS n being pure aluminum ($x = y = 0$), M clusters have Al at the center, on the inner partial dodecahedron and on the icosidodecahedron, and Mn atoms on the icosahedron. M' clusters have Al at the center, on the inner partial dodecahedron and on the icosahedron, and Pd atoms on the icosidodecahedron.

3. Conclusion

The present numerical explorations on plausible sets of perfect unrelaxed chemical models of icosahedral i-AlPdMn structures show that the chemical decoration has a rather limited influence on the global R-factor for models that differ only in the way the chemical species are distributed between the n and n' ASs: this explains why it is so difficult to assert which model is best with respect to chemical distribution.

Irrespective of these restrictions, it is clear that the structural models that are in best agreement with the experimental X-rays results are those with parameters (x, y) close to values $\{(0, 0) — (0, 1)\}$. As a consequence, *models having pure Al plus possibly some Mn on the n AS with Pd confined on n' and bc* are the best candidates for a simple structural description consistent with both the bulk diffraction data and surface analyses as discussed in reference [3].

These results enforce the idea that 5-f termination planes do indeed correspond to actual cuts of the bulk structures, irrespective of the tiny contractions of the interplanar distances between the two last planes. Furthermore, observed local features, as dark stars, in STM images are formed by M, M' clusters since placing Pd on n' eliminates the B clusters being the candidates for them [20].

Acknowledgments

The portion of the work by B.U. and P.T. was supported by the Office of Science, Basic Energy Sciences, Materials Science Division of U.S. Department of Energy (USDOE) under Contract No. DE–AC02–07CH11358.

References

[1] Gierer M, Van Hove M A, Goldman A I, Shen Z, Chang S-L, Jenks C J, Zhang C M and Thiel P A, *Phys. Rev. Lett.* **78**, 467 (1998).
[2] Gierer M, Van Hove M A, Goldman A I, Shen Z, Chang S-L, Pinhero P J, Jenks C J, Anderegg J W, Zhang C M and Thiel P A, *Phys. Rev.*, B **57**, 7628 (1998).
[3] Unal B, Jenks C J and Thiel P A, *Phys. Rev.* B **77** 195419-1–10 (2008).
[4] Quiquandon M and Gratias D, *Phys. Rev.* B **74** 214205-1–9 (2006).
[5] Boudard M, de Boissieu M, Janot C, Heger G, Beeli C, Nissen H-U, Vincent H, Ibberson R, Audier M and Dubois J-M, *J. Phys.: Condens. Matter* **4** 10149–10168 (1992).
[6] Schaub T M, Bürgler D E, Güntherodt H-J, Suck J-B and Audier M, *Appl. Phys.* A **61**, 491 (1995).
[7] McGrath R, Leung L, Barrett S D and Ledieu J, *Proc. Royal Microscopical Society* **40**, 215 (2005).
[8] Jenks C J, Ross A R, Lograsso T A, Whaley J A and Bastasz R, *Surf. Sci.* **521**, 34 (2002).
[9] Dubois J-M, *Useful Quasicrystals.* First ed. World Scientific.Singapore: (2005).
[10] Barbier L, Le Floc'h D, Calvayrac Y and Gratias D, *Phys. Rev. Lett.* **88**, 085506 (2002).
[11] Zheng J-C, Huan C H A, Wee A T S, Van Hove M A, Fadley C S, Shi F J, Rotenberg E, Barman S R, Paggel J J, Horn K, Ebert P, and Urban K, *Phys. Rev.* B **69**, 134107 (2004).
[12] Papadopolos Z, Kasner G, Ledieu J, Cox E J, Richardson N V, Chen Q, Diehl R D, Lograsso T A, Ross A R, and McGrath R, *Phys. Rev.* B **66**, 184207 (2002).
[13] Papadopolos Z, Gröning O, and Widmer R, in *Models, Mysteries and Magic of Molecules*, edited by Boeyens J C A and Ogilvie J F, Springer, New York, (2008).
[14] Krajci M, Hafner J, Ledieu J, and McGrath R, *Phys. Rev.* B **73**, 024202 (2006).
[15] Ledieu J and McGrath R, *J. Phys.: Condens. Matter* **15**, S3113 (2003).
[16] Ledieu J, Unsworth P, Lograsso T A, Ross A R and McGrath R, *Phys. Rev.* B **73**, 012204/1 (2006).
[17] Smerdon J A, Sharma H R, Ledieu J and McGrath R, *J. Phys: Condens. Matter.* **20**, 314005 (2008).
[18] Barbier L and Gratias D, *Progress in Surface Science* **75** 177-189 (2004).
[19] Quiquandon M, Calvayrac Y, Quivy A, Faudot F and Gratias D, in *MRS Symposium Proceedings: Quasicrystals* edited by Dubois J-M, Thiel P A, Tsai A-P and Urban K (MRS, Warrendale, 1999) pp. 95–106 (1999).
[20] Unal B, Jenks C J and Thiel P A, *J. Phys.: Condens. Matter* **21**, 55009 (2009).

Journal of Physics: Conference Series **226** (2010) 012008 doi:10.1088/1742-6596/226/1/012008

Characterization of quasicrystalline Al-Cu-Fe nanoclusters as catalysts for the synthesis of carbon nanotubes

K Kajiwara, S Suzuki, Y Matsui, H Sato and K Hata

Department of Electrical and Electronic Engineering, Mie University, 1577 Kurima-Machiya-cho, Tsu-shi, 514-8507, Japan

E-mail: kajiwara@em.elec.mie-u.ac.jp

Abstract. Quasicrystalline (QC) Al-Cu-Fe nanoclusters as catalysts for the synthesis of carbon nanotubes (CNTs) were characterized in this study. It was shown in previous work that armchair-type CNTs are preferentially synthesized by using Al-Cu-Fe ultrathin films as catalysts in catalytic-chemical vapour deposition. We postulated a plausible mechanism that facets of QC nanoclusters with fivefold symmetry play an important role in the nucleation of a CNT cap with specific six-pentagon-configuration, which gives rise to the selective synthesis of armchair-type CNTs. To verify this hypothesis, the morphology and structure of Al-Cu-Fe nanoclusters were investigated by using transmission electron microscopy and electron diffraction (ED) as well as X-ray diffraction. As a result, it was found that the morphology of their nanoclusters is generally faceted-polyhedra, and the average diameter is ~4 nm, ranging from 1 to 10 nm, which indicates to correlate with the diameter of CNTs synthesized by using these nanoclusters as catalysts. Their ED patterns could be attributed to an icosahedral (*i*) Al-Cu-Fe phase. It was also found that multi-walled CNTs are synthesized successfully by using *i*-Al-Cu-Fe, *i*-Al-Pd-Re, and decagonal-Al-Ni-Co bulk-alloy systems as catalysts, respectively.

1. Introduction

Over the last decade, the synthesis of carbon nanotubes (CNTs), having suitable physical and chemical properties for industrial applications such as field emission cathodes in flat panel displays, has been developed towards the chirality-controlled synthesis [1]. In recent years, it has been reported that the morphology and structure of catalytic metal nanoclusters, for example, Fe-Co [2] and Co-Mo [3], as well as the synthetic conditions play an important role in determining both diameter d_t and chirality (n, m) of CNTs synthesized by catalytic-chemical vapor deposition (CVD) using a carbon source gas supply such as ethanol and CO. Based on the dissolution-diffusion-precipitation model [4] for the synthesis mechanism of CNTs, a hydrocarbon such as ethanol adsorbed on the catalytic cluster surface releases carbon upon decomposition, which dissolves and diffuses into the cluster. When a supersaturated state is reached, carbon precipitates in a crystalline tubular form. It is supposed that (i) the solid solubility of carbon into clusters is a dominant factor rather than chemical reaction on the cluster surface in the dissolution step, and (ii) the template effect of clusters is a dominant factor on the diameter and the tube chirality in the precipitation step.

As shown in previous work [5], we have studied quasicrystalline (QC) Al-Cu-Fe nanoclusters as catalysts for the chirality-selective synthesis of armchair-type, i.e., metallic, CNTs, since it is known that the chirality of CNTs depends on the configuration of six pentagons at the cap in the precipitation step, and an armchair tube is formed if the fullerene C_{60}, having a truncated-icosahedral structure, is bisected normal to a fivefold axis [6]. As described previously [5], we postulated a mechanism of chirality-selectivity using QC nanoclusters as catalysts that a pentagon accompanied with five hexagons is epitaxially formed at the center of a hemi-fullerene cap of the CNT on a facet of icosahedral (*i*) nanoclusters with fivefold symmetry, followed by adding five pentagons and a cylinder of graphite, resulting in an armchair-type CNT due to the catalytic activity of *i*-nanoclusters.

In order to verify this hypothesis, first of all in this exploratory work, we focused on the geometric structure of the clusters, especially on the quasicrystallinity of Al-Cu-Fe nanoclusters. The morphology and structure of Al-Cu-Fe nanoclusters, which are formed by annealing at 723 K for 30 min in Ar ambient [5], were investigated by using transmission electron microscopy (TEM), electron diffraction (ED), and energy-dispersive X-ray spectrometry (EDS), compared with our conventional Al-Co two-layered films [7]. Since Co-based alloys are commonly used for catalysts in the catalytic CVD of CNTs using ethanol as a carbon feedstock [2,3,7-9], and also a few reports on chirality distributions using only Co-based alloys as catalysts are available so far [2,3].

Secondly, whether *i*-Al-Cu-Fe, *i*-Al-Pd-Re, and decagonal (*d*) Al-Ni-Co bulk-alloy systems [10], of which QC phases were confirmed by X-ray diffraction (XRD), have a similar catalytic-activity for the synthesis of CNTs or not were investigated by using scanning electron microscopy (SEM) as well as TEM. Finally, at the current state, a most probable mechanism of chirarity-selectivity by using QC nanoclusters as catalysts is described in this article. However, further work is needed to clarify the catalytic effect of electronic structure of the cluster surface on the synthesis of CNTs.

2. Experimental Details

First, Al (2.1 nm)/Cu (0.6 nm)/Fe (0.3 nm) three-layered films were deposited on rock-salt/SiO_2 (50 nm thick) substrates by sequential electron-beam evaporation, followed by annealing at 723 K for 30 min in Ar ambient in a CVD reactor to form nanoclusters having an icosahedral phase, i.e., *i*-$Al_{62}Cu_{25.5}Fe_{12.5}$ [11-13], as described in detail previously [5]. Subsequently, these annealed Al-Cu-Fe ultrathin films were removed from the substrates, and were put them on TEM grids. Then the morphology of Al-Cu-Fe nanoclusters was observed by TEM, and also both electron diffraction patterns and EDS spectra were obtained from these nanoclusters.

Secondly, in order to compare with the results on Al/Cu/Fe three-layered films, the procedures above-mentioned were performed on conventional Al (2 nm)/Co (1 nm) two-layered films, where the annealing were executed at 973 K for two hours in air ambient [5, 7].

Thirdly, in order to investigate the catalytic activity of QC bulk-alloys, both *i*-$Al_{62.5}Cu_{24.5}Fe_{13}$ and *i*-$Al_{70}Pd_{20}Re_{10}$ alloy pellets, and also a *d*-$Al_{70}Ni_{15}Co_{15}$ alloy pellet were prepared by furnace alloying and subsequent annealing, respectively. Then XRD with Cu Kα radiation was used for evaluation of the quasicrystallinity of these alloy pellets. Both *i*-QC alloy pellets were crushed by grinding with an agate pestle in a mortar, and then these *i*-QC alloy suspensions dispersed by ultrasonication in ethanol were deposited and dried on silicon (100) wafer substrates (10 mm squares and 0.5 mm thick), respectively. These samples were annealed at 723 K for 30 min in Ar ambient in a CVD reactor, followed by synthesizing CNTs at 973 K for 5 min by ethanol catalytic-CVD, i.e., under the same conditions as described previously [5]. Subsequently, as-prepared CNTs were observed by using SEM (S-4000, Hitachi) and TEM (H-800, Hitachi) to characterize the morphology and structure of CNTs.

3. Results and discussion

3.1. The morphology and structure of Al-Cu-Fe nanoclusters

Figures 1(a) and 1(b) show a TEM image and an ED pattern of Al-Cu-Fe nanoclusters, respectively, and the inset in figure 1(a) shows a TEM image of CNTs synthesized by catalytic-CVD using these nanoclusters as catalysts. Figure 2 also shows an EDS spectrum of Al-Cu-Fe nanoclusters. From TEM observation (figure 1(a)), the morphology of most nanoclusters is so-called faceted-polyhedra [14], and the average diameter of nanoclusters, d_c, is about 4.5 nm, ranging from 1 nm to 10 nm, which indicates that d_c correlates to the average diameter of CNTs, $d_t \cong 3$ nm as shown in the inset. From

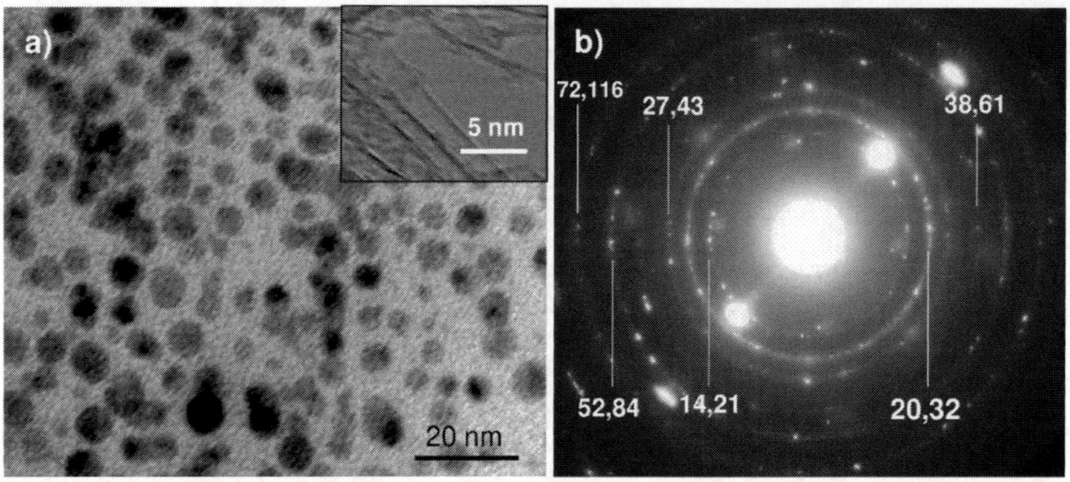

Figure 1. (a) TEM image and (b) ED pattern of Al-Cu-Fe nanoclusters. The inset in (a) shows a TEM image of CNTs synthesized by catalytic-CVD using Al-Cu-Fe nanoclusters as catalysts.

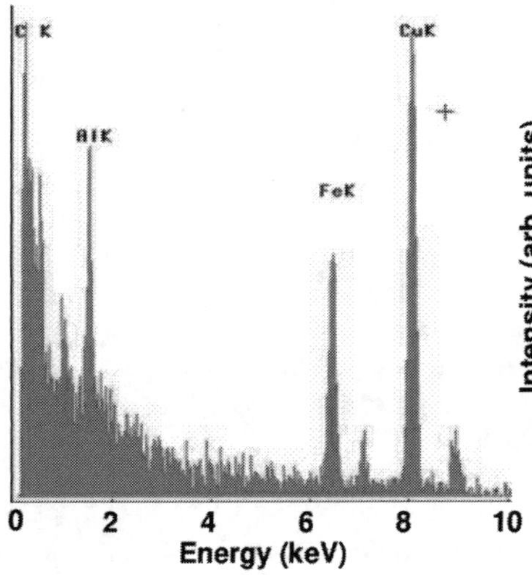

Figure 2. EDS spectrum of Al-Cu-Fe nanoclusters.

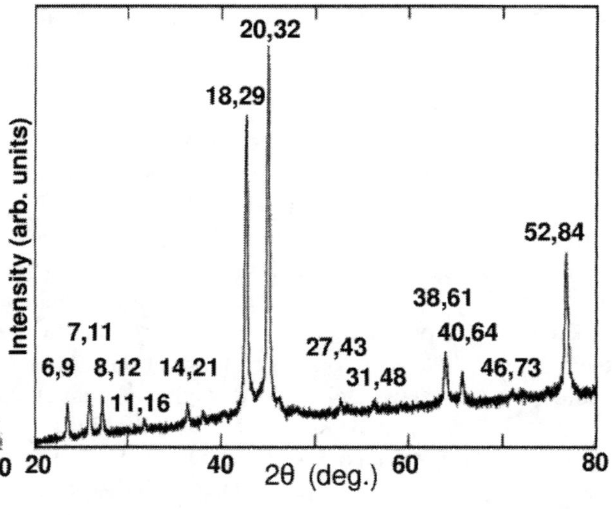

Figure 3. XRD pattern of an *i*-Al-Cu-Fe alloy pellet.

an EDS spectrum (figure 2), as expected, nanoclusters consist of Al, Cu and Fe, while a carbon peak could originate from a TEM grid used here.

Debye rings obtained in an ED pattern (figure 1(b)) give the values of interplanar spacings, d = 0.248, 0.201, 0.175, 0.146, 0.124, and 0.106 nm, respectively, from inner- to outer-ring order. According to the indexing of the XRD pattern obtained from the i-$Al_{62.5}Cu_{24.5}Fe_{13}$ alloy pellet as shown in figure 3, where the (N, M) indexing scheme of Cahn *et al.* [15] is used for a QC phase, these Debye rings could be attributed to the indices of (14, 21), (20, 32), (27, 43), (38, 61), (52, 84) and (72, 116), respectively, in the above order. From these indices, it indicates that the structure of Al-Cu-Fe nanoclusters is confirmed to be of face-centered-icosahedral phase [15].

It is well-known that the formation route of the QC nanoclusters considerably affects the morphology and structure of nanoclusters. In this study, in the case of Al/Cu/Fe sequential deposition, isolated faceted-polyhedral, presumably icosahedral, nanoclusters are formed by an appropriate heat-treatment as shown in figure 1(a), and these nanoclusters have a higher catalytic activity rather than in the case of Al/Fe/Cu sequence (not shown here). It suggests that further study of QC nanoclustes with

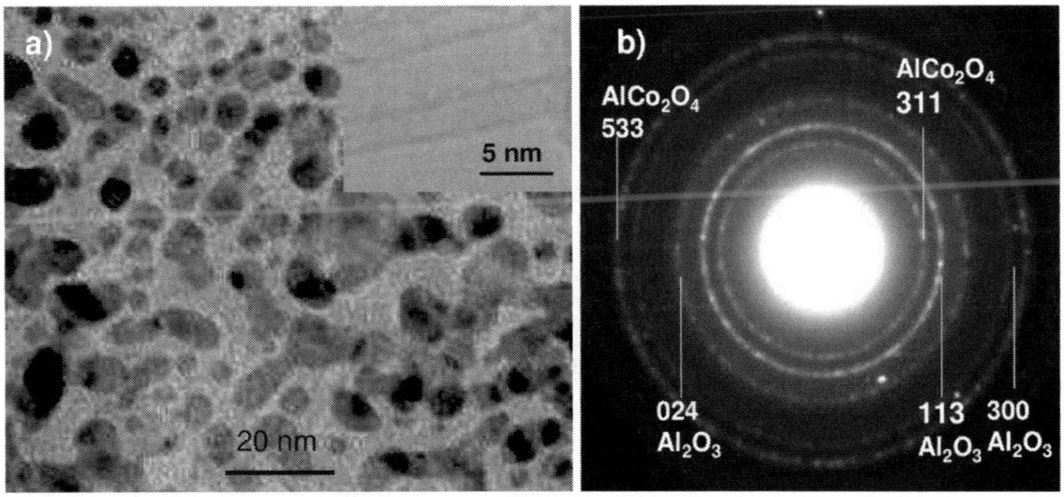

Figure 4. (a) TEM image and (b) ED pattern of Al-Co nanoclusters. The inset in (a) shows a TEM image of CNTs synthesized by catalytic-CVD using Al-Co nanoclusters as catalysts.

more suitable morphology and structure as catalysts is needed for a specific (*n*, *m*) selective-synthesis of CNTs in catalytic-CVD.

3.2. The morphology and structure of conventional Al-Co nanoclusters

Figures 4(a) and 4(b) show a TEM image and an ED pattern of conventional Al-Co nanoclusters, respectively, and the inset in figure 4(a) shows also a TEM image of CNTs synthesized by catalytic-CVD using Al-Co nanoclusters as catalysts, for comparison with the above-mentioned QC Al-Cu-Fe nanoclusters. From TEM observation (figure 4(a)), the morphology of Al-Co nanoclusters is slightly different spherical-, including irregular-shaped, polyhedra from the Al-Cu-Fe nanoclusters, while the average diameter of nanoclusters has a quite similarity to the Al-Cu-Fe nanoclusters. Based on the chirality assignment by radial-breathing mode Raman scattering, we obtained previously [5] a different chirality distribution using Al-Co nanoclusters from that using *i*-Al-Cu-Fe nanoclusters.

Debye rings obtained in an ED pattern (figure 4(b)) give the values of interplanar spacings, d = 0.245, 0.209, 0.173, 0.142, 0.134, and 0.123 nm, respectively, from inner- to outer-ring order. Based on the powder-diffraction database, Al-Co nanoclusters could be attributed to both $AlCo_2O_4$ andα -Al_2O_3 mixed phases, which have a poor crystallinity according to halo-like rings.

It is reported by Liu *et al.* [16] thatα -Al_2O_3 particles less than 3 nm in diameter have a catalytic activity for the synthesis of CNTs, but their chirality distribution of CNTs is different from our results of both Al-Cu-Fe and Al-Co nanoclusters, which were described previously [5]. It was also shown in previous work [5] that CNTs using Al-Co nanoclusters have a broader chirality distribution rather than those using QC Al-Cu-Fe nanoclusters, and the relative abundance of armchair-type CNTs is ~31% for the former, whereas that is ~55% for the latter. It might indicate that the morphology of Al-Co nanoclusters is less preferred to be faceted-polyhedra [17], for example, icosahedra, rather than *i*-Al-Cu-Fe nanoclusters.

3.3. Catalytic activity of quasicrystalline bulk-alloys

Moreover, we have also investigated whether or not CNTs are synthesized by catalytic-CVD using QC bulk-alloys as catalysts. Figures 5(a), 5(b) and 5(c) show SEM images of CNTs synthesized by using (a) *i*-Al-Cu-Fe, (b) *i*-Al-Pd-Re alloy powders, and by using (c) a *d*-Al-Ni-Co alloy pellet as catalysts, respectively. The inset in figure 5(c) shows, as an example, a TEM image of CNTs synthesized by using *d*-Al-Ni-Co alloy pellet as catalysts. From these results, it indicates that multi-walled CNTs (MWNTs) are synthesized successfully by using *i*-Al-Cu-Fe, *i*-Al-Pd-Re, and *d*-Al-Ni-Co alloys, respectively, where their diameters, d_t, ranging from 10 nm to 30 nm, are considerably larger than *i*-

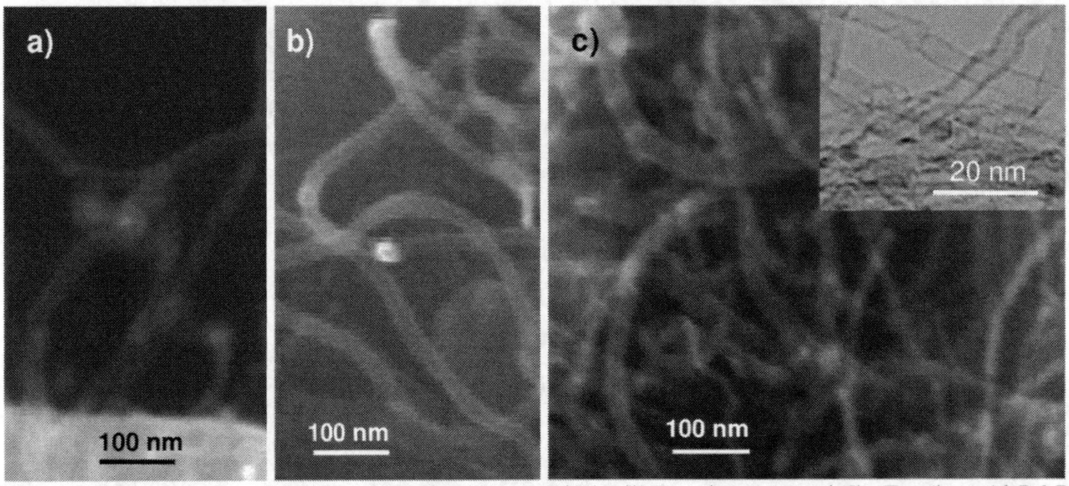

Figure 5. SEM images of CNTs synthesized by catalytic-CVD using (a) *i*-Al-Cu-Fe, (b) *i*-Al-Pd-Re alloy powders, and (c) *d*-Al-Ni-Co alloy pellet as catalysts, respectively. The inset in (c) shows a corresponding TEM image of *d*-Al-Ni-Co alloy catalyst.

Al-Cu-Fe nanoclusters formed from ultrathin films. It is well-known that the diameter of CNTs dominantly depend on that of clusters due to the template effect, and also the number of walls depends on the diameter of CNTs. Based on the energetics of nanotubes, it is suggested that a larger (>3 nm) SWNT tends to collapse unless it is surrounded by neighbouring tubes, for example, as in a MWNT due to surface strain energy.

In the case of *i*-Al-Cu-Fe bulk system, we found the dark precipitates inside the CNTs to be Al-Cu-Fe compounds by means of energy-dispersive X-ray spectra in the TEM images (not shown here). Also, in *d*-Al-Ni-Co bulk system, we obtained ED patterns of clusters inside the CNTs, which could be attributed to the structure of decagonal Al-Ni-Co (not shown here). It is suggested that absorption phenomenon of nonwetting clusters originates from the capillary action by the inner wall of CNTs [18]. According as the bright precipitates inside the nanotubes in SEM images, typically as shown in figure 5(b), are considered to be QC alloy clusters, the average diameter of CNTs, $d_t \cong 30$ nm, is also correlated with that of alloy clusters, d_c. It implies that larger QC clusters also have a significant catalytic activity for the synthesis of MWNTs with larger diameters.

However, this is a preliminary result performed on typical three QC alloy systems under the same synthetic conditions. Further study is still required for finding a more suitable QC alloy system and appropriate synthetic conditions for each alloy system in the field of controlled-synthesis of CNTs.

3.4. Mechanism of chirality-selectivity using quasicrystalline nanoclusters

It is well-known that the fullerenes, typically C_{60} with ~0.7 nm in diameter, have a truncated icosahedral structure with fivefold axis, and also the QC clusters have building-blocks of multi-shell, hierarchical structures, so-called pseudo-Mackay clusters [19]. Therefore, we have hypothesized that a pentagonal-carbon ring is nucleated epitaxially on a facet of QC nanoclusters, subsequently a hemi-fullerene cap with fivefold symmetry is formed around the pentagon, and then an armchair-type nanotube is elongated by the catalytic reaction of the QC nanoclusters.

From a preliminary work [5], in which the considerable preference of armchair-type CNTs by using QC nanoclusters as catalysts is shown, it could indicate that the facets of QC nanoclusters with fivefold symmetry play an important role in the nucleation stage of a CNT cap with a specific six-pentagon-configuration, which gives rise to armchair-type CNTs having a, as an example, (9, 9) chirality, $d_t \cong 1.22$ nm in diameter, corresponding to the diameter of QC nanoclusters.

Based on these evidences obtained from both this study and the previous work, it indicates that the chirality-selectivity of QC nanoclusters originates from the faceted-polyhedral template effect and the surface atomic structure with fivefold symmetry of the QC nanoclusters. That is, the cap structure of CNTs reflects a six-pentagon-configuration in the nucleation stage of CNT-apex due to the interaction between C $2sp^n$ orbitals of carbon chains and Al $3sp^m$ orbitals of QC surfaces. It is suggested that (9, 9) armchair-type CNTs are self-assembled on the facets of pseudo-Mackay icosahedral clusters.

The surface of QC nanoclusters is probably covered with ultrathin aluminium oxide layer. This might influence the synthesis of CNTs, probably the growth rate, the diameter, and the tube chirality. Liu et al. [16] indicated that alumina nanoparticles less than 3 nm in diameter, or Al thin film oxidized in air, have a significant catalytic activity, but give rise to a different chirality distribution from the QC Al-Cu-Fe nanoclusters, i.e., non-armchair type CNTs are preferably synthesized by using alumina nanoparticles. However, further work is needed to clarify the influence of oxidation in the near future.

4. Conclusions

Over the last decade, nanoclusters [17], quasicrystals [10, 12, 19], and CNTs [1, 6] are fascinating fields in nanoscience and nanotechnology. In this article, nanoclusters having QC structures were investigated in applications to the catalysis for chirality-selective synthesis of CNTs.

First, it was found that the morphology of QC Al-Cu-Fe nanoclusters, which is transformed from Al-Cu-Fe ultrathin films by annealing at 723 K for 30 min in Ar ambient, is in generall faceted-polyhedra, and their average diameter d_c is ~4 nm, ranging from 1 to 10 nm, which implies to correlate with the diameter distribution of CNTs synthesized using QC nanoclusters as catalysts. Secondly, their ED patterns could be attributed to a face-centered-icosahedral Al-Cu-Fe phase. It could indicate that the hypothesis mentioned previously [5] is verified from these evidences, that is, the facets of QC nanoclusters with fivefold symmetry play an important role in the nucleation stage of a hemi-fullerene cap with specific six-pentagon-configuration of an armchair-type CNT such as (9, 9) with $d_t \cong 1.2$ nm.

In addition, it was found that multi-walled CNTs are synthesized successfully by using both *i*-Al-Cu-Fe and *i*-Al-Pd-Re alloy powders, as well as by using a *d*-Al-Ni-Co alloy pellet as catalysts for catalytic-CVD using ethanol as a carbon source. It indicates that larger QC faceted-clusters have a significant catalytic activity for the synthesis of multi-walled CNTs with approximately the same diameter as that of QC clusters, $d_t \cong 30$ nm.

Acknowledgments

The authors would like to thank Prof. Y. Saito for his helpful encouragement and also to thank the Venture Business Laboratories in Nagoya University for electron-beam evaporation of Al-Cu-Fe ultrathin films with a thickness-accuracy of ~10 pm. Thanks are due to Prof. T. Takeuchi in Nagoya University for supply of both *i*-Al-Cu-Fe and *i*-Al-Pd-Re alloy samples and their XRD patterns, and also to Prof. A.P. Tsai in Tohoku University for supply of a *d*-Al-Ni-Co alloy sample. This work was supported by the project entitled "Development of controlled-synthesis of carbon nanomaterials with on-demand specifics" at the Venture Business Laboratory in Mie University, Japan.

References

[1] Kajiwara K and Saito Y 2006 Carbon nanotubes for electron emitter devices *Carbon Nanotechnology* ed Liming Dai (Amsterdam: Elsevier) chapter 16 pp 577-609

[2] Miyauchi Y, Chiashi S, Murakami Y, Hayashida Y and Maruyama S 2004 *Chem. Phys. Lett.* **387** 198

[3] Lolli G, Zhang L, Balzano L, Sakulchaicharoen N, Tan Y Q and Resasco D E 2006 *J. Phys. Chem.* B **110** 2108

[4] Meyyappan M 2004 Growth: CVD and PECVD *Carbon Nanotubes Science and Applications* ed M Meyyappan (Boca Raton: CRC Presss) chapter 4 pp 99-116

[5] Kajiwara K, Suzuki S, Sato H and Hata K 2009 *Z. Kristallogr.* **224** 5

[6] Rao C N R, Satishkumar B C, Govindaraj A and Nath M 2001 *CHEMPHYSCHEM* **2** 78

[7] Sato H, Hata K, Hiasa K and Saito Y 2007 *J. Vac. Sci. Technol.* B **25** 579

[8] Murakami Y, Chiashi S, Miyauchi Y, Hu M, Ogura M, Okubo T and Maruyama S 2004 *Chem.*

Phys. Lett. **385** 298
[9] Noda S, Tsuji Y, Murakami Y and Maruyama S 2005 *Appl. Phys. Lett.* **86** 173106
[10] Tsai A P 1999 Metallurgy of Quasicrystals *Physical Properties of Quasicrystals* ed Z M Stadnik (Berlin: Springer) chapter 2 pp 5-50
[11] Klein T and Symko O G 1994 *Appl. Phys. Lett.* **64** 431
[12] Dubois J M 2005 Thin films and nanosized precipitates *Useful Quasicrystals* (Singapore: World Scientific) pp 336-349
[13] Bonasso N, Piageat P, Rouxel D and Weber B 2002 *Thin Solid Films* **409** 165
[14] Balzuweit K, Meekes H, van Tendeloo G and de Boer J L 1993 *Phil. Mag.* B **67** 513
[15] Cahn J W, Shechtman D and Gratias D 1986 *J. Mater. Res.* **1** 13
[16] Liu H-P, Takagi D, Ohno H, Chiashi S, Chokan T and Homma Y 2008 *Appl. Phys. Express* **1** 014001
[17] Baletto F and Ferrando R 2005 *Rev. Mod. Phys.* **77** 371
[18] Schebarchov D and Hendy S C 2008 *Nano Lett.* **8** 2253
[19] Boudard M and de Boissieu M 1999 Experimental Determination of the Structure of Quasicrystals *Physical Properties of Quasicrystals* ed Z M Stadnik (Berlin: Springer) chapter 4 pp 91-126

Evolution and structural transition of quasicrystalline phases in Al-Fe-Cu-Cr and Al-Fe-Cu-Mn alloys during high energy ball milling

Thakur Prasad Yadav[1], Nilay Krishna. Mukhopadhyay[2*], Radhey Shyam Tiwari[1], and Onkar Nath Srivastava[1]

[1]Department of Physics, Banaras Hindu University, Varanasi-221 005 (INDIA)
[2]Department of Metallurgical Engineering, Institute of Technology, Banaras Hindu University, Varanasi-221 005, (INDIA)

*E-mail:mukho_nk@rediffmail.com

Abstract. The main objective of the current work was to investigate the transformation mechanism in quasicrystals during mechanical alloying/milling. The quasicrystalline phases have been synthesized by melting the individual elements as well as by mechanical alloying of the concerned elements using high energy ball milling. The transformation of icosahedral and decagonal quasicrystalline phases to various disordered states in Al-Fe-Cu-Cr and Al-Fe-Cu-Mn alloys during milling of the pre-alloyed material. The milling of a quasicrystalline alloy was carried out using a high energy ball mill by varying milling times up to 40h under liquid hexane medium and at various milling intensity. X-ray diffraction was carried out for evaluating the lattice strain, lattice parameters and crystallite sizes of the mechanically milled samples. The evolution of nano icosahedral and nano decagonal phases as well as crystalline phases was found to occur. The subsequent thermal treatment led to the structural ordering in the concerned phases. The implication of the evolution of various phases, their structural correlations and their relative stability will be discussed.

1. Introduction

Quasicrystals (QC) have well-ordered structures without periodicity, and exhibit noncrystallographic rotational symmetries (e.g. five-fold, eight-fold, ten-fold and twelve- fold) as well as crystallographic rotational symmetries [1-4].

Due to the aperiodicity, these materials are expected to exhibit properties that are very different from conventional metallic materials and these properties can be exploited for industrial applications [5-10]. The stability of quasicrystalline materials during mechanical milling (MM) appears to be important as they can give rise to other stable/metastable phases [11]. The icosahedral phase in Al-Cu-Fe system is of particular interest because it appears to have a very perfect quasicrystalline structure [12]. It is known that QC phase in Al-Cu-Fe undergoes a structural transformation to various types of crystalline phases during MM [13]. The structure of Al-Cu-Fe alloy is sensitive to the additions of other elements such as chromium, manganese which destabilizes the icosahedral phase and promotes the formation of the decagonal and one dimensional quasicrystal [14-15]. It is interesting to emphasize that the stability of deacagonal and one dimensional quasicrystal should be examined during MM. As it is possible to obtain a nanoscale microstructure after MM, which may lead to the formation of equilibrium or stable phases in the Al-Cu-Fe-Cr and Al-Cu-Fe-Mn systems.

The aim of the present investigation is to synthesize a decagonal phase in $Al_{65}Cu_{20}Fe_7Cr_8$ and one dimensional quasicrystal in $Al_{65}Cu_{20}Fe_7Mn_8$ alloy system and to study its phase stability during mechanical milling. An isothermal heat treatment in vacuum and air was also investigated in order to study the possible formation of spinel structure.

2. Experimental Procedure

Alloys with a composition close to $Al_{65}Cu_{20}Fe_7Cr_8$ and $Al_{65}Cu_{20}Fe_7Mn_8$ (Al=99.98%, Cu=99.99%, Fe = 99.98 % Cr=99.97%, Mn = 99.92 % pure) were prepared by melting in an R.F. induction furnace under a dry argon atmosphere. The individual elements were at first mixed in correct stoichiometric proportions and pressed into a cylindrical pellet of 2 cm diameter and 0.75 cm thickness by applying a pressure of $\sim 3 \times 10^4$ N/m^2. The pellet (5 g by weight) was then placed in a silica tube surrounded by an outer pyrex glass jacket. Under continuous flow of argon gas into the silica tube, the pellet was melted using radio frequency induction furnace (18 kW). The ingots formed were re-melted several times to ensure homogeneity. This pre-alloyed as cast ingot was crushed to particles less than 0.5mm in size and placed in an attritor ball mill (Szegvari Attritor) with ball to powder ratio of 40:1. The attritor has a cylindrical stainless steel tank of inner diameter 13 cm. The grinding balls made of stainless steel are of 6 mm in diameter. The speed of the mill was maintained at 400 rpm. The milling operation was conducted for various time ranging from 5h to 40h using hexane as a process control agent (PCA). Powders obtained after 40 h of milling was further annealed at 400, 500, 600 700, 800 and 900 °C for time ranging from 10 to 60h in air ambient. The air annealing has been done for oxidation of milled powders to form spinel phase. Structural and microstructural characterizations of the mechanically activated and annealed powders were performed using scanning electron microscope (SEM – QUANTA 200), X-ray diffraction (XRD) with CuK$_\alpha$ radiation (Philips PW-1710 X-ray diffractometer, λ =1.54026 Å), scanning electron microscopy (Philips XL-20) and transmission electron microscopy (TEM- Techni G^{20} at 200kV).The effective crystallite size and relative strain of mechanically milled powders as well as heat-treated products were calculated based on line broadening of XRD peaks. The use of the Voigt function for the analysis of the integral breadths of broadened X-ray diffraction line profiles forms the basis of a single line method of crystallite-size and strain determination [17].

Figure1. Scanning electron microscope images of $Al_{65}Cu_{20}Fe_7Cr_8$ alloy (a) as cast, (b) 40 h milled; $Al_{65}Cu_{20}Fe_7Mn_8$ alloy (c) as cast & (d) 40 h milled.

3. Results and discussions

Fig.1 represents SEM images showing the morphology of (a) as cast, (b) 40 h milled $Al_{65}Cu_{20}Fe_7Cr_8$ alloy; (c) as cast & (d) 40 h milled $Al_{65}Cu_{20}Fe_7Mn_8$ alloy. In both the as cast alloys the μm size particle was observed and after milling, the powders are markedly agglomerated and the coarse particles are formed on the surface of materials, the size of powders tends to be reduced with increasing the MM treatment time. This is due to the formation of nano quasicrystalline / crystalline grain during MM.

In order to investigate the effect of the Cr substitution in place of Fe in Al-Cu-Fe icosahedral phase, the quaternary alloy $Al_{65}Cu_{20}Fe_7Cr_8$ shows the formation of decagonal phase. Fig. 2(a) shows XRD pattern of the decagonal phase obtained from the as cast $Al_{70}Cu_{20}$ Fe_7Cr_8 alloy. The diffraction pattern corresponding to decagonal phase can be indexed by using six independent indices as proposed by Mukhopadhyay and Lord [18]. All of the peaks are from the decagonal phase, no other phase was identified. Fig. 2(b-e) shows the XRD pattern from different hours (10, 20, 40 and 60 h) MM powder exhibiting broadening of the peaks belonging to the B2 phase after 20 h milling. At the initial stage of milling the formation of icosahedral phase coexisting with decagonal phase have been observed (Fig.2 (b)). It should be noted that the shift of the peaks towards lower θ angle side with increase in milling time indicate the increase in lattice parameter of the B2 phase. Intensity of (110) peak goes on decreasing with increasing milling time and it becomes broadened for milling time of 60 hrs ,which clearly evidences formation of amorphous along with B2 phase (see Fig 2(e)). XRD pattern indicate that the initially sharp diffraction lines are considerably broadened after subsequent ball milling, suggesting that the nano- crystalline phase is present in the milled sample and have evolved during milling. The calculated crystallite size (d) as a function of milling time shows a rapid decrease of the crystallite size to 14 nm with milling time up to 60h.

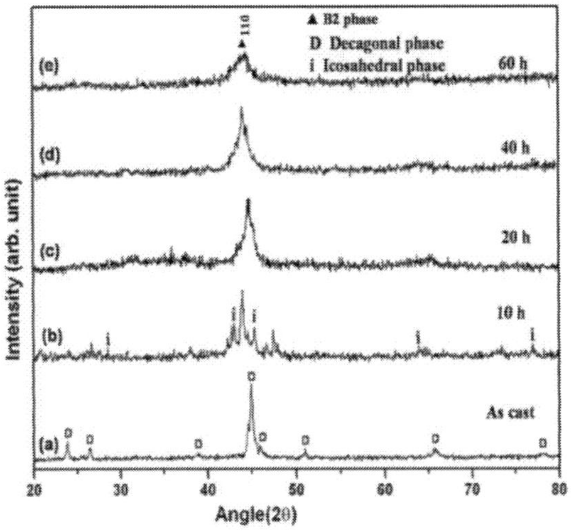

Figure 2. XRD pattern from as cast $Al_{65}Cu_{20}Fe_7Cr_8$ alloy (a) and different hours mechanical milled (10, 20, 40 and 60 h) powder (b-e).

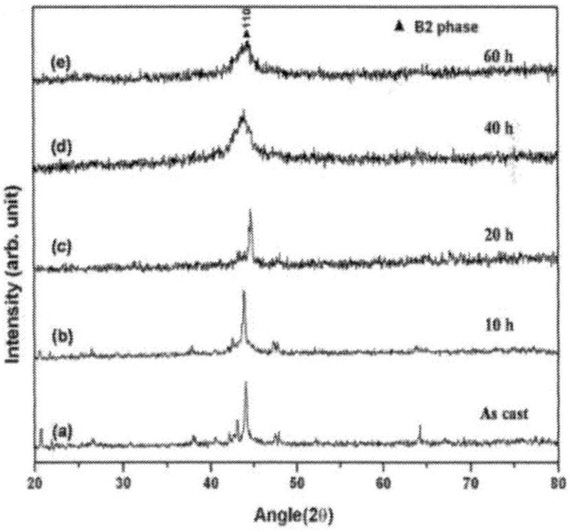

Figure 3. XRD pattern of as cast $Al_{65}Cu_{20}Fe_7Mn_8$ alloy (a) and different hours mechanical milled (10, 20, 40 and 60 h) powder (b-e).

Fig. 3(a) shows XRD pattern of the one dimensional quasicrystal (1DQC) obtained from the as cast $Al_{65}Cu_{20}Fe_7Mn_8$ alloy. Fig. 1(b-e) shows the XRD pattern from mechanical milled (10, 20, 40 and 60 h) powder, exhibiting broadening of the peaks belonging to the B2 phase. From the XRD pattern it is evident that the initially sharp diffraction peaks are considerably broadened after ball milling due to a decrease in the particle size and an increase in internal lattice strains.

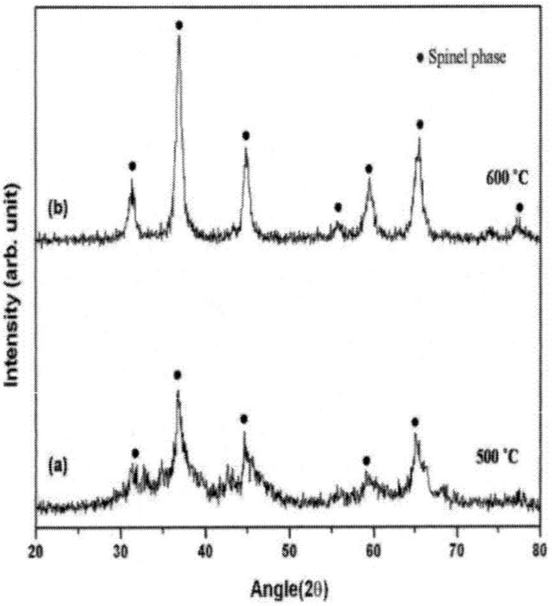

Figure 4. XRD patterns of 40 MM $Al_{65}Cu_{20}Fe_7Cr_8$ powders, air annealing for 60 h at 500 °C (a) and 600 °C (b).

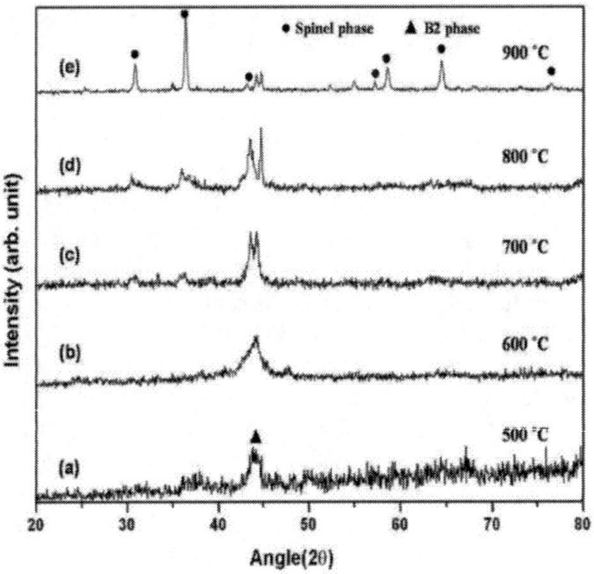

Figure 5. XRD patterns of 40 MM $Al_{65}Cu_{20}Fe_7Mn_8$ powders, air annealing for 60 h at different temperature (a-e).

The investigation on the stability of MM $Al_{65}Cu_{20}Fe_7Cr_8$ alloy has been carried out by annealing in air. In the case of $Al_{65}Cu_{20}Fe_7Cr_8$ MM powder, the formation of single phase spinel has been observed after 60 h of air annealing at 600 °C by 40 h mechanically activated powder. XRD patterns (Fig 4(a-b)) show the peaks corresponding to the spinel structure with the lattice parameter 8.05 ±0.04 Å. It is obvious that the high degree of oxidation has caused the formation of $(Fe,Cr)Al_2O_4$ spinel structures from the MM materials. It is understood that after 40h of milling, the surface and interface areas are increased due to the formation of smaller particles as well as the nanocrystallite. The thermal stability of 40 MM $Al_{65}Cu_{20}Fe_7Mn_8$ powder has been also done at same annealing parameter. The rate of formation of spinel phase in $Al_{65}Cu_{20}Fe_7Mn_8$ powders is very slow and B2 phase is still present along with spinel phase up to 900° C annealing temperature. When the annealing has been done in air, at different temperatures i.e. 500, 600, 700, 800 and 900 °C a significant phase transformation has been found to occur in case of 40 h mechanically activated 1DQC powder. XRD pattern (Fig 5(a-e)) shows the peaks corresponding to the spinel structure with the lattice parameter 8.07 ± 0.04 Å and parent B2 pase with a= 2.9 ± 0.06 Å. It is very interesting ti note that the kinetics of oxidation rate, which is controlled by oxygen diffusion, are not same in both alloy $Al_{65}Cu_{20}Fe_7Cr_8$, $Al_{65}Cu_{20}Fe_7Mn_8$. It should be noted that the oxidation of $Al_{65}Cu_{20}Fe_7Mn_8$ alloy was not insignificant in comparison to that of $Al_{65}Cu_{20}Fe_7Cr_8$ powder after same annealing parameter. This could be due to the higher affinity of Cr for oxygen compared to that of Mn.

The rigorous transmission electron microscopic (TEM) investigation by obtaining selected area diffraction patterns and microstructural features at different stages have been carried out. Fig. 6 shows TEM investigation of 40h MM $Al_{64}Cu_{29}Fe_7Cr_8$ and $Al_{65}Cu_{20}Fe_7Mn_8$ alloys.

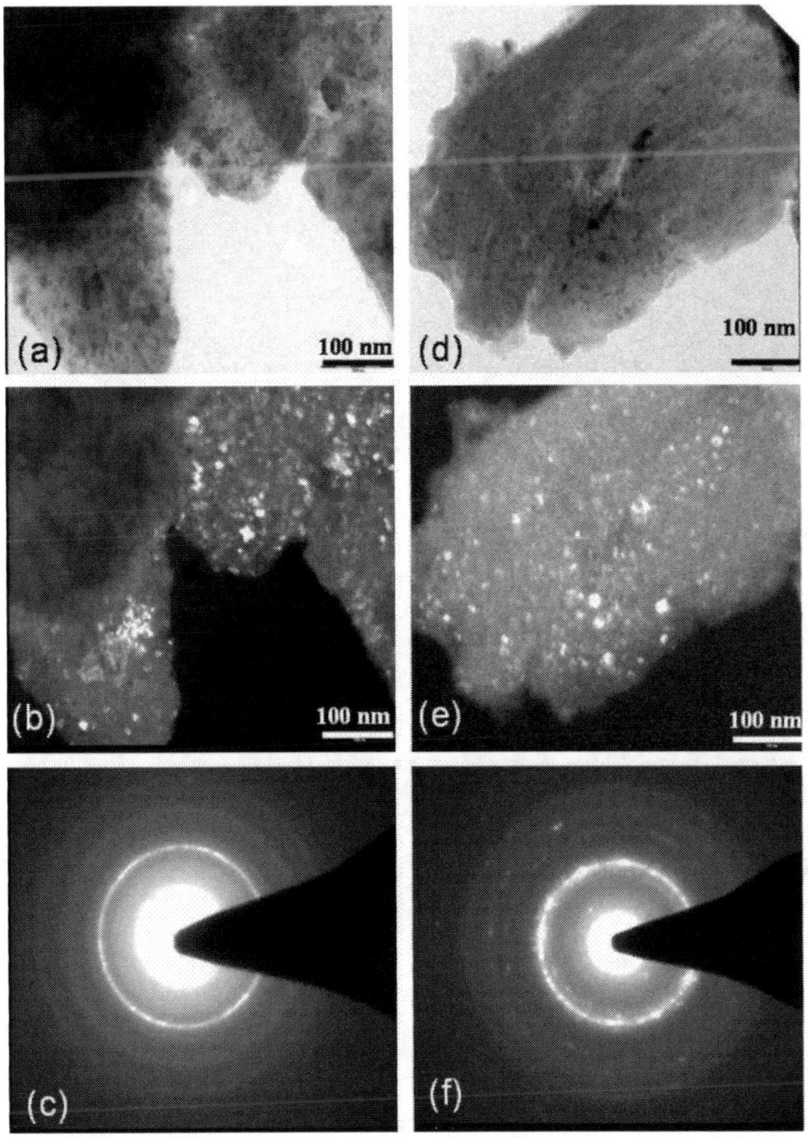

Figure 6. The bright field and dark field TEM microstructure of 40 h MM sample of $Al_{65}Cu_{20}Fe_7Cr_8$ (a-b) and the corresponding electron diffraction pattern in (c). The bright field and dark field TEM microstructure of 40 h MM $Al_{65}Cu_{20}Fe_7Mn_8$ alloy (d-e) and the corresponding electron diffraction patterns in (f).

The bright field and dark field microstructure of 40 h MM sample of $Al_{65}Cu_{20}Fe_7Cr_8$ can be seen in Fig. 6(a-b). Very fine equiaxed particles of the nanometer size (~20 nm) are visible. The corresponding electron diffraction pattern, shown in Fig. 6(c) was indexed due to B2 phase. Fig. 6 (d-e) shows a bright field and dark field TEM micrograph of 40 h MM $Al_{65}Cu_{20}Fe_7Mn_8$ alloy. The large number of small size particles embedded in the big grain in lamellar microstructure can easily be discerned in the Figure 6(d-e). The average particle size is of ~16-24 nm. The corresponding electron diffraction patterns are shown in Fig.6 (f). All the diffraction rings are indexed in the terms of B2 phase with lattice parameter of 2.92 Å. This lattice parameter agrees reasonably well with the XRD results.

HRTEM images of 40 h MM and 60 h annealed $Al_{65}Cu_{20}Fe_7Cr_8$ powder (Figures 7 (a)) clearly show the lattice resolution image of the nanocrystalline spinel phase along (111) plane. The appearance of amorphous like region is due to the presence of formvar used as mounting material in Cu grid. The

HRTEM image of $Al_{65}Cu_{20}Fe_7Mn8$ 40 h MM alloy followed by annealing at 900 °C for 60 h in air (Fig.7 (b)) clearly shows the same type lattice image of spinel phase as mentioned earlier. In this case very sharp fringes along (111) plane indicating the ordering has been observed.

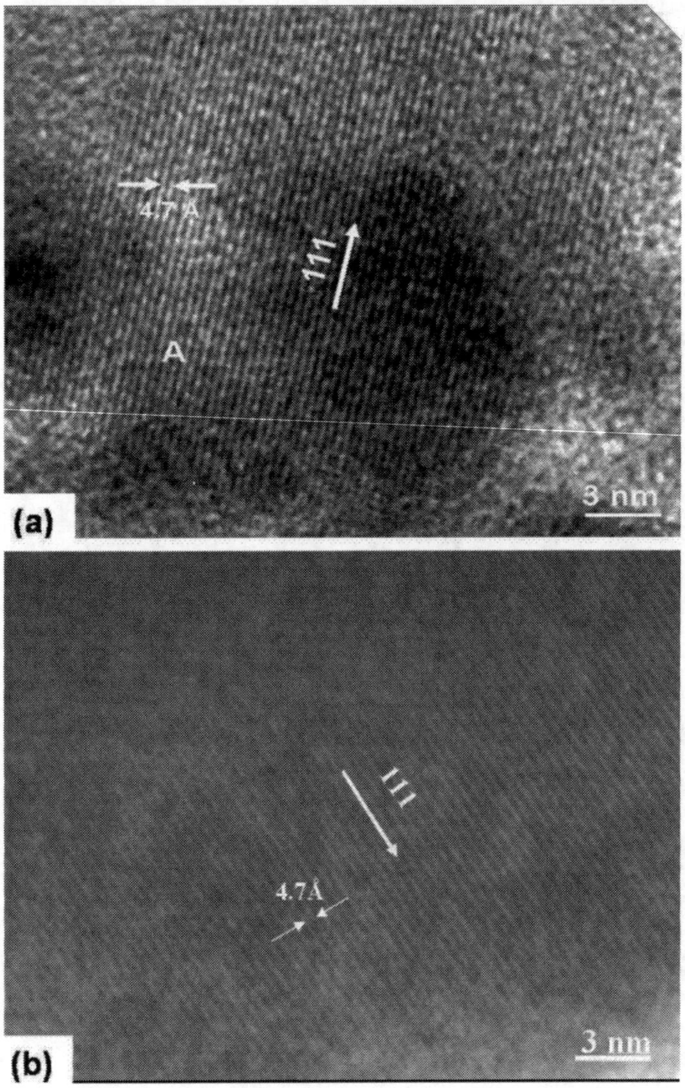

Figure 7. HRTEM images of 40 h MM and 60 h annealed $Al_{65}Cu_{20}Fe_7Cr_8$ powder (a) and $Al_{65}Cu_{20}Fe_7Mn_8$ 40 h MM alloy followed by annealing at 900 oC for 60 h in air (b).

4. Conclusions

It is clear from the present investigation that after milling the decagonal and 1DQC quasicrystal in $Al_{64}Cu_{29}Fe_7Cr_8$ and $Al_{65}Cu_{20}Fe_7Mn_8$ alloy have transformed to the crystalline phase (i.e., disordered B2 phase) with a nanoscale microstructure of 20 nm approximately. At the initial stage of milling (10h) of $Al_{64}Cu_{29}Fe_7Cr_8$ alloy, the decagonal phase transforms to an icosahedral phase which further transformed to B2 phase. In both the cases the evolution of the spinel structure started upon air annealing at 600 °C from the B2 phase formed due to milling of quasicrystalline phases for 40h but $Al_{65}Cu_{20}Fe_7Mn_8$ alloy is more stable compare to $Al_{64}Cu_{29}Fe_7Cr_8$ alloy. From the analysis it appeared that Al-Cu-Fe-Cr decagonal phase exhibits better spinel forming ability compared to that of Al-Cu-Fe-Mn 1DQC. This could be due to the initial structure as well as the higher affinity of Cr for oxygen compared to that of Mn.

Acknowledgement
The authors would like to thank Prof G.V.S. Sastry, Prof. R.K. Mandal, and Dr. M.A.Shaz, for discussions. The authors also gratefully acknowledge the Department of Science and Technology (DST), New Delhi and the Ministry of New-Renewable Energy, New Delhi, India for financial support.

References

[1] Shechtman D, Blench I, Gratias D and Cahn J W 1984 *Phys Rev. Lett*, **53** 1951
[2] Tsai A P, Inouc A, Masumoto T and Kataoka N 1988 *Jpn. J. Appl. Phys.* **27** L2252
[3] Poon S J, Drehman A J and Lawless K R 1985 *Phys. Rev. Lett.* **55** 2324
[4] Tsai A P, Inoue A and Masumoto T 1989 *Mater. Trans. Jpn. Inst. Met.* ***30*** 150
[5] Archambault P and Janot C 1997 *Mater. Res. Soc. Bulletin.***22** 48
[6] Mancinelli Chris, Ko1 J S and Jenks C J 2001 *Mat. Res. Soc. Symp. Proc.* **643** K821
[7] McGrath R, Ledieu J, Erik Cox J and Renee Diehl D 2002 *J. Phys.: Condens. Matter.* ***14*** R119
[8] Besser MF and Eisenhammer T 1999 *Mater.Res. Soc. Bulletin*, ***722*** 59
[9] Urban K, Feuerbacher M and Wollgarten M 1997 *Mater.Res. Soc. Bulletin.* **22** 65
[10] Kelton K F and Gibbons P C 1997 *Mater.Res. Soc. Bulletin.* **22** 69
[11] Tiwari RS,Yadav TP, Mukhopadhyay NK and Srivastava O N 2009 *Z. Kristallogr.* ***224*** 26
[12] Elina Huttunen-Saarivirta 2004 *Journal of Alloys and Compounds* ***363*** 150
[13] Yadav T P, Mukhopadhyay N K, Tiwaria R S and Srivastava O N 2005 *Trans. IIM* ***58*** 1169
[14] Dong C and Dubois J M 1991 Journal of materials science **26** 1647
[15] Tsai A P, Sato A, Yamamoto A,Inoue A and Masumoto T 1992 *J. App Phys* ***31*** L970
[16] Mukhopadhyay N K,Yadav T P,Tiwari R S and Srivastava O N 2008 *Z. Kristallogr.* ***223*** 716
[17] Yadav TP, Mukhopadhyay N K, Tiwari RS and Srivastava O N. 2005 *Mater. Sc. Eng.A* ***393*** 366
[18] Mukhopadhyay NK and Loard E A 2002 *Acta. Cryst. A.* ***58*** 424

Spectroscopic study on chemical ordering of transition metals in decagonal Al-Co-Ni and Al-Co-Cu quasicrystals

M Inukai[1,2], K Soda[1], M Kato[1], S Yagi[1], Y Tezuka[3], Y Yokoyama[4] T Muro[5] and Y Saitoh[6]

[1]Graduate School of Engineering, Nagoya University, Furo-cho, Chikusa-ku, Nagoya 464-8603, Japan

[2]Venture Business Laboratory, Nagoya University, Furo-cho, Chikusa-ku, Nagoya 464-8603, Japan

[3]Faculty of Science and Technology, Hirosaki University, Bunkyo-cho, Hirosaki 036-8561, Japan

[4]Institute of Materials Research, Tohoku University, Katahira, Aoba-ku, Sendai 980-8577, Japan

[5]Japan Synchrotron Radiation Research Institute, Kouto, Sayo-cho, Sayo, Hyogo 679-5198, Japan

[6]Synchrotron Radiation Research Center, Japan Atomic Energy Agency, Kouto, Sayo-cho, Sayo, Hyogo 679-5148, Japan

E-mail: inukai@vbl.mbox.nagoya-u.ac.jp

Abstract. Preferred occupation at the transition metal (TM) sites in the building unit cluster of two-dimensional Al-Co-Ni and Al-Co-Cu quasicrystals has been investigated by comparing their experimental electronic structures with those of the unit clusters calculated by a discrete variational $X\alpha$ method. Analysis of the bonding character and bond energy for local clusters around the TM sites shows that the Al-Co-Cu quasicrystal favours the Cu and Co occupations at the first and second inner TM sites, respectively, and hence the chemical order of these TM's, while the small bond energy difference in the Ni and Co occupations at those TM sites suggests the large chemical disorder of the transition metals in the Al-Co-Ni quasicrystal.

1. Introduction

Decagonal phases of Al-Co-Ni and Al-Co-Cu alloys are typical two-dimensional quasicrystals, where quasiperiodic planes are periodically stacked along a tenfold axis (the *c*-axis) [1]. The atomic arrangement in the quasiperiodic planes is often viewed as a special tiling such as the Penrose one composed of two unit tiles, a skinny rhombus and a fat one, and it may be regarded as consisting of a single decagonal building unit cluster with overlapping rules [1-3]. However, because of the lack of periodicity, it is still challenging to determine the atomic arrangement and clarify the formation mechanism of the unique quasiperiodic arrangement.

For the Al-Co-Ni quasicrystal, high-resolution transmission electron microscopic (HRTEM) studies show that the various structural modifications are formed as a function of the Co/Ni ratio [4], although it is still hard to distinguish the transition metal (TM) elements, Co and Ni, from each other in the microscopic image. For the Co-rich Al-Co-Ni quasicrystal, the X-ray diffraction (XRD) study of a W-

(AlCoNi) approximant $Al_{72.5}Co_{20}Ni_{7.5}$ has brought a new model with fivefold symmetric Al-TM mixed arrangement for the cluster centre [1,5,6], which also describes well the atomic arrangement of decagonal $Al_{64}Co_{14}Cu_{22}$ quasicrystal [7]. Theoretical studies with use of interatomic pair potentials predict such an Al-TM decagonal ring-like arrangement for the cluster centre of the Co-rich Al-Co-Ni approximant (the decagon tile) [8,9], while they show that the decagonal (pentagonal) symmetry of the cluster centre is broken for the Ni-rich Al-Co-Ni approximant (the hexagon-boat-star (HBS) tiling) [9-11]. Electronic band structure calculations for Al-Co-Ni approximants and a recent *ab initio* study of the W-(AlCoNi) approximant show the importance of the energetically favourable Al-Co and Ni-Ni interactions as well as the contribution of the Hume-Rothery mechanism to the structural stabilization [12,13]. It is also pointed out that the chemical ordering between Al and TM may be an origin of the broken symmetry in the Ni-rich Al-Co-Ni quasicrystal [14,15]. However, the chemical ordering is not so clear in particular for the transition metals because of the difficulty in distinguishing them from each other.

In this paper, we have studied the atomic arrangement of the Al-Co-Ni and Al-Co-Cu quasicrystals by comparing the electronic structure obtained by the soft X-ray photoelectron spectroscopy (XPS) and the TM L_α X-ray emission spectroscopy (XES) with those calculated theoretically by the discrete variational $X\alpha$ (DV-$X\alpha$) method for the model unit cluster, based on the XRD and electron microscopic observations, in order to specify the sites occupied by Co, Ni and Cu and clarify the chemical nature bringing its unique atomic arrangement with the various modifications. A part of the present study for the Al-Co-Ni quasicrystal has been already reported in detail elsewhere [16].

2. Experimental and Calculating procedures

Specimens were single-grain decagonal quasicrystals $Al_{72}Co_{16}Ni_{12}$ and $Al_{65}Co_{20}Cu_{15}$ grown by the Czochralski method [17,18]. For the XES measurement, their surfaces were mechanically polished just before the measurement, while clean surfaces for the XPS measurement were prepared by *in situ* fracturing specimens perpendicularly to the tenfold axis under the pressure of 1 - 4 x 10^{-8} Pa.

The XES measurement for the Al-Co-Cu quasicrystal was performed at 100 K with a slit-less spectrometer in the depolarized configuration at the beamline BL27SU of SPring-8, the Japan Synchrotron Radiation Research Institute [19,20]. The photon energy of the spectrometer was calibrated with elastically scattered excitation light, whose energy was determined by measuring photoelectron spectra of the Fermi edge and the $4f$ core level of an evaporated Au film. The excitation photon energy $h\nu$ was tuned at 968.83 eV for the present Co and Cu L_α XES measurements. In order to represent the XES spectra in terms of the binding energy, we have used the Co and Cu $2p_{3/2}$ core level binding energies E_B of 778.53 and 933.89 eV, respectively, obtained from the XPS measurement. The error in the binding energy scale of the XES spectrum is estimated to be within 0.2 eV.

The XPS measurement for the Al-Co-Cu quasicrystal was performed at 20 K with a high-resolution hemispherical analyzer in the angle-integrated mode at the beamline BL23SU of SPring-8 [21]. The total energy resolution and the origin of the binding energy, the Fermi energy E_F, were determined by measuring the Fermi edge of evaporated Au films. The energy resolution was estimated to be 0.27 eV at $h\nu$= 1200 eV. The experimental details for the Al-Co-Ni quasicrystal are described elsewhere [22].

The DV-$X\alpha$ method is one of the real-space first principle molecular orbital calculations, and it is powerful for studying the electronic structure and chemical bonding in the materials including many TM's without the periodicity [23]. In this method, the Hartree-Fock-Slater molecular equation is solved by the self-consistent charge procedure with the Slater's $X\alpha$ potential as the electron exchange-correlation and the random-sampling integration for the matrix elements of Hamiltonian and overlap integrals. For the present calculation, we have modified the available code SCAT [24] for faster calculation with larger cluster than the original one [16].

The model unit cluster has a triple layered structure along the c-axis with two types of quasiperiodic layers, A and B, shown in figure 1, in the A-B-A or B-A-B stacking sequences with the interlayer distance $c/2$ of 0.2000 and 0.2074 nm for the Al-Co-Ni and Al-Co-Cu quasicrystals, respectively. The atomic arrangement in the quasiperiodic layers is based on the model proposed for the W-(AlCoNi) approximant [1,5] and is the same for both the quasicrystals [7]; an underlying Penrose tiling with the fat and skinny rhombi is shown by broken lines in the figure and the quasilattice constant a_q, that is, the edge length of the rhombi is taken as 0.250 and 0.230 nm for the Al-Co-Ni [1] and Al-Co-Cu [25] quasicrystals, respectively. In figure 1, open circles represent Al, and closed circles and diamonds stand for two kinds of transition-metals, TM1 and TM2, respectively, *i.e.* Co or Ni for the Al-Co-Ni quasicrystal and Co or Cu for the Al-Co-Cu one; the total number of TM1 per a pair of A and B layers is set equal to that of TM2, and the occupation of the nearest neighbour TM is avoided for the mixed sites of the proposed model for simplicity. Open squares show hydrogen atoms to reduce the so-called surface effects for the outer TM in the calculation for the Al-Co-Ni quasicrystal, but we could not place these hydrogen atoms for the Al-Co-Cu quasicrystal in the present code. A large thick circle in the figure indicates the size of the unit cluster, ~2 nm, and the nominal composition per a pair of the A and B layers, $Al_{81}TM1_{15}TM2_{15}(H_{50})$, may correspond to $Al_{73.0}TM1_{13.5}TM2_{13.5}$.

In order to avoid the surface effects, the density of states (DOS) for the model cluster is obtained as the sum of the local DOS's of the central layers A and B for the respective B-A-B and A-B-A model clusters. The DOS are shown after convoluting their calculated line spectra with a Gaussian function of the 0.17 eV half width at half maximum.

3. Results and Discussion

Calculated local TM 3*d* DOS's at the various TM sites assigned in figure 1 are shown in figure 2 for the Al-Co-Cu quasicrystal with total and partial DOS's of the model clusters: (a) a Cu centre model, *i.e.* a model with TM1 = Cu and TM2 = Co in figure 1, and (b) a Co centre model, *i.e.* TM1 = Co and TM2 = Cu. As already reported for the Al-Co-Ni quasicrystal [16], the local TM 3*d* DOS's at the sites 2 and 3 for both the model clusters have almost the same energy distribution also in the Al-Co-Cu quasicrystal because of the resemblance in their nearest neighbour arrangement, and they are the lowest in energy due to the large Al-TM interaction. The small difference in their distribution may be caused by the different arrangement of the second nearest neighbours. The 3*d* DOS at the site 4 is distributed over the highest energy region due to the weak Al-TM interaction, while the energy distribution of the 3*d* DOS at the site 5 is similar to that at the site 1. The dependence of the local DOS on the TM site suggests the importance of the Al-TM interactions or the Al coordination. Thus, we will ignore the sites 4 and 5 below, since their local atomic arrangements are incomplete in the present

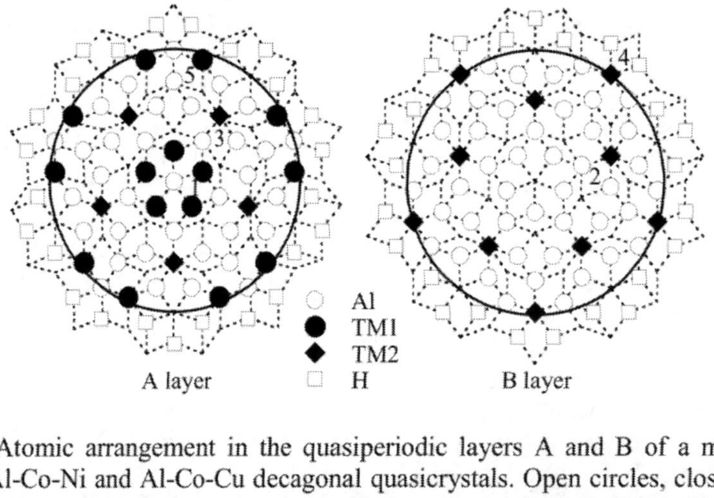

Figure 1. Atomic arrangement in the quasiperiodic layers A and B of a model unit cluster of Al-Co-Ni and Al-Co-Cu decagonal quasicrystals. Open circles, closed circles and diamonds, and open squares represent Al, the transition metals TM1 and TM2, and H, respectively. The non-equivalent tranistion metal sites are assigned with numbers. A large thick circle indicates the size of the unit cluster, ~2 nm.

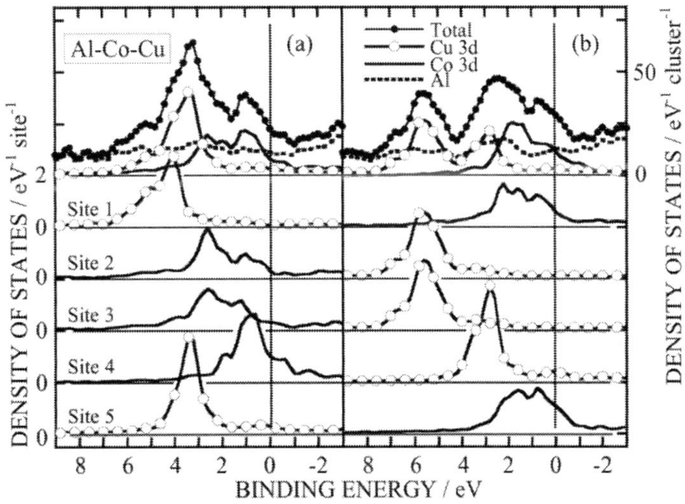

Figure 2. Local density of the Co and Cu $3d$ states at different transition metal sites of the model unit clusters (a) Cu centre model and (b) Co centre model. The transition metal sites, 1 to 5, are assigned in figure 1. Total and partial densities of the states of the unit model clusters are also shown in the top panel.

calculation. Both the Co and Cu $3d$ DOS's at the sites 2 and 3 are located lower in energy than those at the innermost TM site 1, which might imply the preferential occupation at the site 2 (or 3) rather than the site 1. However, the anti-bonding character increases near E_F for the TM-Al bond at the sites 2 and 3 in comparison to the site 1, as observed in the Al-Co-Ni quasicrystal [16]. The energy gain by the Cu occupation at the site 1 may lead to the energy consumption by the Co occupation at the site 2, and *vice versa*. It should be also noted that the Al partial DOS's for both models show a pseudogap near E_F, implying the contribution of the Hume-Rothery mechanism to the stabilization of this quasicrystal.

In figure 3, we compare the TM $3d$ partial DOS, calculated by summing up the local DOS's at the sites 1 to 3, with the experimental XES and XPS spectra for the Al-Co-Ni and Al-Co-Cu quasicrystals. The panels (a) and (b) show the TM $3d$ DOS's for the Ni or Cu centre models and for the Co centre model, respectively. The experimental spectra of the Al-Co-Cu quasicrystal shows well-separated Co and Cu $3d$ bands, while the XPS spectrum of the Al-Co-Ni quasicrystal reveals a single $3d$ band which is composed of the Co and Ni $3d$ bands in the low and high binding energy sides, respectively. In the Co centre model, the Cu $3d$ band is separated largely from the Co $3d$ one. Thus, as a whole, the calculated energy distributions for the Cu centre model agree fairly well with the experimental results, which suggests the preferential occupation of Cu at the site 1. On the other hand, the Co and Ni $3d$ bands are distributed over rather wide region for the Co centre model; a single peaked TM $3d$ band appears for the Ni centre model but the Co $3d$ band is located at higher binding energy than the Ni one, inconsistent with the experimental results. Instead, it is found in the previous study [16] that the $3d$ partial DOS for a model with Co and Ni mixed at the site 1 agrees better with the experimental result than the others. This suggests the chemical disordering may occur in the Al-Co-Ni quasicrystal.

In order to compare quantitatively the chemical stabilities at the TM sites 1 (the innermost TM site of the model cluster) and 2 (the second innermost one), we have calculated the local bond energy at the TM sites and bond order of the Al-TM bonds at the TM sites, because the present DV-$X\alpha$ method does not have enough accuracy, unfortunately, to compare the total energy for the whole system including the deep core levels. The results are summarized in table 1. Here, the local bond energy is defined as the energy gain of the occupied TM band states at the TM site compared to the sum of the relevant occupied orbital energy of an isolated TM atom, and the bond order shows how large the bonding nature in the concerned molecular states [26]. For the Al-Co-Cu quasicrystal, the total bond

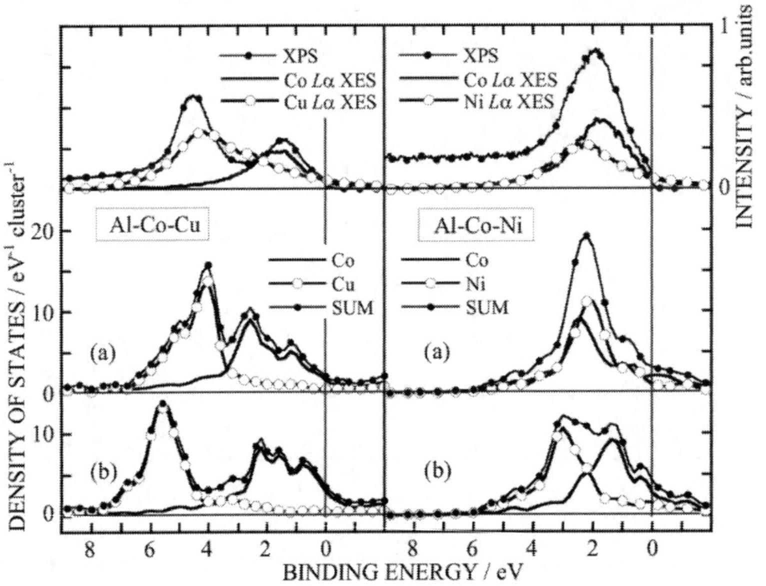

Figure 3. Transition metal (TM) 3*d* band calculated with Co and Cu (or Ni) at two types of the sites, 1 and 2(3) shown in figure 1, for (a) Cu (or Ni) centre model and (b) Co centre model for Al-Co-Cu and Al-Co-Ni quasicystals. The TM *Lα* X-ray emission spectra (XES) and valence-band X-ray photoelectron ones (XPS) of $Al_{65}Co_{20}Cu_{15}$ (left) and $Al_{72}Co_{16}Ni_{12}$ (right) quasicrystals are also shown in the top panel.

energy, which is higher by ~1.6 eV for the Cu centre model than for the Co one, clearly indicates the preferential occupation of Cu at the site 1 and Co at the site 2, and the bond order also may favour the Cu centre model in a sense of almost the same bond strength at the various TM sites. These suggest a highly chemical ordering of Cu and Co in the Al-Co-Cu quasicrystal. For the Al-Co-Ni quasicrystal, the bond order may prefer the Ni centre model to the Co one, but the bond energy is slightly high for the Co centre model. The difference in the bond energy is very small and hence the chemical disordering may be expected to occur easily for the Al-Co-Ni quasicrystal.

Although the peripheral part of the unit cluster is not clear yet in the present study, tendency in the TM occupation may provide some hint to the observed Co/Ni dependence of the modifications in the Al-Co-Ni quasicrystal; Hiraga *et al.* [1,27] have pointed out that the density of unit clusters increases with the Co/Ni ratio, suggesting the increase in the linkage of the unit cluster forming a skinny rhombus, which in turn may contain less Ni than the linkage forming a fat rhombus. The strong Al-Co interaction as well as the chemical disordering might play an important role in forming the various

Table 1. Local bond energy at the transition metal (TM) site and Al-TM bond order.

Model Cluster (TM1, TM2)	Site 1 Bond Energy (eV atom^{-1}) / Bond Order		Site 2 Bond Energy (eV atom^{-1}) / Bond Order		Total Bond Energy (eV atom^{-1})
Al-Co-Ni	Co	Ni	Co	Ni	Co+Ni
(Ni, Co)	–	18.32 / 0.145	19.68 / 0.146	–	38.00
(Co, Ni)	15.90 / 0.155	–	–	22.16 / 0.135	38.06
Al-Co-Cu	Co	Cu	Co	Cu	Co+Cu
(Cu, Co)	–	23.16 / 0.137	18.32 / 0.148	–	41.48
(Co, Cu)	12.69 / 0.154	–	–	27.20 / 0.131	39.89

modifications in a wide composition range.

4. Conclusions

Comparison between the theoretical and experimental electronic structures and analysis of the local bond energy and bond order at the TM sites show that the Al-Co-Cu quasicrystal favours the Cu and Co occupations at the first and second inner TM sites, respectively, and hence the chemical order of Co and Cu, while the small difference in the total of the local bond energy suggests the large chemical disorder of TM's in the Al-Co-Ni quasicrystal, which might bring its various modifications.

Acknowledgements

The XPS and XES measurements were performed under the approval of the Japan Synchrotron Radiation Institute and the Institute of Materials Structure Science. We would appreciate the staff members of the synchrotron radiation facilities, SPring-8 and Photon Factory, for their kind support during the measurements. This work was partly supported by International Union of Crystallography and Venture Business Laboratory of Nagoya University.

References

[1] Hiraga K 2002 *Adv. Imaging and Electron Phys.* **122** 1
[2] Zhang X and Fu X 2006 *Phys. Lett.* A **351** 327
[3] Jeong H-C and Steinhardt P J 1997 *Pys. Rev.* B **55** 3520
[4] Ritsch S, Beeli C, Nissen H-U, Godecke T, Scheffer M and Luck R 1998 *Phil. Mag. Lett.* **78** 67
[5] Sugiyama K, Nishimura S and Hiraga K 2002 *J. Alloys Comp.* **342** 65
[6] Deloudi S, Kobas M and Steurer W 2006 *Phil. Mag.* **86** 581
[7] Taniguchi S and Abe E 2008 *Phil. Mag.* **88** 1949
[8] Krajči M, Hafner J and Mihalkovič M 2006 *Phil. Mag.* **86** 593
[9] Hiramatsu S and Ishii Y 2006 *J. Phys. Soc. Jpn.* **75** 054602
[10] Gu N, Henley C L and Mihalkovič M 2006 *Phil. Mag.* **86** 593
[11] Mihalkovič M, Al-Lehyani I, Cockayne E, Henley C L, Moghadam N, Moriarty J A, Wang Y and Widom M 2002 *Phys. Rev.* B **65** 104205
[12] Krajči M, Hafner J and Mihalkovič M 2000 *Phys. Rev.* B **62** 243
[13] Hassdenteufel K H, Oganov A R, Katrych S and Steurer W 2007 *Phys. Rev.* B **75** 144115
[14] Abe E Saitoh K Takakura H Tsai A P Steinhardt P J and Jeong H -C 2000 *Phys. Rev. Lett.* **84** 4609
[15] Yan, Y and Pennycook S J 2001 *Phys. Rev. Lett.* **86** 1542
[16] Inukai M, Soda K, Kato M, Yagi S and Yokoyama Y 2008 *Z. Kristallogr.* **223** 851
[17] Yokoyama Y, Note R, Kimura S, Inoue A, Fukaura K and Sunada H 1997 *Mater. Trans.* **38** 943
[18] Yokoyama Y, Note R, Yamauchi A, Inoue A, Fukaura K and Sunada H 1999 *Mater. Trans.* **40** 123
[19] Tokushima T, Harada Y, Ohashi H, Senba Y and Shin S 2006 *Rev. Sci. Instrum.* **77** 063107
[20] Ohashi H, Ishiguro E, Okumura H, Hiraya A, Senba Y, Okada K, Saito N, Suzuki I. H, Ueda K, Ibuki T, Nagaoka S, Koyano I and Ishikawa T 2001 *Nucl. Instrum. Methods* A **467** 533
[21] Saitoh Y, Nakatani Y, Matsushita T, Agui A, Yoshigoe A, Teraoka Y and Yokoya A 2001 *Nucl. Instrum. Methods Phys. Res.* A **474** 253
[22] Inukai M, Soda K, Ota S, Miyazaki H, Suzuki T, Kato M, Yagi S, Tezuka Y and Yokoyama Y 2007 *Phil. Mag.* **87** 3003
[23] Adachi H, Tsukuda M and Satoko C 1978 *J. Phys. Soc. Jpn.* **45** 875
[24] http://www.dvxa.org/
[25] Steurer W and Kuo K H 1990 *Acta. Cryst.* B **46** 703
[26] Sutton A P 1996 *Electronic Structure of Materials* (Oxford: Clarendon Press) pp 52-66
[27] Hiraga K, Ohsuna T, Sun W and Sugiyama K 2001 *Mater. Trans.* **42** 2354

Modes vs. modulations: symmetry-mode analysis of commensurate modulated structures compared with the superspace method

J M Perez-Mato, D Orobengoa, M I Aroyo and L Elcoro
Departamento de Física de la Materia Condensada, Facultad de Ciencia y Tecnología, Universidad del País Vasco, Apdo. 644, 48080 Bilbao, Spain

E-mail: jm.perez-mato@ehu.es

Abstract. We know from Landau theory that the natural language to deal with distorted structures is the one of symmetry-adapted modes. A symmetry-mode decomposition is both useful for investigating the mechanisms responsible of these phases and for pure crystallographic purposes. Some symmetry-mode components are usually dominant, introducing a parameter hierarchy or reducing in practice the crystallographic degrees of freedom. Several computer tools are now freely available allowing structure refinements directly using symmetry-mode collective coordinates. Alternatively, the superspace formalism is a very efficient method for similar purposes. By means of several examples we compare the two approaches.

1. Introduction

Structural distortions in distorted structures can be decomposed into contributions from different modes with symmetries given by irreducible representations (irreps) of the parent space group. One can then distinguish primary and secondary distortion components, which have in general quite different weights in the structure, and respond differently to external perturbations. In general, the use of symmetry-adapted modes introduces a physical hierarchy among the structural parameters isolating the actual distortion responsible for the phase. These collective mode parameters are in general better adapted for a controlled refinement of the structure, or for instance, for comparative studies between different materials. Symmetry-mode analyses of commensurately distorted structures have been regularly done since decades (see for instance refs 1-8). But despite their advantages they have remained somehow exceptional. Each mode analysis generally required extensive specific group theoretical calculations, and the mode parameterization was usually done *a posteriori*, and not in the refinement process. This situation has now drastically changed. Freely available programs as AMPLIMODES [9] and ISODISPLACE [10] allow the automatic parameterization with collective symmetry-mode coordinates of any system, and its direct refinement.

On the other hand, the superspace formalism, with a program like JANA2006 [11], has become a very efficient tool for the refinement and analysis of commensurate distorted structure, treated as modulated structures. One or several *primary* modulation wave vectors are defined and the distortion is described as a superposition of harmonics for this set of wave vectors. The superspace symmetry defines the correlations and symmetry restrictions that the atomic displacements must have for each of these harmonics. In a commensurate case, the number of possible harmonics is finite, and a hierarchy between them is usually present.

Symmetry-mode and superspace formalisms are closely related. Displacive irrep modes are in fact modulations and, for instance, the superspace groups resulting from the freezing of a specific irrep distortion mode with incommensurate wave vector have been analysed in detail in the past [12-14]. Now, the possibility of refining any distorted commensurate structure using symmetry-mode coordinates, as an alternative to the superspace description, opens up new perspectives that we explore here. By means of several examples we show the close relation of both approaches, showing their similarities and the differences that make more convenient one or the other method depending on the system.

2. A simple example. Triclinic distortion of NbS_3

NbS_3 belongs to a series of compounds of type MX_3 (M= Nb,Ta, Zr, Ta; X= S, Se, Te) with similar structures. While most of them are monoclinic with $P2_1/m$ symmetry, NbS_3 is triclinic, with space group *P-1*. This is due to a small displacive distortion that multiplies by two the cell parameter along the monoclinic unique axis, and breaks the monoclinic symmetry [15-16]. By comparison with the other compounds of the family, one can easily postulate for NbS_3 a non-distorted *average* parent structure of symmetry $P2_1/m$. The symmetry-mode decomposition of the distortion with respect to this reference, (obtained using AMPLIMODES [9]), is summarized in table 1.

Table 1. Summary of the symmetry-mode decomposition of NbS_3. The *isotropy* subgroup is the space group maintained by the distortion component. For each symmetry-adapted component, the third column indicates the direction in the irrep space and the fifth one the number of independent basis modes.

k-vector	irrep	direction	isotropy subgroup	dim.	amplitude(Å)
(0,0,0)	GM1+	(a)	$P2_1/m$ (**a**,**b**,**c**;0,0,0)	8	0.000(5)
(0,0,0)	GM2+	(a)	*P-1* (**a**,**b**,**c**;0,0,0)	4	0.036(3)
(0,1/2,0)	Z1	(0,a)	*P-1* (**a**,2**b**,**c**;0,0,0)	12	0.520(4)

The NbS_3 distortion can be decomposed into three symmetry components. One, with wave vector (0 ½ 0) (Z point of the Brillouin zone), breaks directly the symmetry to the observed one. Two additional distortions with null wave vector, i.e. at the Γ (or GM) point of the Brillouin zone, and with higher invariance symmetries (isotropy subgroup) are also present. The absolute amplitudes, measured in Å, of these three irrep distortion components clearly show the prevailing weight of the Z1 component, being more than one order magnitude larger than the GM2+ distortion. The GM1+ distortion, fully compatible with the symmetry of the parent structure, depends on the specific parent structure used. It can be made zero if an appropiate average structure is chosen, as done here. One can speak of a *primary* Z1 distortion and a much weaker *secondary* distortion of symmetry GM2+, with a higher symmetry. This is a typical situation, where the structure contains weak distortions compatible with the new symmetry and associated with induced secondary effects, in agreement with Von Neumann's principle, while the strongest distortion is the primary cause of the observed symmetry break and is associated with a single irrep.

The form of the Z1 and GM2+ distortions present in NbS_3 are shown in figure 1. The set of relative atomic displacements represented there, normalized so that the sum of their square modulus is one Ångstrom, constitute the so-called polarization vector, which, multiplied by the amplitudes listed in table 1, yield the actual symmetry-adapted distortion present in the experimental structure. As indicated in table 1, the GM2+ subspace has 4 *basis* modes. The displacements are restricted along the y axis, breaking the mirror plane, while maintaining the lattice and the inversion centre. One independent GM2+ basis mode exists for each of the four 2e atomic orbits within the asymmetric unit of the parent phase. The actual GM2+ distortion observed in the experimental structure can be any linear combination of these four basis modes. Hence four parameters are required to describe it. Similarly, the description of the Z1 distortion requires 12 parameters, as there are 12 independent basis modes. The direction of the displacements is not restricted in this case, and there are three independent basis modes for each 2e orbit. As it should be, summing the 4, 12 and 8 degrees of freedom of the three different symmetry subspaces (including the GM1+) one obtains the same number of positional parameters as a conventional description. The difference is that they are collective coordinates, and have quite different weight in the total distortion.

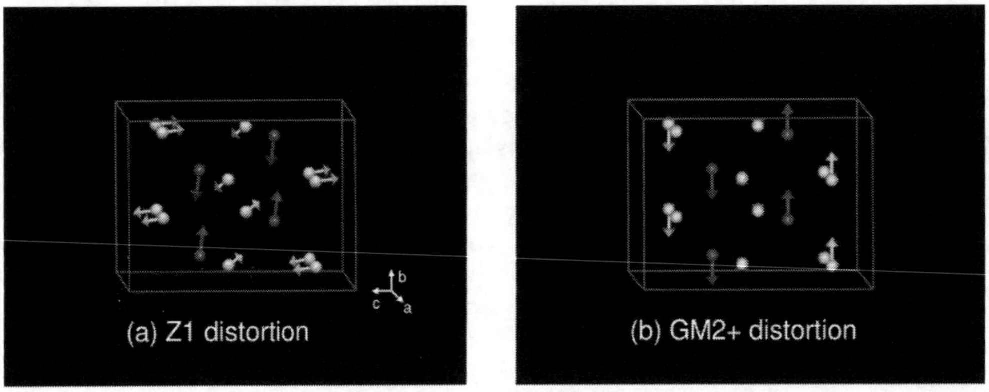

Figure 1. Primary Z1 (a), and secondary GM2+ (b) distortion components present in NbS_3. Blue atoms are the Nb sites. Relative atomic displacements are represented in an arbitrary scale. The atomic positions are those of the parent structure, i.e. four 2e orbits in two consecutive monoclinic cells along the y axis (images from Fullprof Studio [17]).

The extreme weakness of the GM2+ distortion means that a structure refinement limited to the Z1 distortion should yield a very good model, while it would require four positional parameters less than a conventional approach. Note however that the error for the GM2+ amplitude, derived from the reported structure [15], indicates that this secondary distortion is significant, despite its weakness.

NbS_3 was one of the first commensurate modulated structures where it was pointed out that a superspace description could require in practice less parameters than a conventional approach. Van Smaalen [16] showed that it could be described in a very good approximation as a commensurately modulated structure with superspace group $P2_1/m(0\ \beta\ 0)$ with $\beta=1/2$, and a single harmonic for the atomic modulations. The model required four positional parameters less than the conventional description. In fact, this reduction was due to the truncation of the superspace model to a single harmonic, as a complete superspace model allows a second harmonic for the atomic displacements along the y direction. This second harmonic has wave vector (0,1,0) (equivalent to the Γ point), but yields atomic displacements that cannot be described by a "zeroth harmonic". The relation with the symmetry-mode decomposition explained above is then quite obvious. The first harmonic in the atomic modulations describes a Z1 distortion, while the neglected second harmonic modulations correspond to the weak secondary GM2+ distortion. A zeroth harmonic or equivalently the shifts of the average atomic positions represent the GM1+ distortion.

The way the symmetry properties of the atomic displacements change from one irrep to the other, when considering the first and the second harmonic of the modulation is shown in figure 2. One can see that the assumption of the superspace group symmetry, together with the choice of the real space section in superspace, automatically introduces the division into two distinct irrep distortion components, as first and second harmonics of the modulation, implicitly assuming some hierarchy among them.

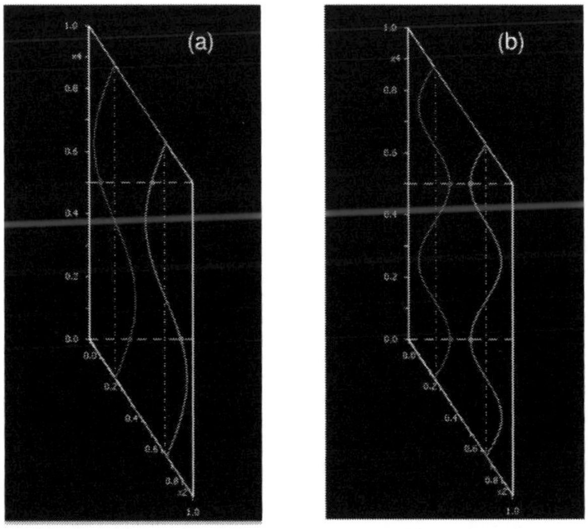

Figure 2. First harmonic (a) and second harmonic (b) modulations on the superspace plane x2x4 (with arbitrary amplitudes) of two symmetry-related atoms of an orbit 2e of the average $P2_1/m$ structure, according to the superspace group $P2_1/m(0\beta0)$ with $\beta=½$. The figure shows how the two atomic displacements within the same cell (at the dashed lines) fulfill the symmetry relation of the Z1 and GM2+ modes, respectively, i.e. they are equal for the first harmonic and opposite for the second one (figures obtained with JANA2006 [11]).

Both approaches are therefore essentially equivalent, but the symmetry-mode description seems somehow closer to the real system. The continuous modulation functions in the superspace, with only two points having physical meaning, seem rather artificial but effective. One must also consider that the superspace group of a commensurate structure is in general not unique. Several choices may be equally possible. In the present case, by choosing as average symmetry the one of the parent structure, the resulting superspace group reproduces the symmetry decomposition of the symmetry-mode analysis. This would not be the case if a triclinic superspace group had been chosen. Note that using a monoclinic superspace group implies to include rotational symmetry operations, which are not necessarily present in the point group symmetry of the diffraction diagram.

As usual, the 3D space group resulting from the superspace construction depends on the section taken along internal space, i.e. on the global phase associated to the modulation. If the section is taken at t= ¼ or ¾ , instead of zero, the resulting 3D structure would have *Pm* symmetry. It was pointed out in [16] that in this case the superspace description with a single harmonic would require 20 parameters, from a total of 28 in a conventional description. It is interesting to see the corresponding mode decomposition, summarized in table 2.

Table 2. Symmetry-mode decomposition of a virtual *Pm* structure resulting from the distortion of the parent $P2_1/m$ structure of NbS_3 with twofold multiplication of the unit cell along the unique axis. Columns as in table 1.

k-vector	irrep	direction	isotropy subgroup	dim.
(0,0,0)	GM1+	(a)	$P2_1/m$ (**a**,**b**,**c**;0,0,0)	8
(0,0,0)	GM2-	(a)	*Pm* (**a**,**b**,**c**;0,1/4,0)	8
(0,1/2,0)	Z1	(-a,a)	*Pm* (**a**,2**b**,**c**;0,1/4,0)	12

Z1 is again the irrep of the primary distortion. The change of section in superspace is reflected in a change of direction within the Z1 two-dimensional irrep space. The system has in this case a secondary distortion with 8 degrees of freedom and with its symmetry given by a different irrep than in the triclinic case. It can easily be checked, as in the previous case, that the superspace construction (with the same superspace group but a different commensurate section) introduces the symmetry constraints in the atomic displacements of the first and second harmonics corresponding to the irreps Z1 (direction (-a,a)) and GM2-, respectively.

3. A distorted perovskite

The example above shows that in simple one-dimensional modulations the mode and superspace descriptions are essentially equivalent. If the appropiate superspace group is chosen, the division of the modulation into harmonics mimics the symmetry-mode decomposition. Although we lack a rigorous proof, probably this formal equivalence is rather general and could be extended to more complex cases. However, it is clear that the superspace approach, compared with a conventional symmetry-mode decomposition (specially now that general computer tools for doing it are available) can be rather unpractical when several irreps with different large commensurate wave vectors are involved.

As an example let us consider the structure of $NaNbO_3$ at room temperature. This is a distorted perovskite with space group *Pbcm*. Its setting is related with the cubic one by the transformation **a-b**, **a+b**,**4c**;1/2,0,1/2 [18,19]. Table 3 summarizes its mode decomposition with respect to the ideal perovskite *Pm-3m* configuration. One can see that there are several wave vectors involved. It is important that the first column of the table only lists a representative of the star of wave vectors of the corresponding irrep. In some cases, this star has more than one wave vector, and those involved in the distortion and particularized by the irrep direction (indicated in the third column) do not necessarily coincide with the one listed in the table, which is only used as a part of the irrep label. These "label" wave vectors are not used explicitly in the symmetry-mode description, which is done defining the atomic displacements of all atoms within the unit cell of the distorted structure.

Table 3. Symmetry-mode decomposition of the *Pbcm* structure of $NaNbO_3$ [19] with respect to the ideal cubic perovskite. Columns as in Table 1

k-vector	irrep	direction	isotropy subgroup	dim.	amplitude (Å)
(0,1/4,0)	DT5	(0,0,0,0,0,0,0,0,a,0,0,-a)	*Cmcm*(**a-b**,**a+b**,**4c**;0,0,1/2)	5	0.55
(1/2,1/2,1/2)	R4+	(0,a,a)	*Imma*(**a+b**,**2c**, **a-b**,;0,0,0)	1	1.38
(1/2,1/2,1/2)	R5+	(0,a,-a)	*Imma*(**a+b**,**2c**, **a-b**,;0,0,0)	2	0.03
(0,1/2,0)	X3-	(0,a,0)	*P4/mmm*(**a**,**b**,**2c**;0,0,1/2)	2	0.08
(1/2,1/2,0)	M5-	(0,0,0,-a,0,0)	*Pmma*(**a+b**,**c**,**a-b**;0,1/2,0)	3	0.16
(1/2,1/2,1/4)	T2	(a,a,0,0,0,0)	*I4/mcm*(**a-b**,**a+b**,**4c**;1/2,1/2,1/2)	1	0.00
(1/2,1/2,1/4)	T4	(a,-a,0,0,0,0)	*I4/mcm*(**a-b**,**a+b**,**4c**;0,0,1/2)	1	1.07

Checking the isotropy subgroups in table 3 one can see that none of the irrep distortion components directly yields the observed *Pbcm* symmetry. Hence there is not an obvious primary distortion. However, the list of amplitudes shows that there are two prevailing distortion components with irreps R4+ and T4. A R4+ distortion (with this irrep direction and isotropy subgroup or with others) is typical in many perovskites due to the instability of three degenerate modes with this symmetry, associated with rigid rotations of the framework of oxygen octahedra. The T4 mode is also a rigid unit distortion of the octahedral framework (see figure 3). The intersection of the isotropy subgroups of these two distortion components is *Pbcm*. Thus, their presence is sufficient to explain the observed symmetry. They can therefore be considered the two primary distortions, while the remaining ones can be taken as induced secondary components. This means from the modulated viewpoint that we can choose (½ ½ ½) and (½ ½ ¼) as the two primary wave vectors of the distortions; while the rest correspond to harmonics and can be obtained as linear combinations of them.

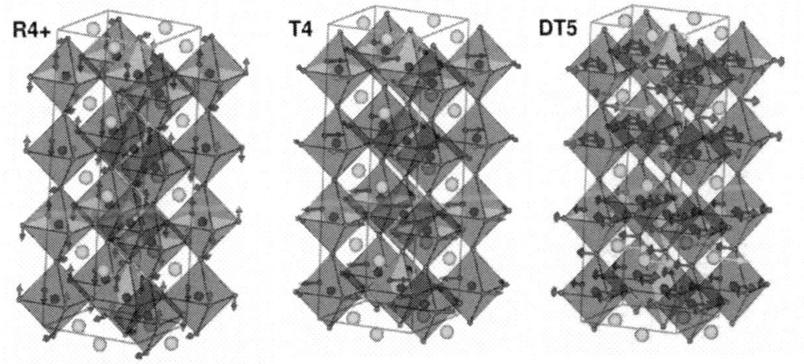

Figure 3. Primary distortion components R4+ and T4, and the secondary one DT5 in $NaNbO_3$ [19]. An arbitrary global scale is used for each case. (images from Fullprof Studio [17])

One could then try to define a (3+2)-dim superspace group with these two primary wave vectors and with the adequate symmetry restrictions on the two first harmonics to make them R4+ and T2 distortions for an appropiate choice of the real space section. The postulated superspace symmetry should be cubic despite the orthorhombic symmetry of the diffraction diagram. However, the choice of primary distortions and primary wave vector is in principle only clear once the symmetry-mode decomposition is done, since a hierarchy among wave vectors and symmetries (extinction rules) will not be easily observable *a priori* in the diffraction diagram. But even if the superspace construction is achieved and its equivalence with the symmetry-mode decomposition could still be maintained, the superspace description would introduce unnecessary complexity and artificial features in the problem. The modulation harmonics describing the atomic displacements would be two dimensional functions, where only a few points correspond to real atomic displacements. In this case, a description in terms of symmetry-modes, with explicit indication of the symmetry constraints of the atomic displacements for each basis mode, seems a soberer and more efficient approach.

4. Modulated phase of gallium under pressure

We finally consider the phase GaII of Ga under pressure [20]. To our knowledge it is the most extreme case in parameter reduction when describing a superstructure within the superspace formalism. This phase, with a very large unit cell and symmetry $C222_1$, has been shown [21] to be a commensurately modulation of a simple *Fddd* structure with only one symmetry-independent atom at the Wyckoff position 8a (1/8, 1/8, 1/8). The superspace group used was *Fddd* $(00\gamma)0s0$, with $\gamma = 9/13$, with a multiplication by 13 of the unit cell with respect to the virtual *Fddd* parent structure. This superspace description allows up to 25 harmonics in the modulation, but three harmonics were sufficient for describing the structure within a very good approximation (see figure 4). This happens despite the quite large atomic displacements (of the order of 0.8 Å). The superspace group description meant a reduction from 38 positional parameters to only four.

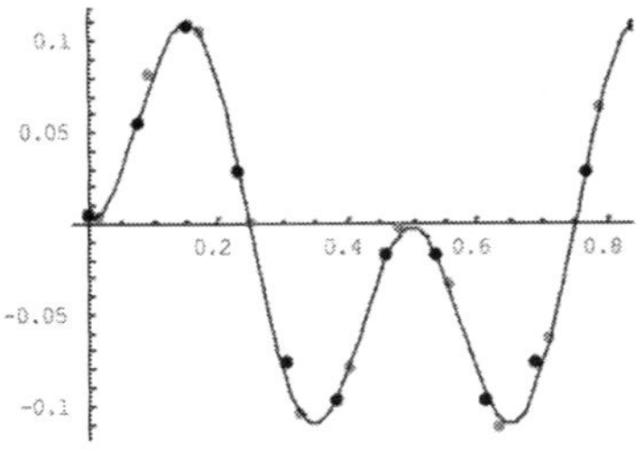

Figure 4. Independent atomic displacements along the x axis of the experimental GaII structure with respect to an average *Fddd* configuration, embedded in a single atomic modulation function when using the superspace group *Fddd*$(00\gamma)0s0$ with $\gamma = 9/13$. (after ref. 21).

Table 4 summarizes the mode decomposition of the experimental $C222_1$ structure with respect to the parent *Fddd* structure. The number of irrep distortion components is 25, as the number of allowed harmonics in the superspace description, but the irreps involved have only 13 possible distinct wave vectors of type n/13 **c*** (n≤13), defined within the first Brillouin zone. All the irrep distortions with n odd have as isotropy subgroup the observed symmetry and could be from this viewpoint possible primary modes, while the distortion components with n even are all secondary, with higher isotropy subgroups. The amplitudes show the dominant weight played by a LD3 distortion. Its wave vector is **q**=9/13**c***, in accordance with the superspace description. This distortion component is clearly the primary one, being more than one order of magnitude larger than the rest, except for two additional ones, which can be identified with a second and a third harmonic of the modulation. Indeed, the LD4[1/13**c***] distortion with a quite large amplitude has a wave vector equivalent to 3**q**, while the third most significant one, LD2[8/13**c***], can be identified with a second harmonic with wave vector 2**q** (see below). These three irrep distortion components correspond with the first three harmonics, which were considered sufficient to describe the structure in [21]. In general, for each wave vector n/13 **c*** two irrep distortion components exist, with symmetries LD3 and LD4 for odd terms and LD1 and LD2 for even terms, except for the case n=13, at the Brillouin zone border with a single irrep.

Table 4. Symmetry-mode decomposition of the experimental structure of phase GaII with respect to its parent *Fddd* configuration. The columns "order harm." indicate the corresponding harmonic in the superspace description. The estimated error of the amplitudes is 0.01 Å in all cases. The first three harmonics are highlighted.

order harm.	k-vector	Irrep	isotropy subgroup	dim	amplitude (Å)	order harm.	k-vector	irrep	Isotropy subgroup	dim	amplitude (Å)
23	(0,0,1/13)	LD3	$C222_1$	2	0.11	21	(0,0,7/13)	LD4	$C222_1$	2	0.11
3	**(0,0,1/13)**	**LD4**	$\mathbf{C222_1}$	**2**	**2.14**	24	(0,0,8/13)	LD1	*Fddd*	1	0.03
6	(0,0,2/13)	LD1	*F222*	1	0.03	**2**	**(0,0,8/13)**	**LD2**	***F222***	**1**	**0.53**
20	(0,0,2/13)	LD2	*Fddd*	1	0.03	**1**	**(0,0,9/13)**	**LD3**	$\mathbf{C222_1}$	**2**	**3.59**
17	(0,0,3/13)	LD3	$C222_1$	2	0.06	25	(0,0,9/13)	LD4	$C222_1$	2	0.05
9	(0,0,3/13)	LD4	$C222_1$	2	0.05	22	(0,0,10/13)	LD1	*F222*	1	0.02
12	(0,0,4/13)	LD1	*Fddd*	1	0.00	4	(0,0,10/13)	LD2	*Fddd*	1	0.00
14	(0,0,4/13)	LD2	*F222*	1	0.03	7	(0,0,11/13)	LD3	$C222_1$	2	0.12
11	(0,0,5/13)	LD3	$C222_1$	2	0.05	19	(0,0,11/13)	LD4	$C222_1$	2	0.05
15	(0,0,5/13)	LD4	$C222_1$	2	0.11	16	(0,0,12/13)	LD1	*Fddd*	1	0.04
18	(0,0,6/13)	LD1	*F222*	1	0.06	10	(0,0,12/13)	LD2	*F222*	1	0.01
8	(0,0,6/13)	LD2	*Fddd*	1	0.01	13	(0,0,1)	Z2	$C222_1$	2	0.07
5	(0,0,7/13)	LD3	$C222_1$	2	0.16						

Each of the irrep distortion components in table 4 corresponds to one of the harmonics used in the superspace description. A modulation harmonic of order m (m=2,...,25) in the superspace description can be demonstrated to be a distortion mode with wave vector $\mathbf{q}_m$= 9m/13 **c*** and with symmetry properties given by the small irrep LD3 or LD1, depending on m being odd or even. But the wave vectors of the irreps in the symmetry-mode decomposition are chosen by convention within the first Brillouin zone. Hence, for a comparison with table 4 one must change each $\mathbf{q}_m$ to an equivalent $\mathbf{q}_n$= n/13 **c*** (n≤13) through some reciprocal lattice translation p2**c*** (p integer): $\mathbf{q}_n = \pm\mathbf{q}_m + 2p\mathbf{c}^*$. This change of wave vector implies in general to change also the corresponding small irrep [22]. For m odd, the LD3 irrep becomes LD4 for p odd, while it remains LD3 for p even. For m even, the label of the small irrep is LD1 or LD2 for p even or odd, respectively. Thus, for instance the LD3[1/13**c***] distortion is the 23rd harmonic of the superspace description, its wave vector being $\mathbf{q}_{23}$= 23 (9/13)**c*** = -(1/13)**c*** + 8(2**c***) with p even, while LD4[1/13**c***] corresponds to the third order harmonic with $\mathbf{q}_3$= 3(9/13)**c***= (1/13)**c***+2**c***, and p odd. Following these rules the correspondance between the 25 harmonics of the superspace description and the 25 irrep components of the mode decomposition can be done, as shown in table 4. Apart from the first three stronger harmonics, it is remarkable that even for the smaller odd harmonics, the underlying hierarchy coming from the primary mode is still observed. Thus, the next two irrep distortion components with the largest amplitudes of 0.16 and 0.12 Å can be identified with the 5th and 7th modulation harmonic.

The equivalence between the two approaches can be found again in this example, but it becomes obvious in this case the higher efficiency of the superspace description. The choice of the superspace group implies a decision about which of the 25 irrep distortions is the primary modulation. But, once a superspace group is assumed, the superspace symmetry takes care of the symmetry properties of each harmonic, and a rough hierarchy among them is implicitly assumed. The mode analysis, on the other hand, does not assume *a priori* any predominance among the 25 irrep components, and one has to define and indicate explicitly one by one the symmetry properties of each possible irrep distortion component present in the structure. Even if a primary distortion could be identified directly from the experiment, the identification of the secondary modes of lowest order among all possible irrep distortion components require non-trivial considerations, as shown above.

5. Conclusions

Symmetry-mode analysis and superspace symmetry are closely connected. Both approaches allow in commensurate distorted structures a division of the configuration space into symmetry-adapted subspaces with very different weight in the total distortion and different responses to perturbations. For superstructures with a small cell multiplication or with several independent wave vectors, the symmetry-mode decomposition is more efficient, and introduces advantages similar to those of the superspace methodology in systems where postulating superspace symmetry becomes rather unpractical.

6. References

[1] Mañes J, Tello M J and Perez-Mato J M 1982 *Phys. Rev.* B **26** 250
[2] Perez-Mato J M, Gaztelua F, Madariaga G and Tello M J 1986 *J. Phys. C: Solid State Phys.* **19**, 1923
[3] Rae A D, Thompson J G and Withers R L 1992 *Acta Cryst.* B **48** 418
[4] Withers R ., Hua G L, Welberry T R and Vincent R 1988 *J. Phys. C: Solid State Phys.* **21** 309
[5] Hatch D M, Artman, J I and Boerio-Goates J 1990 *Phys. and Chem. of Min.* **17** 334
[6] Stokes H T, Hatch D M and Wells J D 1991 Phys. Rev. B **43** 11010
[7] Darlington C N W 2002 *Acta Cryst.* A58, 299
[8] Fennie C J and Rabe K M 2005 *Phys. Rev.* B **72** 100103(R)
[9] Orobengoa D, Aroyo M I and Perez-Mato JM 2009 *J. Appl. Cryst.* in press
[10] Campbell B J, Stokes H T, Tanner D E and Hatch D M 2006. *J. Appl. Cryst.* **39** 607
[11] Petricek V, Dusek M, and Palatinus 2006 *JANA2006* Institute of Physics, Czech Academy of Sciences, Prague, Czech Republic
[12] Janssen T and Janner A 1984 *Physica* (Utrecht) **126** A 163
[13] Perez-Mato J M, Madariaga G and Tello M J 1984. *Phys. Rev.* B **30** 1534
[14] Stokes H T, Campbell B J and Hatch D M 2007 *Acta Cryst.* A **63** 365
[15] Rijnsdorp J and Jellinek F 1978 *J. Solid State Chem.* **25** 325
[16] Van Smaalen S 1988 *Phys. Rev.* B **38** 9594
[17] Chapon L C, Rodriguez-Carvajal J 2008 FullProf Studio
[18] Sakowski-Cowley A C, Lukaszewicz K, Megaw H D 1969 *Acta Cryst* B **25** 851
[19] Hewat A W 1974 *Ferroelectrics* **7** 83
[20] Degtyareva O, McMahon M I, Allan D R, and Nelmes R J 2004 *Phys. Rev. Lett.* **93** 205502
[21] Perez-Mato JM, Elcoro L, Aroyo M I, Katzke H, Toledano P and Izaola, Z 2006 *Phys. Rev. Lett.* **97** 115501
[22] Cracknell A P, Davies B L, Miller S C and Love W F 1979 *Kronecker Product Tables, vol. 1, General Introduction and Tables of Irreducible Representations of Space Groups* (IFI/Plenum, New York)

Structure of incommensurately modulated chromium pyrophosphate studied by Maximum Entropy Method (MEM)

Liang Li, Andreas Schönleber and Sander van Smaalen
Laboratory of Crystallography, University of Bayreuth, 95440 Bayreuth, Germany
E-mail: smash@uni-bayreuth.de

Abstract. A general discussion is given of the Maximum Entropy Method (MEM) as a method to determine the electron density in superspace from X-ray diffraction data. The MEM has been used to determine the electron density in superspace of incommensurately modulated chromium pyrophosphate from X-ray diffraction data published by Palatinus *et al.* [Acta Crystallogr. B **62**, 556]. Chromium pyrophosphate, $Cr_2P_2O_7$, contains ordered regions (83% of the volume) and regions with disorder. Analysis of the MEM density has allowed us to determine the displacive modulation functions within the ordered regions. The disordered regions can be described as the alternate occupation of two conformations of the pyrophosphate unit and two positions of the chromium atom. Both the conformations and the occupational probabilities of them depend continuously on the phase of modulation t.

1. Introduction

A long-standing problem in aperiodic crystallography is the true shapes of modulation functions of modulated crystals. In principle, modulation functions are defined by an infinite number of parameters, but only a limited number of parameters can be varied in structure refinements, because additional parameters would lead to dependencies among them. Usually modulation functions are described by truncated Fourier series with one or a few harmonics or by a special function, like a block-wave or saw-tooth-shaped function. In all these cases the outcome of the refinement is biased by the choice of the model for the modulation functions, and it cannot be guaranteed that the true functions will be found.

A model-independent reconstruction of the electron density in the unit cell can be obtained from the Maximum Entropy Method (MEM) applied to the integrated intensities of Bragg reflections in X-ray diffraction [1]. Local maxima in the density are identified with atomic positions. In this way the crystal structure can be derived from the reconstructed density. For aperiodic crystals the MEM can be used to determine the generalized electron density in the superspace unit cell [2, 3, 4]. Local maxima in t-sections (physical-space sections) of the generalized density then provide the atomic positions in physical space. Positions as a function of t for one period $0 \leq t < 1$ then give the modulated atomic positions from which the modulation functions can be derived.

Chromium pyrophosphate $Cr_2P_2O_7$ has an incommensurately modulated structure at room temperature. The superspace symmetry is $C/m(\sigma_1\, 0\, \sigma_3)\bar{1}s$ with lattice parameters of the basic-structure unit cell of a = 7.019, b = 8.406, c = 4.626 Å and β = 108.61°. The modulation wavevector is $\mathbf{q} = (-0.361, 0, 0.471)$. Structure refinements of a superspace model against X-ray diffraction data have been published in [5]. Palatinus *et al.* [5] have shown that a structure model based on simple modulation functions is insufficient to describe the diffraction data. Instead a model was proposed that involved disorder over two configurations for a small range of t-values (designated model B).

We have recently re-analyzed the diffraction data of $Cr_2P_2O_7$ by the MEM [6]. Based on the interpretation of the generalized MEM-electron density, we have obtained an accurate description of the modulation functions within the ordered regions and we could develop a description of the modulation of the lack of order in the disordered regions. Here we present a summary of these results, after discussing the principles of the application of the MEM in superspace. The final section is devoted to a critical discussion of modulation functions as they can be determined by X-ray diffraction experiments.

2. The Maximum Entropy Method

The generalized electron density $\rho_s(\mathbf{x}_s)$ in $(3+d)$-dimensional superspace is discretized on a $(3+d)$-dimensional grid over the superspace unit cell. This results in $N_{pix} = N_1 \times N_2 \times N_3 \times N_4$ pixels for $d = 1$. The entropy of the discrete generalized electron density is defined as [7]

$$S = -\sum_{k=1}^{N_{pix}} \left\{ \rho_k \log\left[\rho_k/\rho_k^{prior}\right] - \rho_k + \rho_k^{prior} \right\}, \tag{1}$$

where $\rho_k = \rho_s(\mathbf{x}_{sk})$ is the electron density at the grid point with coordinates $\mathbf{x}_{sk}$ and ρ_k^{prior} is the corresponding value of the reference density or prior. The electron density $\{\rho_k\}$ as well as the prior are normalized to reflect the total number of electrons N_e in the unit cell of volume V, according to

$$\frac{V}{N_{pix}} \sum_{k=1}^{N_{pix}} \rho_k = N_e \tag{2}$$

The principle of maximum entropy defines the optimal electron density as the electron density $\{\rho_k\} = \{\rho_k^{MEM}\}$ which maximizes the entropy S (Eq. 1). It is a simple exercise to show that S assumes its maximum value if $\rho_k^{MEM} = \rho_k^{prior}$ for all points k. Information about the electron density can be incorporated into the procedure, if they can be expressed as constraints on the density. Constraints of the form $C = 0$ can be incorporated by the method of undetermined Lagrange multipliers. Then the maximum is determined of

$$Q = S + \lambda\, C \tag{3}$$

for variation of all $\{\rho_k\}$ and the Lagrange multiplier λ. At convergence $C = 0$ is identically fulfilled.

Diffraction data are incorporated in the form of the so-called F constraint, $C_{F^2} = 0$ [8]. The application of the MEM to X-ray diffraction data requires phased structure factors $F_{obs}(\mathbf{H}_i)$ of the $i = 1, \cdots, N_F$ reflections with scattering vectors $\mathbf{H}_i$. It thus relies on a previously solved crystal structure. We have argued that a structure model with a reasonable fit to the diffraction data already gives accurate values for the reflection phases, while the amplitudes of the structure factors come from the experiment [7]. With σ_i the standard uncertainty of $|F_{obs}(\mathbf{H}_i)|$ and $\{F_{MEM}(\mathbf{H}_i)\}$ obtained by (3+1)-dimensional Fourier transform of the trial density $\{\rho_k\}$, the F-constraint is [8]

$$C_{F^2} = -1 + \frac{1}{N_F} \sum_{i=1}^{N_F} w_i \left(\frac{|F_{obs}(\mathbf{H}_i) - F_{MEM}(\mathbf{H}_i)|}{\sigma_i} \right)^2 . \tag{4}$$

Useful weighting schemes include unit weights as well as weights inversely proportional to some power of the length of the scattering vector. The latter method has the effect of increasing the weight for low-angle reflections, which is necessary to obtain a Gaussian distribution of residuals

$$\frac{\Delta F(\mathbf{H}_i)}{\sigma_i} = \frac{[F_{obs}(\mathbf{H}_i) - F_{MEM}(\mathbf{H}_i)]}{\sigma_i} \tag{5}$$

at convergence. It is particularly useful for periodic crystals, because intense reflections have short scattering vectors in those cases. Modulated crystals also have satellite reflections, which can be weak irrespective of the length of the scattering vector. Accordingly, we have found that weights proportional to some power of the structure factor amplitude [9],

$$w_i = |F_{obs}(\mathbf{H}_i)|^n \left(\frac{1}{N_F} \sum_{i=1}^{N_F} |F_{obs}(\mathbf{H}_i)|^n \right)^{-1} \tag{6}$$

provide better results, *i.e.* they lead to the best approximation to a Gaussian distribution of the residuals at convergence. These weights are denoted by Fn; $n = 2$ turned out to be optimal choice for $Cr_2P_2O_7$ [6].

An important issue is the choice of the number of pixels, N_{pix} [7]. The grid needs to reflect the point symmetry of the lattice. For monoclinic symmetry this implies that each of the numbers N_j ($j = 1, \cdots, 4$) must be an even number [7]. Secondly, the number N_j need to be sufficiently large, in order to be able to compute the structure factors $\{F_{MEM}(\mathbf{H}_i)\}$ with sufficient accuracy, and to obtain the atomic positions with sufficient accuracy. The resulting mesh should be smaller than 0.1 Å. Accordingly, we have chosen $72 \times 96 \times 48$ pixels along the first three axes, which correspond to a mesh of $0.097 \times 0.088 \times 0.096$ Å^3 in real space. The number of points along the fourth axes needs special attention. A minimum of four points is required for each expected harmonic of the modulation functions. Sharp features of modulation functions—as presently expected saw-tooth-shaped functions—require many more points. Therefore, we have chosen 48 points along the fourth axis in the present study. Even for this moderate number of points along each of the four axes, the computer program BayMEM [7] uses several Gbyte of RAM.

The MEM of Eqs 1–4 has two fundamentally different applications. With a prior density defined by the independent atom model for the structure, the MEM can be used as an alternative method to the multipole model for charge-density studies [10]. With a uniform prior the resulting MEM-density appears slightly noisy, but it is still of sufficient quality to show local maxima that define the positions of the atoms. Since the present interest is the t-dependent positions of the atoms, we have used the uniform prior for the study on $Cr_2P_2O_7$ [6].

The MEM electron density $\{\rho_k^{MEM}\}$ has be determined with the computer program BayMEM [7] applied to the diffraction data from [5]. Two-dimensional sections of the generalized electron density have been visualized with the computer program Jana2006 [11]. Local maxima and other topological properties of three-dimensional, physical-space sections of the four-dimensional generalized density were obtained with the computer program Edma for each of 100 t-sections in the interval $0 \leq t < 1$. This has resulted in the determination of the modulation functions of all atoms according to the procedure described in Section 1 and [7].

3. MEM-based modulation functions

The MEM was able to reproduce the modulation functions of model B within the ordered regions for all atoms except phosphorus [Fig. 1(a)]. A model M has been developed [6], in which the modulation function of atom P is accurately given by the positions derived from the MEM-density [Fig. 1(b)]. Model M gives a slightly better fit to the diffraction data than model B, and it avoids unphysically short interatomic distances P–O that were found for model B [6].

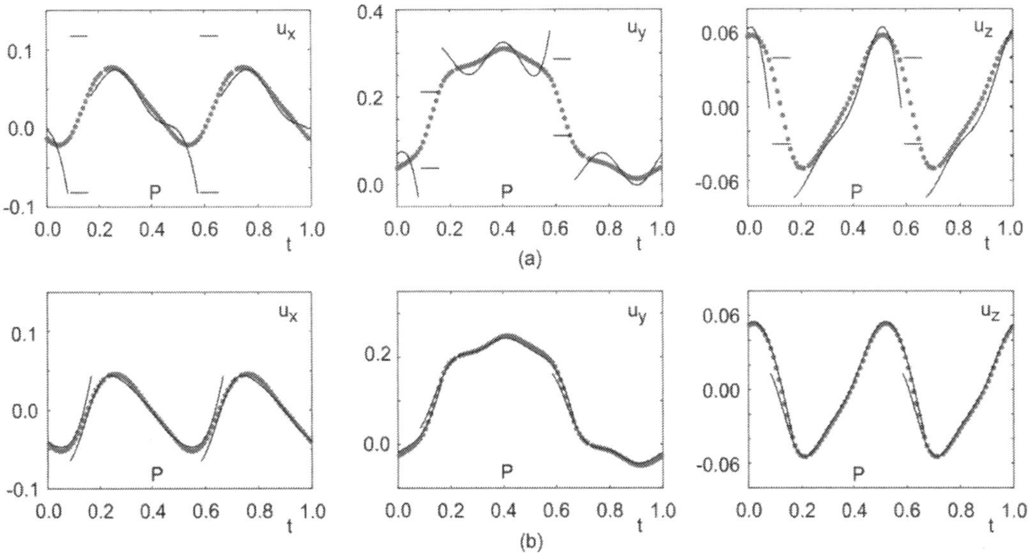

Figure 1. Modulation function of atom P. Displacements along x, y and z are given in Å. Filled circles give the values derived from the MEM-electron density. The full lines in the top panels (a) represent the modulation function of model B [5]. The full lines in the bottom panels (b) have been obtained after refinement of the first harmonic parameters of modulation functions fitted to the MEM-derived points (filled circles). They constitute model M. Figure 1(a) has been reproduced from [6].

The disordered regions involve intervals of width 0.084 in t. In model B the disorder is described by a split-atom model for atoms Cr, O2, O3 and P [*e.g.* see the modulation function around $t \approx 0.12$ and $t \approx 0.62$ in Fig. 1(a)]. The positions within the split-atom regions are not modulated in model B. Atom O1 is described by a continuous, smooth modulation function over the complete interval. The atoms Cr, O2, O3 and P exhibit smeared electron densities within the disordered regions in the MEM-density maps. Two maxima have only been found for atom O2, thus confirming the split-atom model. However, both the positions and the integrated number of electrons that can be assigned to each of the two maxima vary with t. This implies a modulation within the disordered regions that affects the positions of the atoms and concomitantly the occupancies of the two conformations/orientations of the PO_4 units [6]. Although in each t-section of the MEM-density only smeared, asymmetric maxima could be located near the positions of phosphorus, we have modeled the t-dependent position of atom P in the model M for the structure [Fig. 1(b)].

The positions of atoms in crystals are defined as the time-averaged positions of their nuclei. In case of harmonic movements, the time-averaged positions coincide with the local maxima in the electron density (which are time-averaged densities by virtue of the diffraction experiment). However, for anharmonic thermal movements, the local maxima and averaged atomic positions do not coincide. It might then be more accurate to identify the atomic positions with the center of charges of the electron density distribution around each maximum. Figure 2 shows the positions of the centers of charge in comparison to the positions of the local maxima of the MEM-electron density of atoms Cr and O3. They show a good match between local maximum and center of charge within the ordered regions, while the disordered regions are characterized by marked deviations between the two positions. This suggests that the modulated disorder might involve anharmonic motion of the atoms. Accordingly, model M includes third-order ADPs for

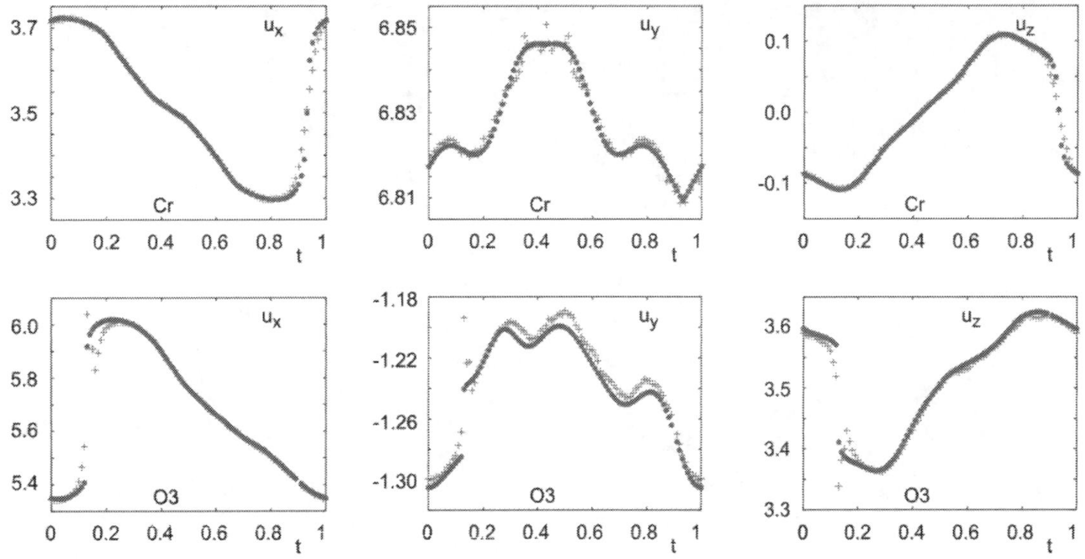

Figure 2. Comparison of the positions (Å) of local maxima (filled circles) and centers-of-charge (crosses) as derived from the MEM-electron density for the atoms Cr and O3.

atom P.

4. Conclusions

The Maximum Entropy Method (MEM) in superspace has been applied to x-ray diffraction data of incommensurately modulated $Cr_2P_2O_7$. The MEM has lead to a description of the structure of $Cr_2P_2O_7$ that differs from the refined structure model B in [5] especially in the disordered regions. Positions of atoms are modulated, such that the CrO_6 and P_2O_7 units attain different positions and orientations, while deformations of these units lead to internal strain. The internal strain is released by the existence of disordered regions, which are characterized by the alternate occupation of one of two conformations/orientations of the CrO_6 and P_2O_7 units. Both the conformations and the occupational probabilities vary with the phase of the modulation.

Model M has been developed on the basis of the MEM-electron density map, but it gives only a slightly better fit to the diffraction data than model B does, while it does not capture the description as developed for the disordered regions of most atoms. Reasons for the failure to find a structure model that captures all features observed in the MEM-density include:

(i) The limited accuracy of ~ 0.01Å of MEM-derived positions together with the impossibility to refine all the parameters that were necessary to construct a model based on $\rho^{MEM}(\mathbf{x}_s)$, due to interdependencies between these parameters. Nevertheless, we believe that the MEM has established the true nature of the modulations in $Cr_2P_2O_7$.

(ii) The failure to identify two positions in the disordered regions of most atoms. This is probably the result of a small distance between the two expected positions, but nevertheless prohibits the accurate determination of these positions.

(iii) The complex nature of the modulations, probably involving modulated anharmonic ADPs, that cannot be refined and neither directly be estimated from the MEM-densities.

References

[1] Gilmore C J 1996 *Acta Crystallogr. A* **52** 561–589

[2] Palatinus L and van Smaalen S 2004 *Z. Kristallogr.* **219** 719–729
[3] van Smaalen S 2007 *Incommensurate Crystallography* (Oxford: Oxford University Press)
[4] van Smaalen S and Li L 2009 *Phys. Scr.* **79** 048305
[5] Palatinus L, Dušek M, Glaum R and El Bali B 2006 *Acta Crystallogr. B* **62** 556–566
[6] Li L, Schönleber A and van Smaalen S 2009 *Acta Crystallogr. B* **submitted**
[7] van Smaalen S, Palatinus L and Schneider M 2003 *Acta Crystallogr. A* **59** 459–469
[8] Sakata M and Sato M 1990 *Acta Crystallogr. A* **46** 263–270
[9] de Vries R Y, Briels W J and Feil D 1994 *Acta Crystallogr. A* **50** 383–391
[10] Hofmann A, Netzel J and van Smaalen S 2007 *Acta Crystallogr. B* **63** 285–295
[11] Petricek V, Dusek M and Palatinus L 2006 *The crystallographic computing system JANA2006* Institute of Physics Praha, Czech Republic

Journal of Physics: Conference Series **226** (2010) 012013 doi:10.1088/1742-6596/226/1/012013

Symmetry, incommensurate magnetism and ferroelectricity: the case of the rare-earth manganites $RMnO_3$

J L Ribeiro

Departamento de Física, Universidade do Minho, 4710-057 Braga, Portugal

jlr@fisica.uminho.pt

Abstract. The complete irreducible co-representations of the paramagnetic space group provide a simple and direct path to explore the symmetry restrictions of magnetically driven ferroelectricity. The method consists of a straightforward generalization of the method commonly used in the case of displacive modulated systems and allows us to determine, in a simple manner, the full magnetic symmetry of a given phase originated from a given magnetic order parameter. The potential ferroic and magneto-electric properties of that phase can then be established and the exact Landau free energy expansions can be derived from general symmetry considerations.

In this work, this method is applied to the case of the orthorhombic rare-earth manganites $RMnO_3$. This example will allow us to stress some specific points, such as the differences between commensurate or incommensurate magnetic phases regarding the ferroic and magnetoelectric properties, the possible stabilization of ferroelectricity by a single irreducible order parameter or the possible onset of a polarization oriented parallel to the magnetic modulation. The specific example of $TbMnO_3$ will be considered in more detail, in order to characterize the role played by the magneto-electric effect in the mechanism for the polarization rotation induced by an external magnetic field.

1. Introduction

In the last few years, there has been a great deal of interest in several metal compounds, where ferroelectricity is induced by a transition to a complex magnetic state. Examples of this class of materials are systems like $RMnO_3$ (R=Gd, Dy or Tb), RMn_2O_5 (R=Tb or Y), $MnWO_4$, $NiVa_2O_8$, $CoCr_2O_4$, $CuFeO_2$, CuO or some complex hexagonal ferrites like $(Ba,Sr)_2Zn_2Fe_{12}O_{22}$ [1-9]. In addition, some of these systems show remarkable effects such as the ability of a magnetic field to rotate or stabilize an electrical polarization, as in the case of $TbMnO_3$ or $GdMnO_3$, respectively [10-12].

In this class of materials, and in contrast with conventional multiferroics like $BiFeO_3$ or $BiMnO_3$, the paramagnetic phase is also paraelectric and the ferroelectric polarization is driven by a modulated magnetic order parameter. This improper ferroelectricity reveals therefore a close resemblance with that originated from lattice modulations, as in the case of systems like $RbZn_2Cl_4$, $(C(NH_3)_4)_2CoCl_4$ or $(CH_3)_3NCH_2COO.CaCl_2.2H_2O$ [13-15].

The first efforts to understand this magnetically driven ferroelectricity have been mainly based on the search for microscopic mechanisms coupling magnetic moments and lattice displacements, such as the symmetric superexchange striction or the anti-symmetric exchange (Dzyaloshinskii-Moriya mechanism [16-17]). Also, phenomenological approaches like the so called "spin current" [18] or the "spiral formulation"[19] have been explored. Here, a spiral magnetic structure with a rotation axis

$\vec{e}$ and modulation wavevector $\vec{k}$ breaks inversion and gives rise to a polarization $\vec{P} // \vec{e} \times \vec{k}$. However, a satisfactory and integrated description of the effect has not yet been achieved.

Establishing the symmetry constrains upon the onset of a polarization vector, toroidal moment or lattice distortion as a result of the stabilization of a given magnetic structure, independently of the microscopic mechanisms involved, is a valuable contribution to further understand the phenomena. Symmetry imposes a rigid framework for the possible interactions between the different degrees of freedom and guarantees the overall consistency of the observed laws. Moreover, given the direct connection between symmetry and phenomenology, via Landau theory, this represents a first step to establish more detailed models and to further elucidate the relevant microscopic mechanisms involved.

The analysis of the symmetry requirements for a magnetically driven ferroelectricity has been pursued in a number of studies which have followed different approaches [20-23]. In this work, the method adopted uses the complete irreducible co-representations (CICR's) of the magnetic space group G of the parent structure, in order to determine the full symmetry of a given modulated magnetic phase. The specific example of the orthorhombic rare-earth manganites $RMnO_3$ will allow us to stress some aspects related to the possible ferroic properties of these compounds and to emphasize several points, such as the role of the magneto-elastic coupling, the essential differences between commensurate and incommensurate modulations regarding the ferroic and magneto-electric properties or the possible onset of an improper polarization due to one irreducible magnetic order parameter. The specific case of $TbMnO_3$ will be further considered, in order to establish the role of the magneto-electric effect in the mechanism for the polarization rotation induced by an external magnetic field.

2. The method for the symmetry analysis: the case of the $RMnO_3$ compounds

The approach of working with the CICR's allows us to use full image of G in the order parameter space instead of considering only the unitary small representation of the vector $\vec{k}$, possibly complemented by the *ad-hoc* inclusion of spatial inversion [20]. This has a number of advantages. In general, this method incorporates in a natural way the physical representations related to $\vec{k}$ and $-\vec{k}$, even if inversion is not a symmetry operation of G. Also, by integrating the full magnetic symmetry, this method allows us to analyze the possible stabilization of the different ferroic parameters, which are precisely distinguished by their different behaviour under spatial inversion and time reversal. Finally, the use of the CICR's allows us to deal with commensurate and incommensurate phases in an unified way, to stress their differences regarding the possible magneto-electric coupling and to use the concept of superspace groups in the case of a magnetic incommensurate order parameter.

Let us consider the example of the orthorhombic rare-earth manganites $RMnO_3$. Here, G is the paramagnetic group $(Pnma)'$, $\vec{k}_1 = \delta(T)\vec{a}^*$ (which corresponds to the Σ line in the reciprocal cell), the star is $\{\vec{k}_1^*\} = \{\vec{k}_1 = \vec{k}; \vec{k}_2 = -\vec{k}\}$ and the small co-representations are one-dimensional. The magnetic modulation in a given ordered phase will be described by the field $\vec{S}(\vec{T},\nu) = \sum_{n,j} S_{nj}\vec{e}_\nu e^{i\vec{k}_n \bullet \vec{T}}$, where $\vec{e}_\nu(\vec{k}_n, j)$ are the eigenvectors of the mode, ν labels the spins in the reference unit cell and $\vec{T}$ denotes a given unit cell of the original structure. The field $\vec{S}$ represents the magnetic order parameter (OP).

The complete co-representation of G in the OP space will be formed by the set of matrices $D_{op}(\vec{k}_1, \{R;\vec{t}\})$ representing the symmetry operations $\{R;\vec{t}\} \subset G$ in the vector space generated by the components S_{nj}. The normal co-ordinates of the active mode, expressed in the same basis, will then be transformed, under the action of $\{R;\vec{t}\}$, as $S'_{mi} = \left[D_{op}(\vec{k}_1, \{R;\vec{t}\})\right]_{mnij} S_{nj}$. This means that the knowledge of the matrices $D_{op}(\vec{k}_1, \{R;\vec{t}\})$ will allow us to characterize how the order parameter is transformed under the different symmetry operations of G and to establish, consequently, the symmetry of the modulated phase.

For the $RMnO_3$ compounds, and in the case of one irreducible OP, the field $\vec{S}(\vec{T},V)$ can be simply written as $\vec{S}(\vec{x}) = \vec{S}e^{i\vec{k}\cdot\vec{x}} + \vec{S}^* e^{-i\vec{k}\cdot\vec{x}}$. Here, $\vec{S} = \vec{S}_0 e^{i\Phi}$ is the eigenvector of the mode and Φ its phase, defined with respect to the underlying discrete lattice. Note that, $\vec{S}(\vec{x})$ is real ($\vec{S}^*_{\vec{k}} = \vec{S}_{-\vec{k}}$) and, given its magnetic nature, odd under time reversal $\theta\vec{S}(\vec{x}) = -\vec{S}(\vec{x})$.

There are 12 irreducible magnetic eigenvectors generated by the four *Mn* spins per unit cell. If one labels the spins located at (o,o,½), (½,o,o), (o,½, ½) and (½, ½, o) as $\vec{S}_1, \vec{S}_2, \vec{S}_3$ and $\vec{S}_4$, respectively, then these eigenvectors $\vec{S}$ correspond to the components of the modes F($\vec{S}_1 + \vec{S}_2 + \vec{S}_3 + \vec{S}_4$), G($\vec{S}_1 - \vec{S}_2 - \vec{S}_3 + \vec{S}_4$), C($\vec{S}_1 - \vec{S}_2 + \vec{S}_3 - \vec{S}_4$) and A($\vec{S}_1 + \vec{S}_2 - \vec{S}_3 - \vec{S}_4$). Each of these eigenvectors is transformed according to one of the four CICR`s of G at $\vec{k} = \delta\vec{a}^*$: $(G_x, C_y, A_z \rightarrow \Delta_1(A_g))$, $(G_z, F_y, A_x \rightarrow \Delta_2(B_{2g}))$, $(F_x, C_z, A_y \rightarrow \Delta_3(B_{3g}))$ and $(G_y, C_x, F_z \rightarrow \Delta_4(B_{1g}))$. Therefore, the set of matrices $\{\hat{D}(g)\}$ specifying how the symmetry operations of G act on the vector space of a particular irreducible OP will correspond to a particular set of CICR matrices. These sets can be readily obtained by following standard methods [24]. For the case of the magnetic space group $(Pnma)'$, the CICR matrices are listed, for example, in the table 1 of [23]. For a reducible OP, the $D_{op}(\vec{k}_1, \{R;\vec{t}\})$ matrices can be constructed from appropriate direct sums of the CICR matrices.

If the modulation wavevector $\vec{k}_1$ is commensurate, the magnetic space group of the ordered phase G′ will be formed by the set of unitary and anti-unitary operations that leave the OP invariant. That is, if $\hat{D}(g)$ and $\hat{D}(\theta g)$ are respectively the matrices induced by the unitary and the anti-unitary operations $\{g;\vec{t}\}$ and $\{\theta g;\vec{t}\}$ of G in the vector space generated by the components of the OP, and if the conditions $\hat{T}\times\hat{D}(g)\times\vec{S} = \vec{S}$ or $\hat{T}\times\hat{D}(\theta g)\times\vec{S}^* = \vec{S}$ hold, then $\{g;\vec{t}+\vec{T}\}$ or $\{\theta g;\vec{t}+\vec{T}\}$ will belong to G'.

If $\vec{k}_1$ is incommensurate, the symmetry of the ordered phase will be described by a magnetic superspace group. The reasoning leading to the notion of a superspace group can be entirely based on the invariance properties of the Landau free energy and it is essentially independent of the fact that the order parameter is of displacive or magnetic nature [25]. Evidently, if $\vec{k}_1$ is incommensurate, translational symmetry is lost along certain spatial directions. Consequently, the symmetry of the incommensurate phase cannot be described by one magnetic space group. However, the incommensurability of the structure in the ordered phase allows us to shift the phase of the order parameter without changing the free energy of the system. If a certain set of phase translations $\{\vec{\tau}\}$ can be combined with a set $\{g;\vec{t}\}$ of symmetry operations of G, in such a way that the generalized operations $\{g;\vec{t},\vec{\tau}\}$ keep the free energy and the OP invariant and, in addition, obey the product rule $\{R_2;\vec{t}_2,\vec{\tau}_2\}\times\{R_1;\vec{t}_1,\vec{\tau}_1\} = \{R_2R_1;\vec{t}_2 + R_2\vec{t}_1, \vec{\tau}_2 + \Re(R_2)\vec{\tau}_1\}$, then these operations can be seen as symmetry elements of the incommensurate structure. The matrix $\Re$ given above is defined by the action of R on the wavevectors of the star $\{\vec{k}_1^*\}$ according to the procedure described in [25].

Similar to the case of a commensurate wavevector, the incommensurate order parameter will be invariant under the action of the unitary $\{g;\vec{t},\vec{\tau}(g)\}$ or the anti-unitary $\{\theta g;\vec{t},\vec{\tau}(\theta g)\}$ operations if

$\hat{\Gamma}(\vec{\tau})\times\hat{D}(g)\times\vec{S}=\vec{S}$ or $\hat{\Gamma}(\vec{\tau})\times\hat{D}(\theta g)\times\vec{S}^*=\vec{S}$. Here, $\hat{\Gamma}(\vec{\tau})$ represents the matrix that expresses, in the space of the order parameter, the phase translation of the magnetic modulation. The set of the symmetry operations that verify these conditions and obey the product rule given above, form the magnetic superspace group of the incommensurate phase.

3. Some results for the RMnO$_3$ compounds

A detailed analysis of the symmetry, ferroic properties and free energy expansions of the magnetically modulated phases allowed in the case of the rare-earth manganites will not be considered here. In the following, we will focus on the results concerning some specific points.

3.1. Ferroelectricity can be induced by an irreducible and commensurate magnetic order parameter.

The application of the method outlined above to the case of one irreducible and incommensurate (INC) order parameter shows that all the symmetry operations of G give rise to generalized symmetry operations of the incommensurate phase. There is, in this case, a one-to-one relationship between CICR`s and magnetic superspace groups: $\Delta_1 \rightarrow P_a(P^{nma}_{\bar{1}11})$, $\Delta_2 \rightarrow P_a(P^{nma}_{\bar{1}1S})$, $\Delta_3 \rightarrow P_a(P^{nma}_{\bar{1}SS})$ and $\Delta_4 \rightarrow P_a(P^{nma}_{\bar{1}S1})$ [1]. In particular, spatial inversion and time reversal are kept as symmetry operations of the incommensurate structure, when complemented by a phase shift of the modulation corresponding to Φ or π, respectively ($\{i;000,\Phi\},\{\theta;000,1/2\}$). This means that, in the case of the RMnO$_3$ compounds, one irreducible and incommensurate phase can never give rise to a polarization *P*, a magnetization *M* or a toroidal moment *T*. Also, the linear magneto-electric tensor must be null. Homogeneous lattice deformations of symmetry A_g are allowed, together with the lattice modulations $e_{11}(2n\vec{k})$, $e_{22}(2n\vec{k})$ or $e_{33}(2n\vec{k})$, with *n* integer. Among these, the more important will be those with $n=1$, which correspond to the elastic modulations with one-half of the wavelength of the magnetic modulation detected experimentally. These lattice modulations are secondary distortions and play no role in the definition of the ferroic properties of a given phase.

However, ferroelectricity can arise in the case of one irreducible OP if $\vec{k}=\delta\vec{a}^*$ is commensurate (C). In this case, the symmetry of the modulated phase will depend on the type of the modulation wavevector (that is, on the odd (even) value of the integers in the fraction describing δ), the CICR of the OP and the global phase Φ. In the case of $\delta=(2l+1)/2m$ and $\Phi=(2l+1)\pi/4m$ (l and m integers), the possible magnetic space groups are $P_a(Pnm2_1)$ and $P_a(Pna2_1)$, if the order parameter has a symmetry Δ_1 or Δ_2 and Δ_3 or Δ_4, respectively. Here, a polarization P_z is allowed. The possibility that a polarization may result from an irreducible magnetic OP and the role played by the global phase Φ of the OP have been often overlooked. If $\delta=(2l+1)/2m$ and $\Phi=0$, the phase will be ferroelastic ($e_{xz}\neq 0$), regardless of the symmetry of the OP. Ferrotoroidal moments and the linear magnetoelectric effect can occur only if $\delta=(2l+1)/(2m+1)$ or $\delta=2l/(2m+1)$ when $\Phi\neq 0$, while ferromagnetism is possible for this type of wavevector only if $\Phi=0$. These conclusions stress the essential difference between C and INC phases regarding the possible secondary parameters.

[1] Here, $P_a(P^{nma}_{\bar{1}S1})$, for example, denotes $P^{nma}_{\bar{1}S1}\times[\{E;000,0\},\{\theta;000,1/2\}]$, where $P^{nma}_{\bar{1}S1}$ is the unitary superspace group.

3.2. The case of a reducible and incommensurate OP

The possible superspace groups obtained for one incommensurate and reducible OP will depend not only on the irreducible representations chosen but also on the phase difference between the irreducible components of the modulation. The different possible symmetries of magnetic phases resulting from the combinations of different pairs of irreducible co-representations show that a polarization can arise in the cases anticipated from the "spiral picture": if the phase difference between the components is $\Delta\Phi = \pi/2$, the magnetic superspace groups allow spontaneous polarizations along the *b*-axis (in the cases $\Delta_1 + \Delta_4$, $P_a(P^{n2_1a}_{\bar{1}s1})$, and $\Delta_2 + \Delta_3$, $P_a(P^{n2_1a}_{\bar{1}ss})$) or along the *c*-axis (in the cases $\Delta_1 + \Delta_2, P_a(P^{nm2_1}_{\bar{1}11})$, and $\Delta_4 + \Delta_3, P_a(P^{nm2_1}_{\bar{1}ss})$). In addition, a polarization parallel to the modulation wavevector can arise, if the modulated phase originates from two components of the same symmetry. For example, the OP $G_x + C_y$ (symmetry $\Delta_1 + \Delta_1$) will give rise to a phase with the symmetry $P_a(P^{2_1ma}_{\bar{1}11})$ and, consequently, to a polarization $\vec{P} // \vec{a}$, if $\Delta\Phi \neq 0$.

The generalized time reversal operation $\{\theta;000,1/2\}$, denoted by the prefix P_a in the labeling of the symmetry groups given above, is always a symmetry operation of this type of magnetic phase. Consequently, homogeneous magnetizations, toroidal moments and linear magneto-electric coefficients are forbidden by symmetry. For example, in the case of the cycloidal phase of $TbMnO_3$ (symmetry $\Delta_2 + \Delta_3$, OP $A_x + A_y$, $\Delta\Phi = \pi/2$), the magnetic superspace group is $P_a\left(P^{n2_1a}_{\bar{1}ss}\right)$. This phase is polar along the *b*-axis ($P_y \neq 0$), as observed, but it is neither magneto-electric nor multi-ferroic. If $\Delta\Phi = 0$, the superspace group would be $P_a(P^{11\frac{2_1}{a}}_{\bar{1}1s})$ and a homogeneous lattice deformation e_{xy} would occur. For a general phase difference between the two components of the order parameter, the superspace group is $P_a(P^{11a}_{\bar{1}1s})$ and, in addition to the secondary modes P_y and e_{xy} previously mentioned, P_x is also possible. Only in the case of a general $\Delta\Phi$, the system is an elasto-electric bi-ferroic, as a ferroeastic deformation and a spontaneous ferroelectric polarization co-exist. However, magneto-electric bi-ferroicity is forbidden.

3.3-Reducible C- phases: a complex landscape

The case of a reducible and commensurate phase is more complex because, in this case, the symmetry of the modulated phase depends both on the relative and global phases of the irreducible components of the OP. Let us consider, as illustrative examples, the cases of commensurate phases with $\delta = 1/4$ resulting from modulations with symmetry $\Delta_2 + \Delta_3$ or $\Delta_2 + \Delta_1$. In the first case ($\Delta_2 + \Delta_3$), the possible magnetic space groups are $P_a(P\bar{1})$ if $\Phi_2 = \Phi_3 = 0$, $P_a(P112_1)$ if $\Phi_2 = \Phi_3 = \pi/8$, $P_a(P12_11)$ if $\Phi_{2(3)} = 0, \Phi_{3(2)} = \pi/2$ or $P_a(P1)$, in the other cases. Note that one OP of this symmetry can stabilize phases that are polar either along the *b*-axis ($P_a(P12_11)$) or along the *c*-axis ($P_a(P112_1)$). Note also that, in the latter case, the components of the OP would be in-phase. In the second example ($\Delta_2 + \Delta_1$), the possible magnetic space groups would be $P_a(P1\frac{2_1}{m}1)$ if $\Phi_2 = \Phi_1 = 0$, $P_a(Pnm2_1)$ if $\Phi_{2(1)} = \pi/8, \Phi_{1(2)} = 5\pi/8$ (or equivalent domains) or $P_a(P1m1)$, in the other cases.

In $TbMnO_3$, the phase stable at zero magnetic field below T_{c2}=28*K* is a $\Delta_2 + \Delta_3$ (mode $A_x + A_y$) incommensurate cycloidal phase with $\Delta\Phi = \pi/2$ [26]. As seen, this phase has symmetry $P_a(P^{n2_1a}_{\bar{1}ss})$

and it is polar along the *b*-axis. Above a critical threshold, an external magnetic field induces a transition to a $\delta = 1/4$ commensurate phase, polar along the *c*-axis. The magnetic structure of this field induced phase has recently been identified as a $\Delta_2 + \Delta_1$ phase (mode $A_x + A_z$), with $\Delta\Phi \approx \pi/2$ [27]. Its symmetry may therefore correspond either to $P_a(P1m1)$ or $P_a(Pnm2_1)$, depending on the global phases adopted. Given that *P* is parallel to the *c*-axis, the latter case must be chosen. Here, the importance of the global phases is again stressed. In such a case, the field induced transition is $P_a(P_{\bar{1}ss}^{n2_1a}) \rightarrow P_a(Pnm2_1)$, which implies a rotation of the plane of the cycloid (in agreement with the "spiral picture") and a change of the symmetry of the order parameter, along with a rotation of the polarization from the *b*-axis to the *c*-axis. However, another possibility could be a transition to the phase $P_a(P112_1)$ (a $A_x + A_y$ phase with $\Phi_2 = \Phi_3 = \pi/8$ (or equivalent phases $(2n+1)\pi/8$)), which would also allow for a polarization strictly directed along the *c*-axis, maintaining the symmetry $\Delta_2 + \Delta_3$ of the OP. This possibility, for example, is not commonly taken into account and can not be anticipated from the usual "spiral picture".

4. Magneto-electric effect and the polarization rotation under field: the example of $TbMnO_3$

In both the cases considered above, a field driven transition from $P_a(P_{\bar{1}ss}^{n2_1a})$ to $P_a(Pnm2_1)$ or $P_a(P112_1)$ involves phases with a symmetry that forbids a linear magneto-electric effect. In these circumstances, can the magneto-electric effect play a relevant role in the mechanism for the field driven rotation of the electric polarization?

The transition from the cycloidal INC phase to the cycloidal C phase, observed in $TbMnO_3$ under a magnetic field, is found to be mediated by an intermediate phase [28]. Therefore, in order to answer the question above, one must first clarify the origin of a phase sequence $P_a(P_{\bar{1}ss}^{n2_1a}) \rightarrow ? \rightarrow P_a(Pnm2_1)$ and elucidate the nature and the magneto-electric properties of the intermediate phase.

The mechanism for the field induced destabilization of the cycloidal $P_a(P_{\bar{1}ss}^{n2_1a})$ phase in $TbMnO_3$ has been recently analyzed [29]. The form of the free energy expansion for the case of this incommensurate phase is entirely dictated by the symmetry of the OP:

$$f_{inc} = \frac{\alpha_2}{2}S_2^2 + \frac{\beta_2}{4}S_2^4 + \frac{\alpha_3}{2}S_3^2 + \frac{\beta_3}{4}S_3^4 + \frac{\gamma_1}{2}S_2^2S_3^2\cos(2\varphi) + \frac{\gamma_2}{4}S_2^4S_3^4\cos^2(2\varphi) + \\ + v_1P_yS_2S_3\sin(\varphi) + v_2e_{xy}S_2S_3\cos(\varphi) + v_3P_xS_2^2S_3^2\sin(2\varphi) + \frac{k_y}{2}P_y^2 + \frac{k_x}{2}P_x^2 + \\ + \frac{k_e}{2}e_{xy}^2 + \frac{\mu}{2}M^2 + \frac{\Delta}{2}M^2S_2^2S_3^2\cos(2\varphi) - MB + ... \tag{1}$$

Here, we have considered the external magnetic field *B* and the induced magnetization *M*, together with a coupling term between *M* and the invariant $S_2^2S_3^2\cos(2\varphi)$, which depends on the relative phase $\Delta\Phi = \varphi$. The equilibrium value of φ will depend on the external field and on the relative values of the expansion coefficients. Let $\bar{\gamma}_1 = \left[\gamma_1 + v_1^2/2k_y - v_2^2/2k_e\right]$ and $\bar{\gamma}_2 = \left[\gamma_2 + 2v_3^2/k_x\right]$. The higher symmetry solutions $\varphi = 0$ or $\varphi = \pi/2$ will be stable if $\left[\bar{\gamma}_1 + \bar{\gamma}_2S_2^2S_3^2\cos(2\varphi) + \Delta B^2\bar{\mu}^{-2}(\varphi)\right] > 0$ or $\left[\bar{\gamma}_1 + \bar{\gamma}_2S_2^2S_3^2\cos(2\varphi) + \Delta B^2\bar{\mu}^{-2}(\varphi)\right] > 0$, respectively. At zero field, the latter condition is verified experimentally, implying that $\left[\bar{\gamma}_1 - \bar{\gamma}_2S_2^2S_3^2\right] > 0$. Also, the destabilization of the cycloidal polar phase by the external magnetic field requires that $\Delta < 0$. Under these conditions, the INC phase is destabilized by a magnetic field $B > B_{c1}$, where B_{c1} is given by:

$$B_{c1} = (\mu + |\Delta| S_2^2 S_3^2)\sqrt{\frac{\bar{\gamma}_1 - \bar{\gamma}_2 S_2^2 S_3^2}{|\Delta|}} \tag{2}$$

These results have been established in [29]. One can, however, explore a little further this mechanism. Let us first note that the phase $\varphi = 0$ is potentially stabilized for fields $B > B_{c2}$, where:

$$B_{c2} = (\mu - |\Delta| S_2^2 S_3^2)\sqrt{\frac{\bar{\gamma}_1 + \bar{\gamma}_2 S_2^2 S_3^2}{|\Delta|}} \tag{3}$$

If, for simplicity, we assume that $\mu >> |\Delta|$, so that $\bar{\mu}(\varphi) \approx \mu$, then $B_{c2}^2 \approx B_{c1}^2 + 2\mu^2 \bar{\gamma}_2 S_2^2 S_3^2 |\Delta|^{-1}$. Since $\bar{\gamma}_2 > 0$, $B_{c2} > B_{c1}$. This means that, for the intermediate field range $B_{c1} < B < B_{c2}$, the phases corresponding to $\varphi = \pi/2$ and $\varphi = 0$ will be both unstable and there must necessarily exist an intermediate phase of lower symmetry. The stability of this intermediate phase requires that:

$$\cos(2\varphi) \approx \frac{1}{\bar{\gamma}_2 S_2^2 S_3^2}\left[\frac{\lfloor\Delta\rfloor}{\mu^2} B^2 - \bar{\gamma}_1\right] \tag{4}$$

If $B=B_{c1}$, this condition imposes $\varphi = \pi/2$. If $B=B_{c2}$, then $\varphi = 0$. That is, above B_{c1}, the field drives a continuous variation of the relative phase of the two components of the spin modulation, from $\varphi = \pi/2$ to $\varphi = 0$. This drift of φ reduces, as seen, the symmetry of the phase from $P_a(P_{\bar{1}ss}^{n2_1a})$ to $P_a(P_{\bar{1}1s}^{11a})$, if we neglect the breaking of $\{\theta;000,1/2\}$ and mirror symmetry induced by the external magnetic field. This reduction of symmetry costs energy and allows the onset of an effective and non-linear magnetoelectric coupling. The increase in energy with respect to the polar cycloidal phase $\varphi = \pi/2$ can be expressed as a function of the magnetic field as:

$$\Delta f_{inc} = f_{inc}(\varphi) - f_{inc}(\frac{\pi}{2}) = \frac{1}{4\bar{\gamma}_2}\left[\frac{|\Delta|^2}{\mu^4} B^4 - \left[\bar{\gamma}_1 - \bar{\gamma}_2 S_2^2 S_3^2\right]^2\right] = \frac{1}{4\bar{\gamma}_2}\frac{|\Delta|^2}{\mu^4}\left[B^4 - B_{c1}^4\right] \tag{5}$$

Also, in the intermediate field range, the effective magneto-electric and magneto-elastic coupling can be explicitly established from the field dependence of the secondary parameters:

$$P_y = -\frac{\upsilon_1}{k_y}\sqrt{\frac{|\Delta|}{2\bar{\gamma}_2\mu^2}\left(B_{c2}^2 - B^2\right)} \tag{6a}$$

$$P_x = -\frac{\nu_3|\Delta|}{k_x\bar{\gamma}_2\mu^2}\sqrt{B^2\left(B_{c1}^2 + B_{c2}^2\right) - B_{c1}^2 B_{c2}^2 - B^4} \tag{6b}$$

$$e_{xy} = -\frac{\upsilon_2}{k_e}\sqrt{\frac{|\Delta|}{2\bar{\gamma}_2\mu^2}\left(B^2 - B_{c1}^2\right)} \tag{6c}$$

The increase of the energy of the intermediate $P_a(P_{\bar{1}1S}^{11a})$ phase, given by (5), is driven by the magnetic field via non-linear magneto-electric and magneto-elastic effects. This reduces the relative stability of the phase with respect to potential competing phases. In this process, the magneto-electric coupling induces a rotation of the polarization but solely in the *ab*-plane. The rotation of the polarization from the *ab*-plane to the *c*-axis results from a separate mechanism: the stabilization, via a first order transition, of a different competing (commensurate) phase.

References

[1] Kimura T, Goto T, Shintani H, Ishizaka K, Arima T and Tokura Y 2003 *Nature* **426** 55
[2] Hur N, Park S, Sharma P A, Ahn J S, Guha S and Cheong S W, 2004 *Nature* **429** 392
[3] Lawes G, Harris A B, Kimura T, Rogado N, Cava R J, Aharony A, Entin-Wohlman O, Yildirim T, Kenzelmann M, Broholm C and Ramirez A P 2005 *Phys. Rev. Lett.* **95** 087205
[4] Kimura T, Lashley J C and Ramirez A P 2006 *Phys. Rev. B* **73** 220401(R)
[5] Yamasaki Y 2006 *Phys. Rev. Lett.* **96** 207204
[6] Taniguchi K, Abe N, Takenobu T, Iwasa Y and Arima T 2006 *Phys. Rev. Lett.* **97** 097203
[7] Heyer O, Hollmann N, Klassen I, Jodlank S, Bohatý L, Becker P, Mydosh J A, Lorenz T and Khomskii D 2006 *J. Phys.:Cond. Matter* **18** L471
[8] Kimura T, Lawes G and Ramirez A P 2005 *Phys. Rev. Lett.* **94** 137201
[9] Cheong S W and Mostovoy M 2007 *Nature Materials* **6** 13
[10] Teague J R, Gerson R and James W J 1970 *Solid State Commun.* **8** 1073
[11] Bertaut E F, in *Magnetism*, edited by G. T. Rado, H. Suhl (Academic Press, New York, 1963), Vol. III 149
[12] Kimura T, Kawamoto S, Yamada I, Azuma M, Takano M and Tokura Y 2003 *Phys. Rev. B* **67** 180401
[13] Mashiyama H 1980 *J. Phys. Soc. Jpn* **49** 2270
[14] Gesi K, 1986 *Ferroelectrics* **66** 269
[15] Perez-Mato J M 1988 *Solid State Commun.* **67** 1145
[16] Sergienko I A, Sen C, Dagotto E 2006 *Phys. Rev. Lett.* **97** 227204
[17] Sergienko I A, Dagotto E 2006 *Phys. Rev. B.* **73** 094434
[18] Katsura H, Nagaosa N, Balatsky A V 2005 *Phys. Rev. Lett.* **95**, 057205
[19] Mostovoy M 2006 *Phys. Rev. Lett.* **96** 067601
[20] Harris A B 2007 *Phys. Rev. B* **76** 054447
[21] Schweizer J 2005 *C. R. Phys.* **6** 375
[22] Radaelli P G, Chapon L C 2007 *Phys. Rev. B* **76** 054428
[23] Ribeiro J L 2007 *Phys. Rev. B* **76** 144417
[24] Kovalev O V, *Representations of Crystallographic Space Groups: Irreducible Representations, Induced Representations and Corepresentations* 1993 Gordon and Breach, Amesterdam
[25] Pérez-Mato J M, Madariaga G, Tello M J 1984 *Phys. Rev. B* **30** 1534
[26] Kenzelmann M, Harris A B, Jonas S, Broholm C, Schefer J, Kim S B, Zhang C L, Cheong S W, Vajk O P and Lynn J W 2005 *Phys. Rev. Lett.* **95** 087206
[27] Aliouane N, Schmalzl K, Senff D, Maljuk A, Prokes K, Braden M, Agyriuo D N 2009 *Phys. Rev. Lett.* **102** 207205
[28] Aliouane N, Argyriou D N, Strempfer J, Zegkinoglou I, Landsgesell S, von Zimmermann M 2006 *Phys. Rev. B* **73** 020102(R)
[29] Tolédano P 2009 *Phys. Rev. B* **79** 094416

Journal of Physics: Conference Series **226** (2010) 012014 doi:10.1088/1742-6596/226/1/012014

Advances in solution of modulated structures reflected by Jana system

M Dušek, V Petříček and L Palatinus

Institute of Physics of the ASCR, v.v.i., Na Slovance 2, 182 21 Prague 8, Czech Republic

dusek@fzu.cz

Abstract. The latest development in the x-ray diffractometers makes modulated structures more frequent and accessible than ever before. The last version of Jana software, Jana2006, contains improvements facilitating solution of modulated structures for non-specialists. The initial structure model, including modulations, is now created in communication with the charge flipping program Superflip. Combination of discontinuous functions with position modulation does not require any more orthogonalization procedure because position modulation is newly expressed by Legendre polynomials. For difficult tasks Jana2006 adds possibility to combine data sources, including powder and single crystal data, regardless on the source of diffraction (x-ray, synchrotron, neutron). This feature helps in refinement of magnetic structures. They have obtained full support in Jana2006, including tools for testing magnetic symmetry and representative analysis. Both nuclear and magnetic structures can be modulated and the can be jointly refined against one or more data sets.

1. Introduction

In the last few years we have seen rapid progress in home laboratory instruments for x-ray diffraction. With sensitive area detectors and new x-ray optics overlooking of satellites has become almost impossible and number of known modulated structures has been growing rapidly. Programs for modulated structures are now used by larger number of crystallographers than ever before and many of them are not specialists for aperiodic structures. The new demand on the computing tools is accessibility and friendliness retaining at the same time all possibilities for complicated problems. This direction is reflected in Jana2006 system [1] by including the charge flipping method for ab initio solution of modulated structures and by better support of often used discontinuous functions.

On the other hand, the progress in the experimental possibilities enables solution of more complicated structures. During the last years there was growing interest among the users of Jana2006 to refine jointly neutron diffraction data and synchrotron data, especially for the purpose of the magnetic structures refinement. Jana2006 reflects also this kind of requests by introducing possibility to jointly refine a structure against almost any combination of diffraction data including combination of powder and single crystal data. The support for magnetic structures has been added recently, too.

2. Integration of charge flipping

Charge flipping [2] has become a new alternative for solution of the phase problem. The input for charge flipping is diffraction data only; neither symmetry information nor chemical composition is required. Charge flipping yields three or more (depending on number of indices) dimensional electron density map. In order to get the structure model the electron density map must be interpreted. For standard structures it may reveal atomic positions, disorder and anharmonic features, for modulated structures the map contains atomic domains which can be described in terms of basic atomic positions and modulation functions.

The importance of charge flipping in the field of modulated structures is much larger than for standard structures because this is the only well established method which can solve a modulated structure without preliminary knowledge of the average structure. This is especially useful for strongly modulated structures where finding of the average structure may be difficult and a refinement of modulation functions from arbitrary small displacements may be misleading.

Jana2006 communicates with the charge flipping program Superflip [3] by the way that it prepares input files, executes Superflip as an external program and interprets the resulting electron density map in terms of basic atomic positions and modulation functions. For structures where harmonic position modulation dominates one can immediately start refinement of the structure found from the charge flipping. For discontinuous functions, however, the starting parameters (centre and width) must be found manually using sections through the electron density map from charge flipping.

Superflip can also search for symmetry in the resulting electron density map. This can be used as an independent symmetry check in addition to the classical methods based on R_{int} and systematic extinctions.

3. Discontinuous functions in Jana2006

Six of the ten latest modulated structures published in Acta Crystallographica B [4-13] use discontinuous functions, either crenel or sawtooth, for the description of the structure model. Apparently discontinuous functions are no longer reserved for special cases. They should be understood as a basic tool and receive a corresponding support that would allow their utilization by non-specialists.

Discontinuous functions were introduced in 1990 (sawtooth function, [14]) and in 1995 (crenel function, [15]) in order to facilitate refinement of structures with abrupt modulation changes of position and occupation. Crenel function describes modulation of occupancy by defining an interval in x_4 where the corresponding atom exists. It is defined by the width and the centre, both of them can be refined. The sawtooth function combines crenel modulation of occupancy with linear modulation of position and it is therefore defined by the width, the centre and the slope of the line along x_1, x_2 and x_3 axes of the basic unit cell. Both crenel and sawtooth functions can be combined with harmonic position modulation.

In previous versions of Jana software usage of crenel or sawtooth function required a lot of user interaction. The centre, width and (for sawtooth) slope had to be found in the Fourier sections x_1-x_4, x_2-x_4 and x_3-x_4, respectively. The combination of crenel or sawtooth functions with harmonic modulation usually required to use orthogonalization procedure [15] in order to make the modulation functions mutually orthogonal in the restricted interval of x_4 corresponding to the width. This semi-automatic procedure calculated – based on the centre and width of crenel – selection rules defining which sine and cosine terms of the harmonic modulation function can be used in the refinement. The orthogonalization procedure had to be repeated after any significant change of the width or the centre of crenel or sawtooth function. The change in selection rules often destroyed the structure model. Finally, since there is no CIF definition for selection rules, refinement could not be repeated from the publication CIF files.

Instead of harmonic modulation functions, Jana2006 combines crenel functions with Legendre polynomials [16]. The Legendre polynomial $P_n(z)$ can be defined by equation

$$P_n(z)=\frac{1}{2^n n!}\frac{d^n}{d^n x}\left(x^2-1\right) \quad (1)$$

The Legendre polynomials are orthogonal over the interval (-1,1), i.e. they satisfy the equation

$$\int_{-1}^{1} P_n(x)P_m(x)dx = \frac{2}{2n+1}\delta_{mn} \tag{2}$$

The original crenel interval $(x_{40}-\Delta/2, x_{40}+\Delta/2)$ can be transformed to the interval (-1,1) by

$$\xi = 2(x_4 - x_{40})/\Delta\,, \tag{3}$$

where x_{40} and Δ are the centre and width of the crenel function, respectively. For a full interval $\Delta = 1$, i.e. $(x_{40}-1/2, x_{40}+1/2)$, this approach can be used, too.

Thus, the Legendre polynomials can combined with crenel function without any additional orthogonalization procedure which facilitates usage of discontinuous functions and leads to more stable and less correlated refinement. For the description of modulation the odd and even polynomials are coupled, P_1 and P_2 forming the first modulation wave, P_3 and P_4 the second wave etc. By this way an analogy is achieved with the original expressions for modulations by harmonic waves. For instance, for position modulation we get

$$\mathbf{r}_\nu = \mathbf{r}_{\nu,0} + \sum_n \mathbf{U}^s_{\nu n}\sin(2\pi n x_4) + \mathbf{U}^c_{\nu n}\cos(2\pi n x_4) \tag{4}$$

when harmonic waves are used, and

$$\mathbf{r}_\nu = \mathbf{r}_{\nu,0} + \sum_n \mathbf{S}^o_{\nu n}P^o_n[2(x_4 - x_{40})/\Delta] + \mathbf{S}^e_{\nu n}P^e_n[2(x_4 - x_{40})/\Delta] \tag{5}$$

with Legendre polynomials. In the equation (4),(5) $\mathbf{r}_{\nu,0}$ is the basic position of the atom ν; $\mathbf{U}^s_{\nu n}$ and $\mathbf{U}^c_{\nu n}$ are Fourier amplitudes of the sine and cosine term, respectively, of the nth modulation wave of the atom ν; "o" and "e" superscript stands for odd and even member of the Legendre polynomial forming the modulation wave.

The first order term P_1 defines a simple linear function at interval $(x_{40}-\Delta/2, x_{40}+\Delta/2)$, which corresponds to a sawtooth modulation. Thus there is no more reason for treating sawtooth and crenel separately; crenel combined with the first Legendre modulation wave acts always like a sawtooth function with inclination encoded in the first three modulation parameters. For example, a sawtooth function defined over the whole interval of x_{40} can be expressed as a crenel function defined over the the same full interval and combined with Legendre modulation wave (figure 1). An article explaining in detail the topic of Legendre polynomials used for description of modulation is under preparation.

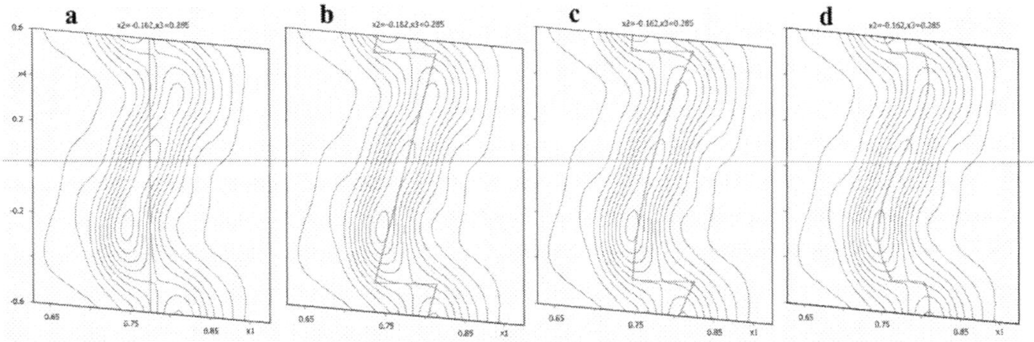

Figure 1. Sawtooth function created like combination of crenel function (x_{40} =0.04, Δ =1) with positional modulation expressed by Legendre polynomials. (a) No combination; (b) 1st order polynomial, (c) 1st and 2nd order polynomial; (d) 1st, 2nd, 3rd and 4th order polynomial. The resulting modulation function is overlaid with the corresponding Fourier maximum. The horizontal dotted line indicates position of the centre of the crenel function.

4. Magnetic structures and joint refinement

At very low temperatures some magnetic materials undergo magnetic phase transition due to ordering of magnetic moments. This ordering leads to the presence of magnetic reflections that can be observed in the diffraction pattern with help of neutron diffraction. The magnetic structure factor can be described like

$$\vec{\mathbf{F}}_M(\mathbf{h}) = p\sum_i f_i(h)T_i(\mathbf{h})\vec{\mathbf{S}}_i \exp(2\pi i\mathbf{h}\cdot\mathbf{r}_i), \tag{6}$$

where f_i is magnetic form factor, T_i is temperature factor, $\vec{\mathbf{S}}_i$ is magnetic moment and $\mathbf{r}_i$ is an atomic position. The reason for the growing interest among users of Jana2006 to use this program for refinement of magnetic structures is due to the fact that the distribution of magnetic moment over the nuclear structure can be described by modulation wave

$$\mathbf{S}_i(x_4) = \mathbf{S}_{i0} + \sum_{n=1}^{N}[\mathbf{S}_{ins}\sin(2\pi nx_4) + \mathbf{S}_{ikc}\cos(2\pi nx_4)], \tag{7}$$

where $\mathbf{S}_{i0}$ is the basic magnetic moment, $\mathbf{S}_{ins}$ and $\mathbf{S}_{ikc}$ are amplitudes of the sin and cos term of the n-th modulation wave, respectively. Thus the tools developed for modulated structures could be generalized and used also for magnetic structures.

The magnetic reflections contribute to the diffraction pattern by the way that their intensities are summed with intensities of reflections corresponding to nuclear structure. From the calculation point of view we can consider a magnetic sample to be a multiphase consisting of nuclear and magnetic phase. From the experimental point of view the data measured under the temperature of the magnetic phase transition are often short strips of neutron powder diffraction data that nicely show the magnetic reflections but their quality is not sufficient for precise determination of the nuclear structure. Therefore a combination seems to be very practical of the powder neutron data with a complete measurement, not necessarily using neutrons, done for the purpose of the exact determination of the crystal structure. Apparently introducing of magnetic structures into the Jana system should be accompanied with a new possibility of joint refinement of the same structure against different diffraction data, of a multiphase structure against the same diffraction data or – generally – of a multiphase structure against several diffraction data.

The latest version of Jana2006 contains complete set of tools for investigation and determination of magnetic structures. Magnetic symmetry can be defined with magnetic space or superspace group symbols using comma for the time inversion. A tool for testing magnetic symmetry allows calculating global magnetic moment pertinent to the selected symmetry symbol and immediate comparing the extinction rules for magnetic reflections (following from the chosen symmetry) with the experimental diffraction pattern. Representative analysis can be generated for arbitrary dimension and the resulted representations can be unambiguously assigned with the magnetic symmetry symbol. The later required some theoretical investigation which will be soon published in a separate article. The magnetic structure, either modulated or three dimensional, either incommensurate or commensurate, can be refined together with nuclear structure which can be also modulated, using the same or different diffraction data.

The impulse for introducing joint refinement into Jana system came from the field of magnetic structures but finally the method works for any combination except electron diffraction data (figure 2). The most useful combination remains joint refinement against neutron and synchrotron or x-ray data. The minimized function for n single crystal phases and m powder phases can be expressed like

$$S=\sum_{\mathbf{h}} w_{\mathbf{h},1}\left(F_{obs,1}(\mathbf{h})-F_{calc,1}(\mathbf{h})\right)^2+\ldots+\sum_{\mathbf{h}} w_{\mathbf{h},n}\left(F_{obs,n}(\mathbf{h})-F_{calc,n}(\mathbf{h})\right)^2+ \\ \sum_i w_{i,1}\left(I_{obs,i,1}-I_{calc,i,1}\right)^2+\ldots+\sum_i w_{i,m}\left(I_{obs,i,m}-I_{calc,i,m}\right)^2 \quad , \qquad (8)$$

where $F_{obs}(\mathbf{h})$ and $F_{calc}(\mathbf{h})$ are observed and calculated structure factors while $I_{obs,i}$ and $I_{calc,i}$ are observed and calculated intensities of the powder profile point i. For joint refinement against combination of powder and single crystal data the weight w of a structure factor or of a profile point is of critical importance because one measurement should not be over weighted by another measurement. The partial experiments should be done carefully and the weighting scheme must be strictly derived from the counting statistics.

Another problem in joint refinement comes from different meaning of "temperature" parameters for neutron and x-ray diffraction data. Jana2006 introduces an overall atomic displacement parameter ADP which scales the ADP for partial data sets.

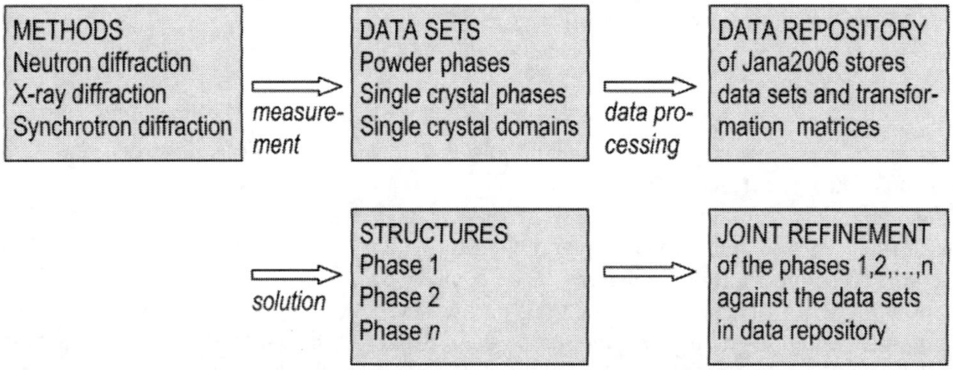

Figure 2. The flowchart illustrating possibilities of joint refinement with Jana2006. Single crystal domains have the same structure while single crystal phases have different structures. Magnetic structure is understood as a special kind of phase.

Journal of Physics: Conference Series **226** (2010) 012014 doi:10.1088/1742-6596/226/1/012014

Up to now only a few examples exists of successful application of joint refinement of powder and single crystal data but they have not been published yet. The same situation is for magnetic structures. Refinement of coexisting single crystal phases with jana2006 has been reported in [17].

References

[1] Petříček V, Dušek M and Palatinus L 2006 Jana2006 - the crystallographic computing system. Institute of Physics of the ASCR, Prague, Czech Republic, www-xray.fzu.cz/jana
[2] Ozslányi G and Süto A 2004 *Acta Cryst.* **A60** 134-41
[3] Palatinus L and Chapuis G 2007 *J.Appl.Cryst.* **40** 786-90
[4] Arakcheeva A and Chapuis G 2008 *Acta Cryst.* **B64**, 12-65
[5] Muguerra H, Grebille D. and Bourée F 2008 *Acta Cryst.* **B64** 144-53
[6] Arakcheeva A, Pattison P, Chapuis G, Rossell M, Filaretov A, Morozov V and Van Tendeloo G 2008 *Acta Cryst.* **B64** 160-171
[7] Bereciartua P J, Zuniga F J and Breclewski T 2008 *Acta Cryst.* **B64** 405-16
[8] Lazic B, Krüger H, Kahlenberg V, Konzett J and Kaindl R 2008 *Acta Cryst.* **B64** 417-25
[9] van Smaalen S, Dinnebier R, Sofin M and Jansen M 2007 *Acta Cryst.* **B63** 17-25
[10] Guiblin N, Fuhrer C A, Häner R, Stoeckli-Evans H, Schenk K and Chapuis G 2006 *Acta Cryst.* **B62** 506-12
[11] Palatinus L, Dušek M, Glaum R and El Bali B 2006 *Acta Cryst.* **B62** 556-66
[12] Evain M, Petříček V, Coué V, Dessapt R, Bujoli-Doeuff M and Jobic S 2006 *Acta Cryst.* **B62** 790-97
[13] Rodriguez S M B, Palatinus L, Petříček V and Chapuis G 2006 *Acta Cryst.* **B62** 1043-50
[14] Petříček V, Gao Y, Lee P and Coppens P 1990 *Phys. Rev.* B **42** 387-92
[15] Petříček V, van der Lee A and Evain M 1995 *Acta Cryst.* **A51** 529-35
[16] Arfken G 1985 Mathematical Methods for Physicists, 3rd ed. Orlando, FL: Academic press, 416
[17] Krüger H, Kahlenberg V, Petříček V, Fritz P and Wertl W 2009 *J. Solid State Chem.* **182** 1515-23

Journal of Physics: Conference Series **226** (2010) 012015 doi:10.1088/1742-6596/226/1/012015

A careful phase analysis and TEM investigation of the incommensurately modulated, $(Bi_{1-x}M^{II})Fe^{III}O_{3-x/2}$ (M = Ca and Sr), solid solution phases

RL Withers[1], J Schiemer[1], L Bourgeois[2], L Norén[1] and Y Liu[1]

[1] Research School of Chemistry, Australian National University, Canberra, A.C.T, 0200, Australia
[2] Monash Centre for Electron Microscopy and Department of Materials Engineering, Monash University, Melbourne, Australia

withers@rsc.anu.edu.au

Abstract. A careful phase analysis and TEM investigation of the $(Bi_{1-x}Ca_x)Fe^{III}O_{3-x/2}V_{x/2}$ (V for vacancy), $\sim 0.20 \le x \le \sim 0.50$, and $(Bi_{1-x}Sr_x)Fe^{3+}O_{3-x/2}V_{x/2}$, $0.2 \le x \le 0.67$, perovskite related solid solution fields has been carried out. Both solid solution fields are, in general, (3+1)-d incommensurately modulated with an incommensurate primary modulation wave-vector magnitude that varies systematically with composition. High Resolution (HR)TEM imaging is used to show the presence of at least 6-fold twinning on a sometimes rather fine scale, the fine scale twinning being more prevalent in the M = Sr than the M = Ca system. The (3+1)-d superspace group symmetry is suggested to be $Cmmm(01\gamma)0s0$ or one of the non-centrosymmetric sub-groups thereof. The magnetic and electrical properties of the systems are briefly discussed and a tentative structural model proposed.

1. Introduction

Bismuth ferrite ($BiFeO_3$) exhibits multiferroic properties at room temperature and is a promising lead-free piezoelectric material. A major drawback with $BiFeO_3$, however, is its high current leakage. Attempts to improve its electrical properties have been made by substitutional doping with a range of rare earth as well as alkaline earth elements on the Bi, or perovskite A, site. In this paper, the results of a careful phase analysis, Transmission Electron Microscope (TEM) and physical properties investigation of the alkaline earth substituted, wide range non-stoichiometric $(Bi_{1-x}Ca_x)Fe^{III}O_{3-x/2}V_{x/2}$ (V for vacancy), $\sim 0.2 \le x \le \sim 0.5$, and isomorphous $(Bi_{1-x}Sr_x)Fe^{III}O_{3-x/2}V_{x/2}$, $0.2 \le x \le 0.67$, perovskite related solid solution phases are reported.

2. Synthesis and phase analysis

The M = Sr sample was prepared by conventional solid state reaction (see Withers *et al* [1]). The M = Ca samples were prepared by a rapid two-stage reaction method using high purity Bi_2O_3, $CaCO_3$ and Fe_2O_3 powders, following Khomchenko *et al* [2]. The starting powders were initially homogenized for ~ 20 minutes in an agate mortar before a first reaction at 850 °C for 20 minutes. The resultant samples were then re-ground for a further 20 min, pressed into pellets at 480 MPa in a steel die and sintered for a further 20 minutes according to composition. Samples with $x < 0.3$ were sintered at 940 °C while samples with x > 0.3 were sintered at 990 °C. Use of sintering temperatures above these caused melting of the samples. Despite the surprisingly short reaction times, extremely well-crystallized reaction products were obtained via this process.

The signature of a continuously variable solid solution phase, namely composition-dependent, continuously variable, unit cell dimensions, is observed in both systems. The metric symmetry of the well-defined, perovskite-related parent (subscript p below), or average, structure of the M = Ca phase was tetragonal (with the c_p dimension always slightly larger than the $a_p = b_p$ dimension, see Fig.1) but very close to metrically cubic while the metric symmetry of the underlying average structure of the M = Sr sample was cubic (see also [3]). No additional satellite reflections were apparent in powder XRD patterns from either solid solution phase. Fig.1 shows a plot of the powder XRD refined metrically tetragonal unit cell dimensions of the underlying perovskite-related, parent structure of the M = Ca phase as a function of the Ca content, x. Note that the lines drawn through the data points in Fig.1 are a guide to the eye only. Samples synthesized outside the shaded composition range were clearly two phase and enabled the composition range of existence of the M = Ca phase to be determined to run from $\sim 0.20 \le x \le \sim 0.50$. In the case of the M = Sr phase, the corresponding composition range of existence is somewhat broader, namely from $\sim 0.20 \le x \le \sim 0.67$ [3].

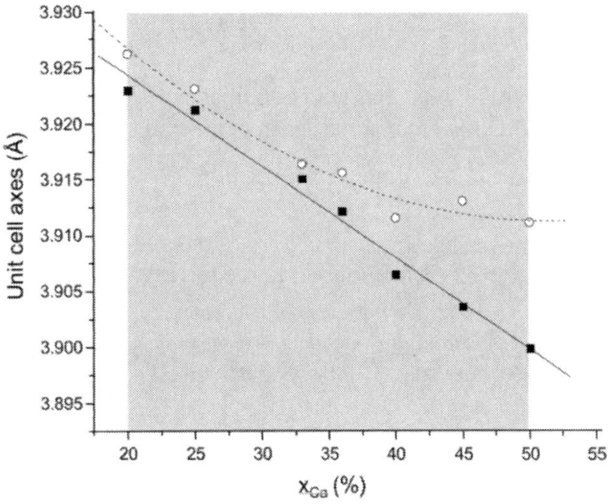

Fig.1 Shows the XRD refined, metrically tetragonal unit cell dimensions, $a_p = b_p$ (solid squares) and c_p (open circles), of the parent structure of the M = Ca phase as a function of Ca content, x.

3. HRTEM imaging and twinning

Despite the metric symmetries, both solid solution phases are, in general, (3+1)-d incommensurately modulated with a primitive primary modulation wave-vector $\mathbf{q}_{prim} = {}^1/_2\mathbf{b}_p{}^* + \gamma\mathbf{c}_p{}^*$ that varies systematically with composition [4]. The local symmetry is thus at most orthorhombic. The higher apparent symmetry of the underlying average structures arises from inherent fine scale twinning, particularly in the case of the M = Sr phase. Fig.2a, for example, shows a typical $<100>_p$ zone axis HRTEM image of an x = 0.46, M = Sr sample along with the corresponding Fast Fourier Transform (FFT) of the whole image. Incommensurate 'superlattice fringes' of periodicity $a_p/(1-2\gamma) \sim 3.47\ a_p \sim 13.7$ Å, running largely along only one or other of the two $<001>_p{}^*$ directions of reciprocal space excited in any one such $<100>_p$ type image (and corresponding to the incommensurate satellite reflections apparent in the FFT, see Fig.2b), are clearly apparent. In other areas, no obvious superlattice fringes are apparent at all suggesting that the incommensurate component of the modulation in such regions is running along the third $<001>_p{}^*$ direction of reciprocal space *i.e.* along the projection direction. Thus the M = Sr phase is at least triply twinned on quite a fine, ~ 5-10 nm, scale. It is thus clear that the incommensurate component of the primitive primary modulation wave-vector runs along only one of the three possible $<001>_p^*$ directions in any one local region.

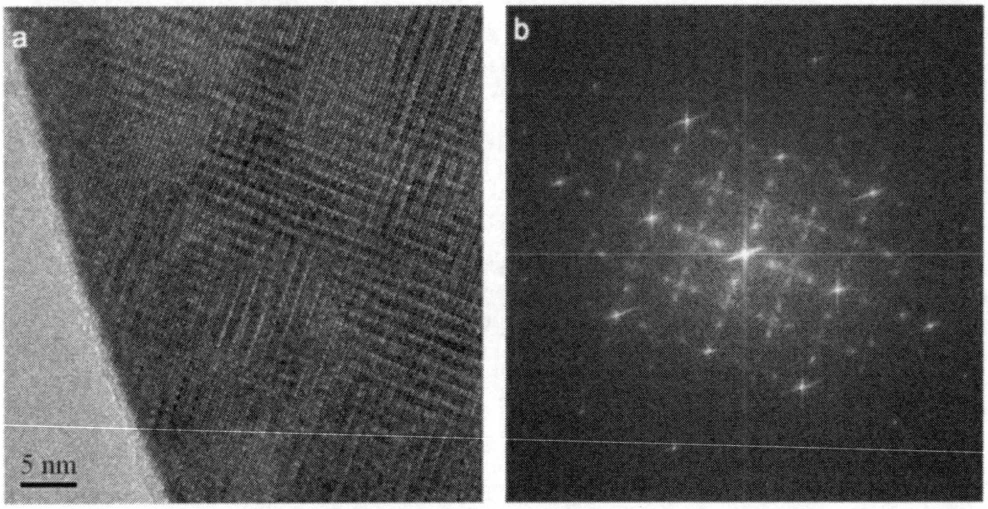

Fig.2. Shows a typical (a) <100> zone axis HRTEM image of an x = 0.46, $\gamma = 0.356(1)$, M = Sr sample along with (b) the corresponding Fast Fourier Transform (FFT) of the whole image.

In addition to this form of twinning, a rather more subtle form of twinning occurs in both phases involving a switch in the primitive primary modulation wave-vector from $[0, {}^1/_2, \gamma]_p^*$ to $[{}^1/_2, 0, \gamma]_p^*$. Such twinning is apparent in the change in periodicity along the superlattice fringes themselves from b_p to $2b_p$ as can be seen in going from the bottom left to the top right hand side of the $<100>_p$ zone axis HRTEM image of an x = 0.36, $\gamma = 0.4$, M = Ca grain shown in Fig.3 below.

Fig.3. A $<100>_p$ zone axis HRTEM image of an x = 0.36, $\gamma = 0.4$, M = Ca grain. Note the change in periodicity along the superlattice fringes from the bottom left to the top right hand side of the image.

4. Electron diffraction and superspace group symmetry.

Given the prevalence of twinning even in the M = Ca case, it is not surprising that selected area electron diffraction patterns (EDP's) are almost invariably twinned (see, for example, the typical

$<001>_p$ type zone axis EDP's shown in Figs.4a to c below) making it rather difficult to unambiguously determine the corresponding (3+1)-d superspace group symmetry.

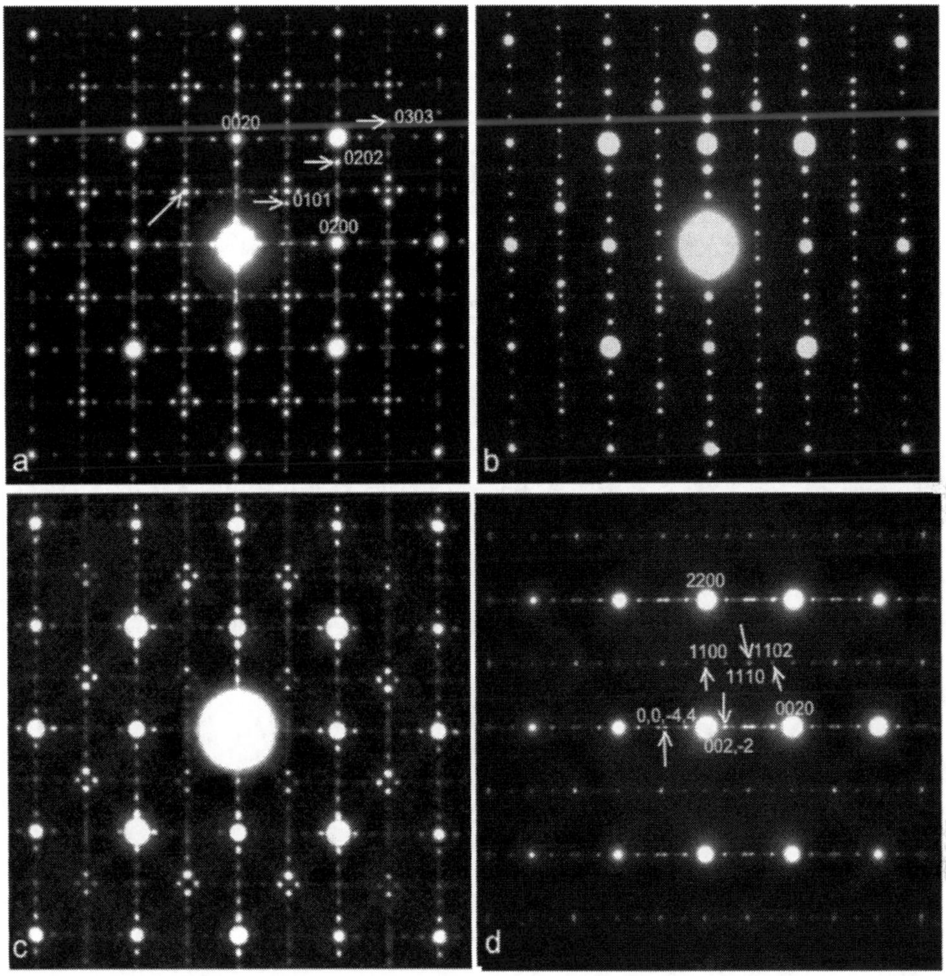

Fig.4. Shows typical $<001>_p$ type zone axis EDP's with γ values of (a) 0.386(1), (b) 0.375(1) and (c) 0.417(1) found in x = 0.333, 0.375 and 0.25, *M* = Ca samples. (d) shows a typical $<\text{-}110>_p$ zone axis EDP of an x = 0.2 sample with γ = 0.417(1). Indexation in all cases is with respect to the (3+1)-d reciprocal lattice basis vector set $M^* = \{{}^1/_2\,\mathbf{a}_p^*, {}^1/_2\,\mathbf{b}_p^*, {}^1/_2\,\mathbf{c}_p^*, \mathbf{q} = \gamma\mathbf{c}_p^*\}$.

In addition to the strong Bragg reflections of the underlying perovskite-related parent structure, labelled **G** hereafter, there are a host of additional (in general, incommensurate) satellite reflections characterized by the composition-dependent, primitive primary modulation wave-vector $\mathbf{q}_{prim} = {}^1/_2\mathbf{b}_p^* + \gamma\mathbf{c}_p^*$ (see *e.g.* the reflections labelled 0101, 0202 *etc.* in Fig.4a). These satellite reflections arise from coupled O/V (V an oxygen vacancy) and Bi/Ca ordering and associated structural relaxation [1,5]. In addition to these $\mathbf{G} \pm m\mathbf{q}_{prim}$ (m an integer) satellite reflections, $\mathbf{G} \pm {}^1/_2 <110>_p^*$ (arrowed but not labelled in Fig.4a) as well as $\mathbf{G} \pm {}^1/_2\,[111]_p^*$ (see *e.g.* the reflection labelled 1110 in Fig.4d) Bragg reflections are also, in general, present. The same is also true for the *M* = Sr solid solution. Indexation in Fig.4 is thus with respect to the (3+1)-d reciprocal lattice basis set $M^* = \{{}^1/_2\,\mathbf{a}_p^*, {}^1/_2\,\mathbf{b}_p^*, \mathbf{c}_p^*, \mathbf{q} = \gamma\mathbf{c}_p^*\}$. The presence of the $\mathbf{G} \pm {}^1/_2\,[111]_p^*$ reflections is presumably either a weak structural response to the G-type anti-ferromagnetic ordering of the Fe moments associated with the wave-vector ${}^1/_2\,[111]_p^*$ (see *e.g.* [6]) or are associated with polyhedral rotations. The

G ± $^1/_2$ <110>$_p$* reflections are presumably associated with polyhedral rotations and associated structural relaxation. Weak incommensurate satellite reflections of even order are observed around the **G** ± $^1/_2$ <110>$_p$* reflections (see *e.g.* the reflection labelled 1102 in Fig.4d) but no satellite reflections at all are observed around the **G** ± $^1/_2$ [111]$_p$* reflections (see also Fig.4d).

The absence of the **G** ± $^1/_2$ <110>$_p$* Bragg reflections in the twinned <100>$_p$ type zone axis EDP shown in Fig.4c suggests that these reflections are of **G** ± $^1/_2$ [110]$_p$* type and only exist in the Zero Order Laue Zone at the [001]$_p$ zone axis orientation. This is also consistent with the apparently single domain [001] zone axis FFT of an x = 0.667, *M* = Sr sample reported in Fig.3b of [5]. The systematic extinction conditions characteristic of both solid solution phases are then F(*HKLM*) = 0 unless *H*+*K*+*M* is even and F(*H*0*LM*) = 0 unless *H*, *M* both even implying the superspace group symmetry *Cmmm*(01γ)0s0 (a non-standard setting of *Cmmm*(10γ)s00, see [7]) or one of the non-centrosymmetric sub-groups thereof. Note that ferroelectric ordering would require one of the latter sub-group symmetries and rule out the former. The *Pmmm*(0$^1/_2$γ) superspace symmetry previously suggested for the *M* = Sr solid solution [1] excluded the presence of the weak **G** ± $^1/_2$ [110]$_p$* and **G** ± $^1/_2$ [111]$_p$* reflections. When these reflections are included the superspace group symmetry *Cmmm*(01γ)0s0 as described above is obtained *i.e.* the *M* = Ca and Sr solid solutions are isomorphous. Finally, note that the anti-ferromagnetic ordering associated with the modulation wave-vector $^1/_2$ [111]$_p$* (see *e.g.* [6]) is not described by the above superspace group symmetries.

5. Properties measurements

The measured room temperature magnetizations of *M* = Ca, x = 0.20, 0.36 and 0.50, samples (in Bohr magnetons/formula unit) as a function of applied field are shown in Fig.5.

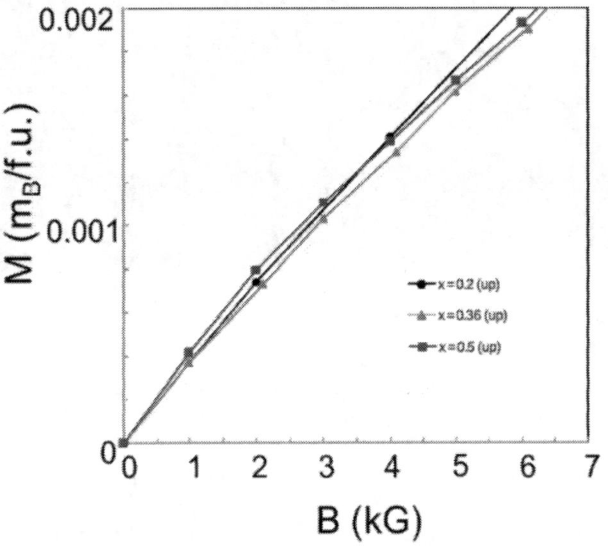

Fig.5. The measured room temperature magnetization of *M* = Ca, x = 0.20, 0.36 and 0.50, samples (in Bohr magnetons/formula unit) as a function of applied field.

The consistent measured magnetization for these samples up to 7 kG indicates that there is no change in iron valence state throughout the solid solution range, as also supported by Mössbauer measurements. Ferroelectric hysteresis loops of the *M* = Ca samples could not be obtained at room temperature and zero magnetic field as a result of the intrinsic semiconductivity of the materials. This does not, however, necessarily preclude the existence of co-ordinated ferroic type ion displacements. Kundys *et al* [7], for example, have reported magnetic field induced ferroelectric behavior for the *M* =

Sr phase. Likewise, bond valence sum calculations on the structural model proposed below suggest a need for off-centre Bi ion displacements in the central $BiFeO_3$-like part of the structure.

6. A suggested structural model

A previously suggested structural model for the solid solution phase [1] involving octahedral as well as square pyramidal Fe co-ordination was found not to fit to the observed HRTEM images of the *M* = Ca samples or to the three distinct Fe co-ordination environments required by Mossbauer results [4]. An alternative disordered variant of the two symmetry-equivalent model structures shown in Fig.6 below, however, does fit quite well with the observed HRTEM images as well as being consistent with the overall stoichiometry and the Mossbauer results [4].

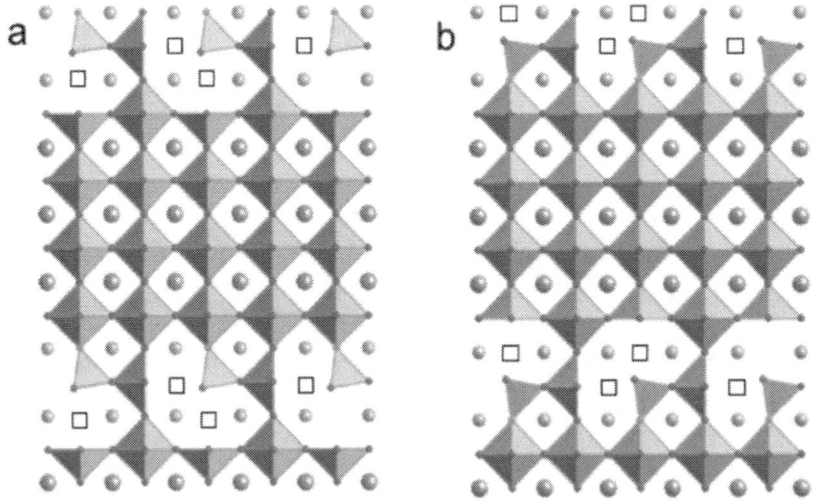

Fig.6. Shows [100] projections of various models for $\gamma = 2/5$, corresponding to a 5 x superlattice along c_p (vertical) and to HRTEM images such as that shown in Fig.3. The oxygen vacancies are represented by the open squares, the Bi ions by the large balls, the Ca ions by the medium sized balls and the oxygen ions by the smallest sized balls, green for half occupied and red for fully occupied.

References

[1] RL Withers, L Bourgeois, K Balamurugan, N Harish Kumar, PN Santosh and PM Woodward, *J. Solid State Chem.* **182** (2009) 2176-2184.

[2] VA Khomchenko, DA Kiselev, JM Vieira, L Jian, AL Kholkin, AM Lopes, YG Pogorelov, JP Araujo, M Maglione, *Journal of Applied Physics* **10** (2008) 024105-6.

[3] J Li, Y Duan, H De and D Song, *J. Alloys and Cmpds*. **315** (2001) 259-264.

[4] J Schiemer, RL Withers, L Norén, Y Liu, L Bourgeois and G Stewart, *Chem. Materials* **21** (2009) 4223-4232.

[5] C Lepoittevin, S Malo, N Barrier, N Nguyen, G van Tendeloo and M Hervieu, *J. Solid State Chem.* **181** (2008) 2601-2609.

[6] W Chen, AJ Williams, L Ortega-San-Martin, M Li, DC Sinclair, W Zhou and JP Attfield, *Chem. Materials* **21** (2009) 2085-2093.

[7] T Janssen, A Janner, A Looijenga-Vos and PM de Wolff. Incommensurate and commensurate modulated structures. In: Wilson AJC, ed. International Tables for Crystallography, Vol C. Dordrecht: Kluwer Academic Publishers (1995) 797-835.

[8] B Kundys, A Maignan, C Martin, N Nguyen and C Simon, *Appl. Phys. Letters* **92** (2008) 112905-3.

Diffraction effects due to modulation of orthorhombic crystals by aperiodically arranged twin boundaries

A Ustinov[1,2] , L Olikhovska[2], N Glavatska[2] and I Glavatskyy[2]
[1]E.O.Paton Electric Welding Institute of NASU, Kyiv, 03680 Ukraine
[2]G.V.Kurdyumov Institute for Metal Physics of NASU, Kyiv, Ukraine

E-mail: ustinov3@yandex.ru

Abstract. A number of the orthorhombic crystals formed due to diffusionless transformations, of which Ni_2MnGa crystal is an example, show diffraction patterns suggesting a modulation along the <110> directions. Nanotwined substructures with aperiodically arranged twin boundaries (TBs) on the (110) planes are found in such crystals. In the work, peculiarities of the diffraction intensity distributions for orthorhombic crystals with the aperiodic TBs arrangement are analyzed. It is shown that the intensity distributions for the twin-modulated structures formed in given orthorhombic crystals differ from those resulted from a wave-modulation of the crystals.

1. Introduction

The crystal structure of Ni-Mn-Ca alloys has been the subject of numerous studies with special attention on the structure of the so-called modulated martensites. For martensitic structures, the neutron, X-ray and electron diffraction measurements have revealed the presence of extra-reflections that suggested ordering in the stacking sequence of the atomic planes along <110> directions.

Nature of such periodic displacement is still a question to be clarified. Two models are today discussed on equal terms to interpret the layered structures in Ni-Mn-Ga system: as a result of a periodic displacement of the atomic planes [1-4] or a periodic stacking of these planes [3] similar to what was defined for Ni-Al alloys [5,6].

Recent HRTEM studies have shown that nanotwinned substructures are characteristic of Ni-Mn-Ga modulated martensites [7]; the TBs on the (110) plane are not arranged periodically throughout an extended area so that the number of atomic layers in the twin-related lamella only predominantly equals 4 and 2; 5 and 2; 5 and 5; 7 and 5-layers for martensites are characterized by modulation periods of $6\mathbf{d}_{110}$-, $7\mathbf{d}_{110}$ -, $10\mathbf{d}_{110}$ -, and $12\mathbf{d}_{110}$, respectively. Comparison between the SAED patterns and the HRTEM images gave evidence that the number of extra-spots correlates to the sum of the most probable thicknesses of twin-related lamella.

The number of layers in these lamellas (distances between nearest (TBs)) is not strictly fixed values of l_1 and l_2; these values are just the predominant thicknesses and there are domains of other thicknesses. Deviation from these values allows an assumption to be made that arrangement of the TBs in the crystals is rather of a *short-range order* than of a long-range order.

Effect of the wave-modulations on the diffraction patterns is well studied [8]. The manner in which aperiodic arrangement of the TBs affects the diffraction pattern of an orthorhombic crystal should be found to provide a means of distinguishing between the wave-modulated and twin-modulated structures. So, the diffraction intensity distributions from orthorhombic crystals containing aperiodicaly arranged TBs on (110) plane have been calculated and a comparison has been made between the diffraction features caused by these two types of modulations.

2. XRD pattern simulation

Assuming that twinning only changes the stacking sequence of the atomic layers without changing the composition and structure of these layers, the crystal structure as a whole can be given as a stacking of the layers parallel to plane (110).

It is convenient to choose a new coordinate system ($\mathbf{A}_1,\mathbf{A}_2,\mathbf{A}_3$) instead the coordinate system ($\mathbf{a},\mathbf{b},\mathbf{c}$) derived from the axes of orthorhombic phase (Figure 1). The relation between $\mathbf{A}_1,\mathbf{A}_2,\mathbf{A}_3$ and $\mathbf{a},\mathbf{b},\mathbf{c}$ is seen to be the following:

$$\mathbf{A}_1=\mathbf{c};\mathbf{A}_2=\frac{1}{2}(\mathbf{b}-\mathbf{a});\mathbf{A}_3=\frac{\eta^2\cdot\mathbf{a}}{(\eta^2+1)}+\frac{\mathbf{b}}{(\eta^2+1)} \quad (1)$$

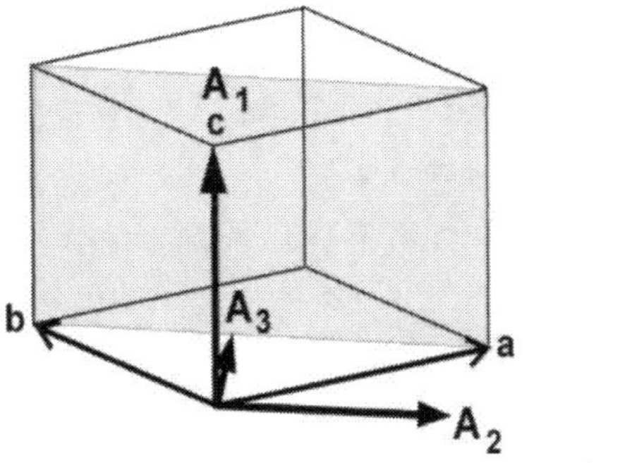

Figure 1. A cell of orthorhombic structure. The shaded plane is a crystallographic plane (110).

where $\eta=b/a$. So in the new coordinate system, the close-packed planes are planes (001), and the plane shear direction is [010]. Further on the indexation of the crystallographic planes and the associated planes diffraction peaks will correspond to the new coordinate system.

The position of the nth layer (001) relatively origin in a twinned crystal can be given as

$$r_n=\frac{1}{2}\mathbf{A}_3 n+\frac{1}{2}\mathbf{A}_2 n+\frac{1}{2}\mathbf{A}_1 n+\boldsymbol{\delta}_0 n-2k_n\boldsymbol{\delta}_0 \quad (2)$$

where $\boldsymbol{\delta}_0=\frac{\eta^2-1}{2(\eta^2+1)}\mathbf{A}_2$; k_n is the number of the layers being in the twinned orientations, as compared with the origin layer, between the origin and the nth layers. According to the kinematic diffraction theory [9], distribution of the diffraction intensity from a twined crystal composed of N identical layers along [001] can be written as in the following form

$$I(HK\zeta)=C^2\sum_{m=-N}^{m=N}(N-|m|)exp[\pi i(l+\zeta)m]exp(\pi iK\Delta_m), \quad (3)$$

where m is a distance between two atomic layers; $\Delta_m=\left[\frac{\eta^2 m}{\eta^2+1}-\frac{(\eta^2-1)k_m}{\eta^2+1}\right]$; k_m is the number of the layers being in the twin orientation between two atomic layers are spaced at interval m atomic layers; ζ is the variable along [001]. It is seen from Eq. (3) that the term $\langle exp(\pi iK\Delta_m)\rangle$ is a constant if K = 0. So in this case the stacking of the layers does not affect the intensity distribution.

It follows from (3) that for the intensity to be calculated value of the factor $\langle exp(\pi iK\Delta_m)\rangle$ must be determined; the value is dependent on the TBs arrangement in the crystal. In the framework of the models, crystals containing the TBs of which arrangement has been given by different statistical lows were generated using the Monte Carlo procedure. The «grown» in such a way crystals were comprised of 10000 layers, and the $\langle exp(\pi iK\Delta_m)\rangle$ term was calculated over 300 layers. For this crystal intensity $I(HK\xi)$ was calculated by Eq. (3) along the reciprocal lattice rods with integer indices H and K ($K\neq 0$). The intensity distributions were calculated for a constant ratio between the lattice constants a and b, $b/a = 0.94$.

3. Results of XRD pattern simulation

Let us describe TB arrangement the crystal using a statistical approach based on assigning *a priori* probability *P(l)* of a given TB location at l interplanar distance from the TB nearest to it. Such a model of TB arrangement in the crystal follows from the assumption of establishment of an interaction between the nearest TB which depends on the distance between them only and does not depend on the arrangement of other TBs (*short-range order in TBs arrangement).*

In the simplest case if a "repulsion" between nearest TBs takes place at the distances smaller than l_i layers, each next TB must not be formed closer to the previous one than at a given minimal distance, l_i ; it occurs at other distances, l, with a certain constant probability α:

$$P(l)=\begin{cases}0, & if\ l<l_i\\ \alpha\ , & if\ l\geq l_i.\end{cases} \tag{4}$$

With such a arrangement of the nearest TB, the probability of detection of two TBs at distance of L – layers from each other can be expressed as $w(L)=\alpha(1-\alpha)^{L-1}$. Allowing for (4) at $L<l_i$ $w(L)=0$, and at $L\geq l_i$ - $w(L)=\alpha(1-\alpha)^{L-l_i}$. Therefore, the most probable distance, at which two nearest TB can occur, will be at $L=l_i$. It is seen that if parameter $\alpha\to 1$, then $w(l_i)\to 1$, i.e. a long-range order in TB arrangement is established.

Considering that the nearest TBs, essentially, are boundaries of a twin domain, at crossing which the crystal orientation changes to a twin orientation, it may be assumed that the TB interaction will depend on orientation of the domain separating these TBs. Two models of the TBs distribution in the crystal, namely a single-mode and a bimodal, are possible in a general case. The single-mode distribution results from independence of the probability *P(l)* on the twin lamellae orientation. If such dependence takes place, the parameters characterizing the TBs arrangement at various points of the crystal should be different so that a bimodal TBs distribution will be observed. In general case, the domain orientation could affect not only the most probable distance between two neighboring TBs, but also the probability of the TB occurrence. In the case that the probability distributions (4) are taken for both modes, the difference between the modes is given by different values $l_1\neq l_2;\alpha_1\neq\alpha_2$.

3.1. Single-mode arrangement TBs.

Let us the probability of the occurrence for each next TB is independent on the orientation of the twin-lamellae located just ahead the TB. This means that $l_1=l_2=l;\alpha_1=\alpha_2=\alpha$. A change of the diffraction pattern depending on the probability α is illustrated below by the example of a structure characterized by value l_i = 5. Figure 2 gives the intensity distributions calculated for the structures formed at various α.

As at small values α (for instance at α = 0.1), the crystal consists of alternation of relatively thick twin domains (average lamel thickness in this case will be equal to $\bar{L}=l_i/\alpha$ layers) the reciprocal lattice identity period will show two diffraction peaks near the positions related to b/a = 0.94. With increasing α to 0.3 the main diffraction peaks become progressively broader, and the intensity distribution as a whole is inhomogeneous. Notice that extra peaks appear at the positions related to a tetragonal structure ($a = b$). Intensities of these extra peaks increases as α increases and, for a range of the α values, intensity distribution along a reciprocal lattice rod appears as a set of triplets composed of central diffraction line and two strong lines. New extra peaks form as α increases still further (curves for α = 0.5-0.7 in Figure 2). It is evident from comparison of the curves that all the peaks change their positions as α changes. The peaks are shifted towards both smaller and higher values of coordinate ζ. In any case, as α increases every peak approaches the position of the closest peak inherent in the ordered structure

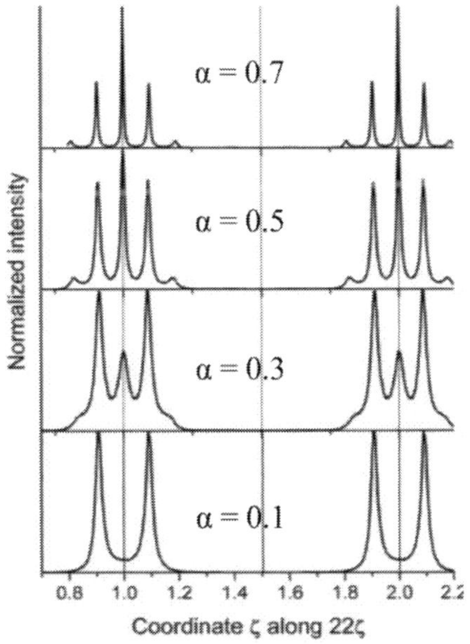

Figure 2. The intensity distribution simulated at various values of α for the crystal with single-mode arrangement of TB at *l* = *5*.

(5). Therefore, higher values α give the intensity distributions showing a series of almost regularly spaced peaks as it is shown in Figure 2 for α = 0.7. The positions of the strongest peaks are worthy of special attention in that the peaks are at the positions corresponded to a crystal with $a = b$. So, a single-mode distribution of the TBs in an *orthorhombic* crystal causes the diffraction pattern along <110> rods that resembles the diffraction pattern from a modulated crystal of a tetragonal symmetry with period of $\mathbf{10d_{110}}$. In the case that a tetragonal crystal is modulated by commensurate wave of the period $\mathbf{10d_{110}}$ the related diffraction pattern should show the same series of regularly spaced peaks [8]. So, for $\alpha > 0.7$ the diffraction intensity from the aperiodic twin-modulated structure is very similar to that from the wave-modulated structure of the same modulation period. Such a similarity leads to problems in an effort to find symmetry (orthorhombic or tetragonal) of the crystal and nature of the crystal modulation (aperiodic twin-modulated or wave-modulated. On the other hand, it is should noted that unlike the diffraction patterns of a crystal with wave modulation, the diffraction pattern of a crystal aperiodically modulated by twin boundaries, has its special features. This is manifested, primarily, in that the most intensive peaks in the identity period (base peaks), the position and shape of which are determined by the parameters and symmetry of the crystal lattice and degree of its perfection, do not change their characteristics at wave modulation of the crystal, irrespective of the type of its wave modulation [8, 10]. The given example shows that the shape of the most intensive peaks, which can be identified as the base ones, i.e. inherent to the crystal proper, depends on the parameters of TB aperiodic arrangement. Therefore, comparison of the profiles of base diffraction peaks, located on reciprocal lattice rods with K = 0 and K ≠ 0 allows establishing the type of crystal modulation without using HRTEM methods.

3.2. *Bimodal arrangement TBs.*

We will consider as an example the case when a bimodal arrangement of the TBs in an orthorhombic crystal takes place assuming that (4) holds for both modes with $l_1 = 2$ and $l_2 = 3$. We have restricted ourselves to the case when $\alpha_1 = \alpha_2 = \alpha$, although the difference between α_1 and α_2 should affect the diffraction pattern as well.

Figure 3 gives the intensity distributions calculated for structures at various α. At small values (α = 0.1) the diffraction pattern is similar to the one shown in Fig. 3 in the case of single-mode modulation of the crystal. However, as α increases to 0.3, the diffraction peaks tend to change both their profiles and positions in a different way. A redistribution of the intensity also occurs so that one of the peaks becomes much stronger than the other. As α increases to 0.3, a weak diffuse extra peak tends to form between the *a*- and *b*-diffraction components. The extra peak is only close to the position related a structure with $a = b$, as opposed to the case of $l_1 = l_2$ where such a peak is exactly in this position. New extra-peaks form as α increases still further (curves for α = 0.5-0.7 in Figure 3). It is evident from comparison of the curves that all the peaks change their positions as α changes. In any case, as α increases every peak approaches the position of the closest peak inherent

Journal of Physics: Conference Series **226** (2010) 012016 doi:10.1088/1742-6596/226/1/012016

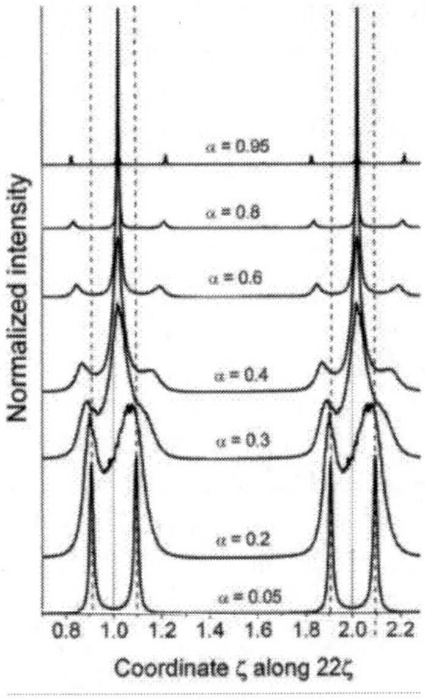

Figure 3. The intensity distribution simulated at various values of α for the crystal containing the twin-related domains of the most probable thickness 3 and 2 layers. The dashed lines show the positions related to the ratio *b*/*a*=1.1.

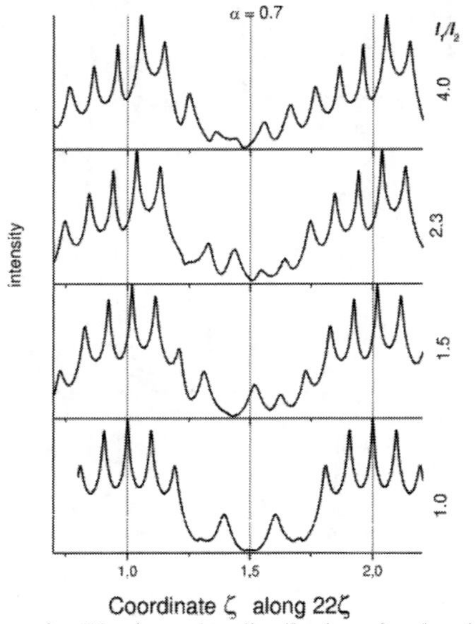

Figure 4. The intensity distribution simulated at various values of l_1/l_2 at conditions that $l_1 + l_2 = 10$ and $\alpha_1 = \alpha_2 = \alpha = 0{,}7$. The intensity in the right part of the figure is given in *logarithmic* scale to show all very weak extra-peaks.

in the ordered structure $\langle 2\bar{2}\rangle$. Therefore, higher values α give the intensity distributions showing a series of almost regularly spaced peaks as it is shown in Figure 3 for α = 0.95.

Assuming that the intensity distributions given in Figure 3 are experimental diffraction patterns, let us analyze these in an effort to find structural characteristics of the crystal. A conclusion of a modulated structure follows from the presence of extra peaks. Number of the inter-peak intervals between two strongest peaks gives a modulation period of $5\mathbf{d}_{110}$, and this agrees with the sum of the most probable distances between TBs, $l_1 = 3$, $l_2 = 2$.

As it is seen from Figure 4, the intensity distributions for structures of the same period ($M = l_1 + l_2$) but constructed using various values l_1 and l_2 show the same number of the peaks. Bimodal nature of the crystal twin modulation is manifested in the absence of symmetry in the distribution of intensity relative to the middle point in the reciprocal lattice identity period. The greater is the difference between l_1 and l_2, the more considerable asymmetry is observed. The asymmetry manifests itself as an inhomogeneous broadening and shift of the 'orthorhombic' components, if value α is relatively low, similar to what is shown in Figure 3. When α value is high enough for formation of extra peaks, the peak positions and intensities are asymmetrical with respect to the midpoint in between the strongest peaks.

Positions of strongest peaks with respect to integer-valued ζ give an evidence of an orthorhombic symmetry of the crystal. However, the strongest peaks are markedly shifted from the positions related to the given cell orthorhombicity *b*/*a* = 0.94. The shift is 0.035 (in the units of coordinate ζ), whereas the distance from the peak position to the nearest integer value of coordinate ζ is only 0.025. Thus, at bimodal ($l_1 \neq l_2$) modulation of orthorhombic crystal the position of base peaks is similar to the crystal with orthorhombic symmetry, however, their position in the identity period, as in the case of single-mode modulation, will not correspond to the true crystalline lattice parameters. In this case, these parameters should be included into the range of varied parameters and should be determined from the experimental data along side the parameters determining the statistics of TB arrangement in the crystal.

5. Conclusions

The results of the calculations demonstrate that aperiodic TBs on the {110} planes in orthorhombic crystals cause the presence of extra peaks in the diffraction patterns along <110> directions of the crystals. The peaks tend to be regularly spaced as the order in the TBs arrangement increases, and their take an δ-shape. The position of the most intensive (base) peaks in the identity period and their shape depend not only on the inherent values of lattice parameters, its symmetry and degree of perfection, but also on its type (single- or bimodal) and degree of TB ordering. Thus, the qualitative similarity of diffraction effects which are due to wave modulation, and aperiodic arrangement of TB in the crystal, makes analysis of experimental data more complicated, both in terms of determination of the symmetry and parameters of the lattice, and the nature of its modulation. In such cases it is necessary to perform quantitative analysis of satellite arrangement in the identity period and shape of base diffraction peaks. In the case of TB aperiodic arrangement, the arrangement of satellites in the identify period will not be equidistant, and the width of basic diffraction peaks located on rods with $K \neq 0$ will be greater that that for peaks with $K = 0$. It should be noted that these differences will be the greater the less ordered are the TB.

Despite the diffraction features resulted from aperiodic TBs arrangement are found to orthorhombic and tetragonal crystals, these would be used to analyze crystals where the formation of TBs occurs due to low twinning shear.

Acknowledgements

The authors are grateful to INTAS and NSAU for partial financial support of the project (Ref. No : 06-1000024-9396).

References

[1] Chernenko V A, Seguí C, Cesari E, Pons J. and Kokorin V V 1998 *Phys. Rev.* **B57** 2659
[2] MartynovV V and Kokorin V V 1992 J. *Phys. III* (France) **2** 739
[3] Pons J, Chernenko V A, Santamarta R and Cesari E 2000 *Acta Mater.* **48** 3027
[4] Ullakko K, Huang J K, Kantner C, Kokorin V V and O'Handley R C 1996 *Appl. Phys. Lett.* **69** 1996
[5] Martynov VY, Enami K, Khandros LG, Nenno S and Tkachenko A V 1983 *Phys. Met. Metallogr.* (USSR) **55** 136
[6] Noda Y, Shapiro S M, Shirane G, Yamada Y and Tanner L E 1990 *Phys. Rev.* **B42** 10397
[7] Pons J, Santamarta R, Chernenko V A and Cesari E 2006 *Mater. Sci. Eng.* A **438-440** 931
[8] S Van Smaalen 2007 *Incommensurate Crystallography* (Oxford: University Press)
[9] Wilson A J C 1942 *Proc. R. Soc. A* **180** 277
[10] Ustinov A I, Olikhovskaya L A and Shmyt'ko I M 2004 *Crystall. Reports* **45** 365 and 374

Study of cracks and phase transitions in cadmium iodide crystals using X-Ray diffraction

S K Chaudhary & H Kaur
Dept of Physics, M.D. University, Rohtak-124001 (India)

E-mail:sunilkc2001in@yahoo.com

Abstract. Polytypism has been observed in a large number of materials where the nearest. Neighbour relationship between identical two-dimensional layers of atoms can be satisfied in more than one way. The phenomenon has posed interesting problem for the Scientists, since the nature of force that causes ordering over the scale ranging from few angstrom to few thousands of angstrom units is not known. The Cadmium iodide has been purified and single crystal grown using a zone-refining system. The frequent appearance of cracks in the grown crystals when exposed to N_2 laser at room temperature has been explained on the structural considerations of CdI_2. The crystals of cadmium iodide have also been grown from solution in two batches using unpurified and purified material. A comparative study has been made on these batches for polytypism using x-ray diffraction technique. Formation of small period polytype 2H is governed by both temperature and impurities contained in the starting material. 4H is the most stable polytype. Higher occurrence of unidentified polytype in crystals of purified material has been attributed to free movement of edge dislocations during growth. The results have been examined, keeping in view the empirical considerations of earlier investigations.

1. Introduction

There are many physical factors associated with the growth of polytypes and a number of solutions have been put forward by the researchers to account for the phenomenon. These have been based on considerations like temperature; rate of crystallization and presence of impurities. Of these the presence of impurities is prominent one. Salje [1] in his large and wide studies has emphasized that the polytype growth is kinetic in nature and supports nucleation and growth mechanism of polytypes. While some specific studies on the effect of impurities on polytype formation have been carried out on SiC, ZnS, and PbI_2 crystals [2, 3, and 4]. Little work has been done on CdI_2 that are richly polytypic in nature. Till date, several hundreds of CdI_2 polytypes have been discovered and nearly 200 structures have been worked out. In a combined optical, SEM and STM study of cadmium iodide, precise information about the growth features on the crystal surface and mode of growth of polytype 4H by screw dislocation have been reported [5]. Further, Kumar has studied the electrical, optical and dielectric properties of pure and doped melt grown CdI_2 and PbI_2 crystals. In the doped crystals no new polytypes have been observed and were found to be harder than the undoped ones [6]. Earlier Jain and Trigunayat [7] have studied the effect of impurities on the nature of growth of polytypes in solution grown crystals of cadmium iodide. Since the zone-refining enables one to exercise control on the degree of purification, so it was hoped that this effect could be better studied in a semi-quantitative way by growing crystals of cadmium iodide under controlled conditions of temperature and humidity using purified material and to examine as how their polytypism differed w.r.t. the crystals of unpurified material grown under identical conditions.

2. Experimental details

The cadmium iodide has been purified by using zone refining technique and crystals have been grown by two methods.

2.1. From melt (using zone –refining system)

The details of purification of material and growth of crystals from melt have been described earlier [8]. The complete purification necessitates 20 zone passes (manufacturers of Analar grade cadmium iodide have mentioned the presence of alkalis in the raw material. Atomic absorption spectroscopy tests show only 1ppm of K and Na.). The grown ingot measured nearly 10 cms in length and 0.5 cm in thickness. Small pieces were cleaved from the ingot for the study of X-Ray diffraction and exposure to laser beam.

2.2. From solution

The crystals have been grown by simple method [9] of slow evaporation of aqueous solution of cadmium iodide. For controlling the conditions of growth, a humidity cabinet was employed in which humidity and temperature could be separately controlled. The door of the cabinet had a glass window through which the crystals could be observed during the growth.

10 grams of analar grade/purified cadmium iodide was dissolved in 20 cc of doubly distilled water and filtered through a fine quality filter paper. The filtered solution was transferred into a well-cleaned shallow flat-bottomed crystallizing dish that was kept in humidity cabinet. The values of humidity and temperature were maintained at 70% and 38°C respectively. The dish was kept inside the humidity cabinet until a number of well-developed platy crystals were formed. The crystals were hexagonal in shape and measured nearly 1-3 mm across and 50-500 micrometer thickness with shining (0001) basal faces.

The crystals were carefully transferred to a well-cleaned microscope slide for the optical examination. The perfect ones were chosen for further investigation. The well-developed morphology of the crystals helped in mounting them in a desired orientation on the X-ray camera. 15^0 oscillation photographs of the crystals were taken with crystal so positioned that a long succession of 10.l reflections, that are suited for the polytypic identification were recorded on zero layer line.

From the viewpoint of obtaining reliable statistical results it was proposed to carry out the X-ray diffraction study of nearly 90 crystals each of

(1) Crystals grown from the raw material.
(2) Crystals grown from the purified material.

Separate X-Ray diffraction photographs were obtained from the faces of each crystal.

3. Results and Discussion

3.1. Melt grown crystals

All the crystals grown by zone refining system have the structure 4H. In the melt growth, the temperature of growing crystal is just below the melting point of the substance and any metastable structure, if at all formed must transform into a structure which has minimum energy configuration at that temperature which in fact is 4H (Figure 1)

Figure 1. An a-axis 15°-oscillation photograph of a CdI_2 crystal, showing the reflections of most common polytype 4H. CuKα radiation; 3 cm camera

3.2. Cracking under laser and storage

During the optical gain studies of perfect cadmium iodide crystals with N_2 laser at room temperature, some cracks were observed in the crystals. [10]

Further it is to be reported that when the crystals of cadmium iodide (Figure 2) were stored in an isolated place for a long period (~ 5 years) cracks and many defects appeared on the crystals and transparency also decreased considerably as shown in Figure 3. Similar is the situation for another crystal (Figure 4, Figure 5)

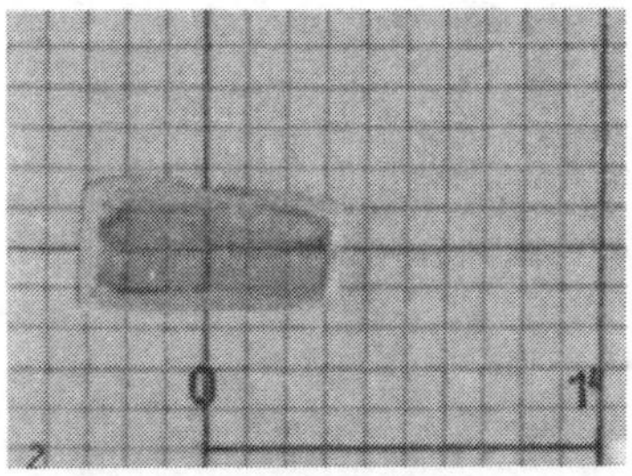

Figure 2. Photograph showing a CdI_2 Crystal cleaved from the ingot.

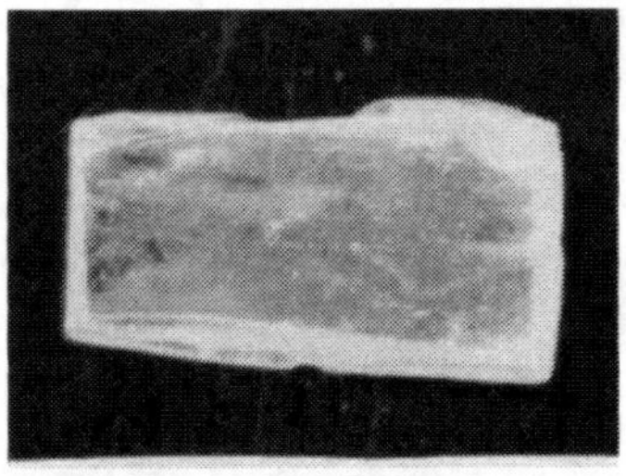

Figure 3. Same crystal (as in Fig.2) after Being stored for a period of ~ 5 years.

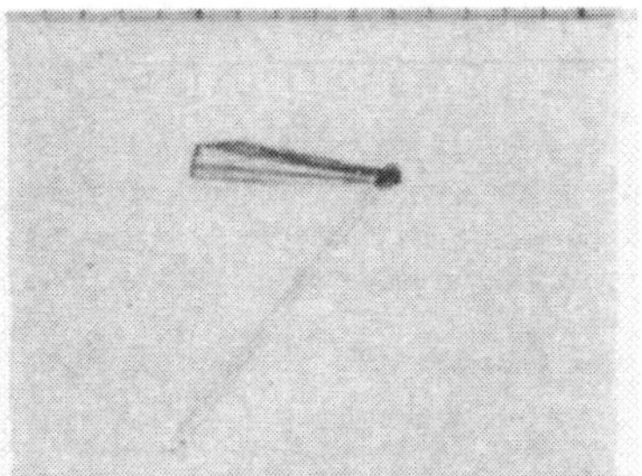

Figure 4. A CdI_2 crystal mounted on the glass fiber with one crack only.

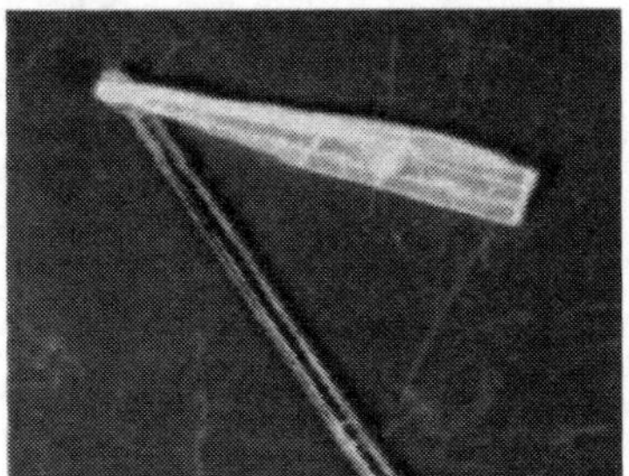

Figure 5. Same crystal (as in Figure 4) after being stored for a period of ~5 years. (Many a cracks are visible)

The CdI_2 structure consists of various stackings of CdI_2 sheets in each of which a layer of Cd ions is sandwiched between two close packed layers of iodine ions. The binding within a molecular sheet, believed to be largely ionic, is quite strong but two adjacent sheets are bonded together with weak van der walls forces. Wahab and Trigunayat [11] have shown that there is an increase in I-I separation of CdI_2 molecules with rise in temperature, thus implying the weakening of I-I bonds.

When exposed to N_2 laser, the necessary energy for the formation of cracks is provided by the laser pulse. The laser pulse has enough energy to weaken the bonding between various sandwiches, leading to appearance of cracks. The appearance of cracks has been explained in terms of movement of edge dislocations and their obstruction against some impurity particle. It may be added here that during purification by zone- refining system, the impurities with $K_0 = (C_s/C_l) < 1$ can be removed. The role of impurities with $K_0 > 1$ is passive as these impurities are to be accommodated by the host lattice.

The steady state dislocation velocity in a material is usually related to the net shear stress acting on the dislocation σ_d by a power law equation of the form $Vss \propto \sigma_d^m$ where m varies between 5-40 for metals but for ionic compounds like CdI_2 the dislocation velocity is approximately proportional to the stress i.e. m=1 [12]. From the above fact it is very clear that when exposed to the laser, there is a very strong stress due to thermal heating and very fast movement of dislocations that leads to formation of cracks instantaneously.. Whereas, in the case of CdI_2 stored for longer times the stresses available are due to impurities (of the order of few ppm) which are very small and development of cracks takes as long as many years. However a detailed microscopic work would be considered.

3.3. Solution grown crystals

Table 1 shows the frequency of occurrence of various polytypes in solution grown crystals from raw and purified material. It is clear that the most common polytype 4H has a dominating occurrence viz. nearly 50% of the total in the case of raw material and 32%in the case of purified material. The percentage of disordered or unidentified structures is 32% of the total in the raw material and nearly 51% in the case of purified material.

Table 1. Frequencies of occurrence of various cadmium iodide polytypes in solution grown crystals from raw and purified material.

TOTAL NUMBER OF CRYSTALS	POLYTYPES			
	2H	4H	Higher	Unidentified
(Raw material) 90	1	90	46	58
(Purified material) 90	7	57	43	93

The higher occurrence of 4H in raw material may be attributed to the fact that the presence of impurities play a dual role in the case of polytypic crystals: (!) producing local stresses in the crystal structure leading to generation of dislocations and stacking faults (!!) providing hindrance to the movement of dislocations and consequential reduction /elimination of stacking faults[13].

In the case of 4H, nuclei preferably are formed around some impurity particle for subsequent growth of 4H. Since the crystals of above category have the maximum impurities, they have the highest percentage of 4H in them. The dislocations originating during the crystal growth are hindered due to the presence of impurities thus eliminating the stacking faults, pinning down transformation front for other structures and leading to the formation of 4H polytype. By the same token, the lesser percentage of 4H in the pure material can be attributed to the fact that in this case the dislocations can move relatively more freely and there are greater chances of creation of random stacking faults leading to lesser incidence of 4H type. The above discussion also explains the lesser frequency of occurrence of the unidentified in the raw material and higher occurrence in the purified material.

In the present work the number of higher polytypes obtained in the crystals grown from both pure and raw material are nearly equal.

The polytype 2H has been observed only once in the crystals grown from raw material whereas in the case of crystals grown from pure material it has been observed 7 times. Earlier Jain and Trigunayat [7] have reported that temperature of growth has marked effect on the formation of 2H. Higher the temperature, lower is the frequency of 2H and it comes down to naught at 40°C. In our case all the crystals have been grown at 38°C indicating that it is not only the temperature that affects the formation of 2H but also the impurities contained in the starting material.

The crystals in general largely owe the regularity of their internal structure to the existence of atomic forces of attraction and repulsion, which exercise their influence over a short range ~20 Å. The observed higher polytypes with cell heights upto thousands of angstrom units implies certain external agencies like (dust particles, lattice vibrations, vibrational entropy, electron density) are also operational in their creation.

The frequency of occurrence of 2H only once in the crystals grown from raw material and 7 times in the crystals grown from pure material indicates that action of atomic forces dominates in the case of formation of 2H in the purified material.

From the above discussion it becomes an open question and interesting problem that if 2H polytype will have the highest probability of occurrence when grown from the pure material under dust free atmosphere?

In the above studies only some specific conclusions could be made for only small period polytypes. However to study the effect of purification/impurities on the distribution of higher polytypes we need to study a very large number of crystals.

References

[1] Salje E, Palosz B and Wruck B 1987 *J .Phys. C: Solid State Physics* **20** 4077
[2] Lundqvist D 1948 *Acta. Chem. Scand.* **2** 177
[3] Kozielski M J 1976 *Bulletin de l Academie Polonaise des Sciences* **24** 367
[4] Chand M and Trigunayat G C 1977 *J. Crystal Growth* **39** 299
[5] Singh R, Samanta S B, Narlikar A V and Trigunayat G C 2000 *Bull. Mater. Sci.* **23** No. 2, 131
[6] Kumar S 2004 *X-Ray studies, Electrical, Optical and Dielectric studies of* PbI_2xA_1*-x*CdI_2xA_1*-x,* PhD Theses, Jamia Milia Islamia New Delhi
[7] Jain P C and Trigunayat G C 1980 *J. Crystal Growth* **48** 107
[8] Chaudhary S K and Trigunayat G C 1982 *J. Crystal Growth* **57** 558
[9] Mitchell R S 1956 *Z. Krist.* **108** 296
[10] Chaudhary S K, Arun Gaur and Trigunayat G C 1990 *Cryst. Res. Technol.* **25** K87
[11] Wahab M A and Trigunayat G C 1981 *Solid state communications* **36** 885
[12] Kanninen M F and Rosenfield A R 1969 *Phil. Mag.* **20** 569
[13] Singh K and Trigunayat G C 1989 *Phase Transitions* **16/17** 529

Modulated structure of β-brass CuZn compressed to 90 GPa

Olga Degtyareva[1,2] **and Valentina F Degtyareva**[2,3]
[1]School of Physics and Centre for Science at Extreme Conditions, University of Edinburgh, Mayfield Road, Edinburgh EH9 3JZ, UK
[2]Geophysical Laboratory, Carnegie Institution of Washington, 5251 Broad Branch Road North West, Washington D.C. 20015, USA
[3]Institute of Solid State Physics, Russian Academy of Sciences, Chernogolovka, Moscow district,142432 Russia

E-mail: degtyar@issp.ac.ru (Valentina F Degtyareva)

Abstract. Cu–Zn is a classic example of an alloy system displaying a sequence of phases along an alloy composition, called Hume-Rothery phases. The crystal structure of these phases is determined by valence electron concentration (that is the average number of valence electrons per atom), and the lowering of the electronic energy is considered the key factor for the structure stabilization. Using powder x-ray diffraction, we study the β-phase of an equiatomic CuZn alloy in its body-centered cubic (bcc) phase in the pressure range up to 90 GPa and find a transformation to a modulated trigonal structure at around 40 GPa. We analyze the structural distortion of bcc CuZn by looking at the configuration of the Brillouin zone of the bcc and the trigonal structures and their interaction with the Fermi surface. We demonstrate that the stabilization of the complex high-pressure structure can be explained with the Hume-Rothery effect.

1. Introduction

Several complex crystal structures have been reported recently in elemental metals under pressure (see reviews [1-3]), including incommensurately modulated and host-guest structures. The traditional explanation of the pressure-induced complexity of the alkali and alkali-earth metals suggests an s-d electron transfer (see review [4]). However, the observation of a large number of structural transformations and a variety of structural types in alkali metals under pressure prompted consideration of other factors that might be responsible for the formation of these phases [2]. Thus, for the complex cubic structure of Li with 16 atoms in the unit cell that appears at around 40 GPa, as well as the long period superstructures of Rb and Cs with 52 and 84 atoms in the unit cell, respectively, a Hume-Rothery mechanism has been suggested to be responsible for the stability of these structures [2]. Formation of a Brillouin-Jones zone with many planes in a close contact with the Fermi surface leads to an opening of an energy pseudo-gap and lowering of the electronic energy, which provides minimization of the lattice energy.

Hume-Rothery effect has been known to stabilize complex phases in some binary alloys (for example Cu–Zn) at ambient pressure. Cu–Zn is a classic example of an alloy system displaying a sequence of phases along the alloy composition, called Hume-Rothery phases. The crystal structure of these phases and their boundaries of stability along the alloy composition are determined by valence electron concentration (that is the average number of valence electrons per atom) [5]:

$$\alpha(\text{fcc}) \rightarrow \beta(\text{bcc}) \rightarrow \gamma\,(\text{complex cubic}) \rightarrow \varepsilon(\text{hcp})$$
$$1.35 \rightarrow 1.5 \rightarrow 1.62 \rightarrow 1.75\ (\text{valence electron/atom}).$$

Here, in the first row, the sequence of the Hume-Rothery phases is shown as observed along the alloy composition, where fcc stands for face-centered cubic, bcc – for body-centered cubic, and hcp – for hexagonal close-packed structures. In the second row, the upper boundaries of stability of these phases are shown using valence electron concentration that is directly related to the alloy composition, with the number of valence electrons equal to 1 for Cu and 2 for Zn. The stability of these phases is attributed to the Brilluoin zone - Fermi surface interaction, i.e. it is caused by the Hume-Rothery effect. This interaction between the Brillouin zone and Fermi surface increases with pressure. On compression, the electronic contribution to the energy dominates over the electrostatic contribution (the former goes as $V^{-2/3}$, the latter as $V^{-1/3}$, with V denoting the volume), and the electronic energy is lowered by the appearance of new Bragg planes near the Fermi surface as the system adopts structures of lower symmetry. Thus, this stabilization factor becomes more important on compression.

In this respect, it is interesting to study the stability of Hume-Rothery phases in Cu-Zn under pressure. One can expect a phase transition in the α- and β-phases with fcc and bcc structures, respectively, to more complex phases due to an enhancement of Hume-Rothery effect on compression. Indeed, our recent preliminary results on phase stability of Cu–Zn alloys to 50 GPa showed that α-phase started to transform to a more complex phase at 17 GPa, β-phase also started to transform at 37 GPa, while γ-phase with a complex cubic structure remained stable to the maximum pressure of that study [6].

In subsequent work [7], we studied pressure effects on the β-phase (bcc structure), collecting diffraction data with improved quality and also to much higher pressure of 90 GPa. We observe diffraction patterns corresponding to a structural distortion of the bcc phase above 40 GPa, that show splitting of the main bcc reflections and appearance of the additional weak reflections. Here we report a tentative structure solution of this high-pressure phase on the basis of a trigonal modulated structure and attribute its formation and stability to the enhanced Hume-Rothery effect.

2. Experimental details

The β-phase of a Cu–Zn alloy with the composition Cu – 50at%Zn (further denoted as CuZn) was prepared by melting the proper amounts of Cu and Zn of 4N purity in evacuated silica tubes. It has a body-centered cubic (bcc) structure. Whether or not this alloy is chemically ordered, cannot be established from x-ray diffraction, as Cu and Zn have nearly identical atomic scattering factors for an x-ray beam. The alloy was studied under pressure using angle-dispersive x-ray diffraction with synchrotron radiation (beam line 16-ID-B (HPCAT) at the APS in the Argonne National Laboratory. The sample was loaded in a symmetrical Mao-Bell cell in a sample chamber of the rhenium gasket together with nitrogen as a pressure-transmitting medium. To determine the pressure, we used *in situ* fluorescence measurements of ruby chips loaded in the sample chamber [8]. After closing the cell with the sample and nitrogen, the pressure in the cell was measured as 44.5 GPa. Diffraction patterns from CuZn were collected from 44.5 GPa to 90.0 GPa, the maximum pressure reached in this study. A focused monochromatic beam of wavelength $\lambda = 0.3678$ Å was used and the data were recorded on a MAR image plate. Diffraction data were integrated azimuthally using FIT2D [9], and structural information was extracted from a Lebail fit to the full profile using JANA2000 [10]. Cell parameters and the value of the modulation wavevector were obtained from the least-square fit of main and modulation reflections using the U-FIT [11].

3. Results and Discussion

A body-centered cubic (bcc) phase was observed in CuZn up to 37 GPa in our previous studies [6], before the splitting of the main reflections started to be evident. In the present experiment, the splitting of the (110), (211) and (220) bcc reflections is observed from 44.5 GPa, the lowest pressure of this experiment, while the (200) does not split during the compression (Fig. 1).

On pressure increase to 46 GPa, additional reflections with weak intensity start to appear. This is at slightly higher pressure than observed in our previous experiment [6], where the additional reflections appeared at 37 GPa. The difference probably arises from the fact that we used a different pressure transmitting medium (liquid nitrogen) in this experiment which is known to give a better quasi-hydrostatic pressure transmitting medium in comparison to the ethanol-methanol mixture used in previous experiment [6]. In the present experiment, one extra reflection appears above 46 GPa (Fig. 1a). Above 66 GPa, many more extra lines are visible, which could be interpreted as a transition to a more complex phase or that it is still the same phase and the extra reflections were too weak to be observed at lower pressure. The splitting of the main reflections and weak additional reflections are clearly visible on the 2D diffraction image of CuZn at 90 GPa (Fig. 1b). The lines from nitrogen, used as pressure transmitting medium, can be tracked using data from [12]. ε-N_2 transforms at 62 GPa to zeta-phase, where a strong line appears at around 1.88 Å, overlapping with the first group of strong reflections of CuZn at 62 GPa. It shifts to 1.83 Å at 90 GPa and can explain a weak reflection observed in our data at 1.83 Å. There are no diffraction peaks from rhenium gasket in any of the spectra.

Summarising our present results and combining them with those from [6], we conclude that the bcc CuZn phase transforms at 41(4) GPa to a more complex high-pressure phase, where the main reflections split and the extra weak reflections appear.

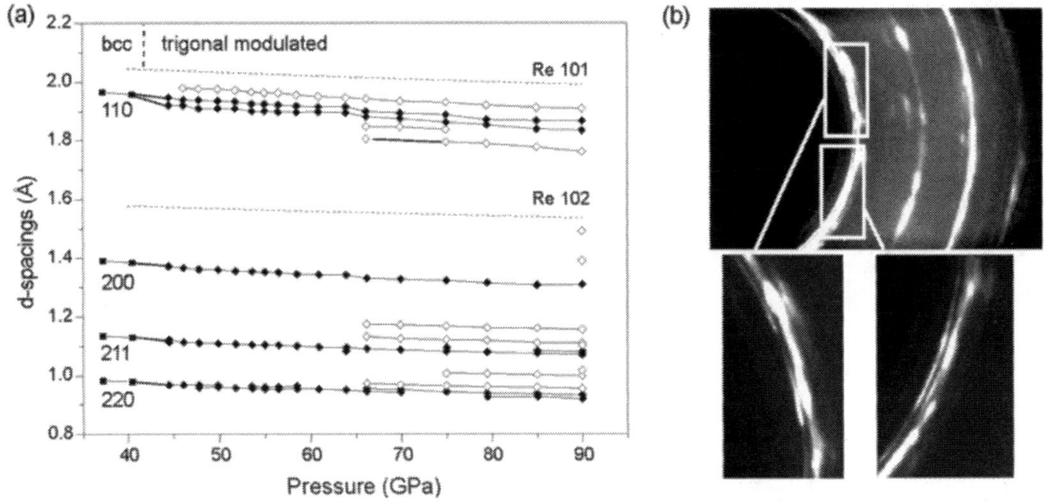

Figure 1. (a) Pressure dependence of the d-spacings observed for the CuZn alloy in its bcc phase and the high-pressure phase, which is shown to be trigonal modulated. The d-spacings of the bcc phase are shown with solid squares (data are taken from [6]) and their indexing is given; the main reflections of the trigonal modulated phase are shown with solid diamonds and the weak superstructure reflections – with open diamonds. Dotted lines show the calculated positions of Re, used as a gasket material, to demonstrate that these positions do not overlap with the sample reflections. (b) 2D diffraction image from CuZn alloy at pressure of 90 GPa, with two insets showing an enlarged view on the first strong group of reflections that consists of a splitted main peak (corresponding to the (110) peak of bcc) and superstructure reflections.

First, we describe our attempt to account for the splitting of the bcc reflections. As the positions and the intensities of the main peaks of the high-pressure phase are close to those of bcc phase, we have searched for a structure that would be a slight distortion of the bcc structure. A possible cell that describes the splitting is a rhombohedral distortion of the bcc with $a_r = 2.270$ Å and $\alpha = 109.28^{\circ}$ at 90 GPa (close to the ideal value for bcc $\alpha = 109.47^{\circ}$) with one atom in the unit cell in the position

1*a* (0 0 0) of the space group R-3m. In the hexagonal setting, the lattice parameters are a = 3.703(3) Å, c = 2.292(3) Å with three atoms in the unit cell in the position 3*a* (0 0 0), (1/3 2/3 2/3) and (2/3 1/3 1/3). A similar rhombohedral distortion has been proposed for the bcc structure in elemental vanadium under pressure [13], with no additional reflections observed. If we now search for a superstructure to describe the additional weak reflections of CuZn, however, no suitable solution can be found. We search for a solution creating a modulated superstructure using (3+1)-dimensional superspace groups [14]. The corresponding (3+1)D superspace group R-3m(00γ) does not give any suitable solutions for any value of the modulation vector γ. This is why we need to choose a different basic structure.

A trigonal structure with the same lattice parameters but with lower space group P-3m and atoms in the positions 1*a* (0 0 0) and 2*d* (1/3 2/3 1/3) and (2/3 1/3 2/3) (inset in Fig. 2) also describes the positions of the main reflections (see upper row of tick marks in Fig. 2). Similar distortion of the bcc structure is known for Ti and Zr at the β to ω phase transition (see [1] and references therein). This solution predicts, however, several reflections with zero intensity that are not observed in the experiment. In attempt to describe the positions of superstructure reflections we construct a modulated structure with the corresponding (3+1)D superspace group P-3m1(00γ). The value of $\gamma \approx 0.2$ predicts positions of the modulated reflections that describe the observed additional weak reflections in CuZn at 90 GPa (see lower row of tick marks in Fig. 2). The refined value of the modulation vector is γ = 0.202(2) at 90 GPa, which was obtained from a least-square fit of main and modulation reflections using the U-FIT program [11]. One unsplit main reflection ((200) in the bcc setting) was used for the refinement together with several non-overlapping modulation reflections. Analysis of the superspace group symmetry shows that the modulation wave has a non-zero component along the trigonal axis, and the atoms can be shifted only along the trigonal axis.

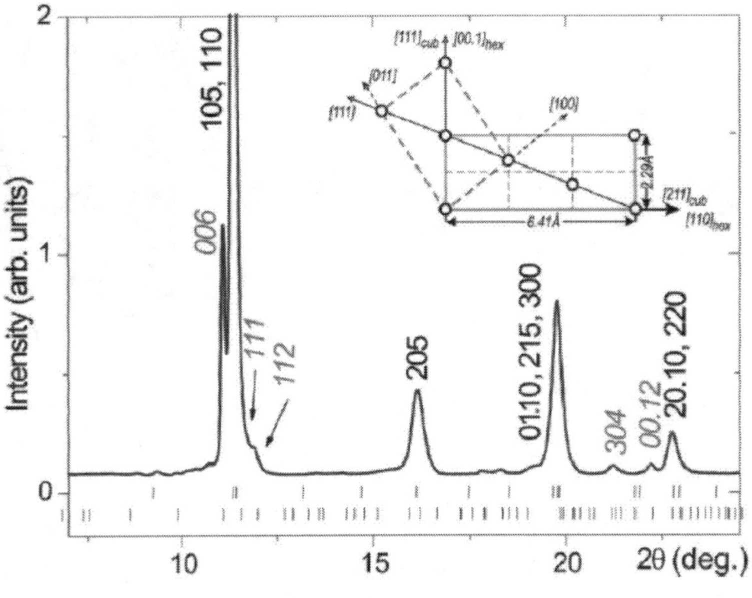

Figure 2. Integrated diffraction pattern of the CuZn alloy sample in the high-pressure phase at 90 GPa. Indexing on the basis of the long-period supercell with the five-fold c-axis is shown. The indices in black correspond to the main reflections, while the indices in italic grey correspond to the superstructure reflections. The upper row of reflections shows the calculated positions for the main reflections, and the lower row shows the superstructure (or modulated) reflections. (Inset) The basic trigonal structure of the high-pressure phase of CuZn with the cell outlined with solid lines, shown in relation to the bcc structure outlined with dashed lines.

At 90 GPa, the refinement of the structural parameters of CuZn for the trigonal solution in the P-3m1(00γ) superspace group are: a = 3.703(3) Å, c = 2.292(3) Å, γ = 0.202(2). The value of the modulation vector equal 1/5 within the error, and that means that this modulated structure can be described by a long-period cell with the 5-fold c-axis. The resulting lattice parameters are a = 3.703 Å and c = 11.460 Å. All observed reflections can be now re-indexed with three integer hkl's (space group P3), as shown in Fig. 2.

The reason for the structural distortion of the bcc phase and appearance and stability of the modulated trigonal structure can be understood within the Hume-Rothery mechanism. We analyze the structural distortion of the bcc phase CuZn by looking at the configuration of the Brillouin zone of the bcc and the trigonal structures and their interaction with the Fermi surface using the program BRIZ [15]. The number of planes in the Brillouin zone increases (Fig. 3); zone filling with electronic states increases from 75% (bcc) to 81% (trigonal structure). This brings more of the lower-energy electrons in states of higher density and the overall electronic energy is expected to be reduced. These factors indicate an enhancement of the Hume-Rothery mechanism under pressure.

We note, that a known example of a formation of a long-period superlattice in an alloy is the CuAu-II phase [16,17], and its stability has also been described using the Hume-Rothery effect. There are, however, differences to the present case of the high-pressure modulated structure of CuZn. CuAu has one valence electron per atom forming a tetragonal structure, and its superlattice is stabilized by adjusting an existing strong plane of the Brillouin zone to the Fermi surface to minimise the electronic energy. CuZn has on average valence electron concentration of 1.5 forming a bcc phase (Hume-Rothery phase). The displacive modulation found in the present work under pressure leads to a formation of new planes in the Brillouin zone (Fig. 3) that causes further reduction of the electronic energy.

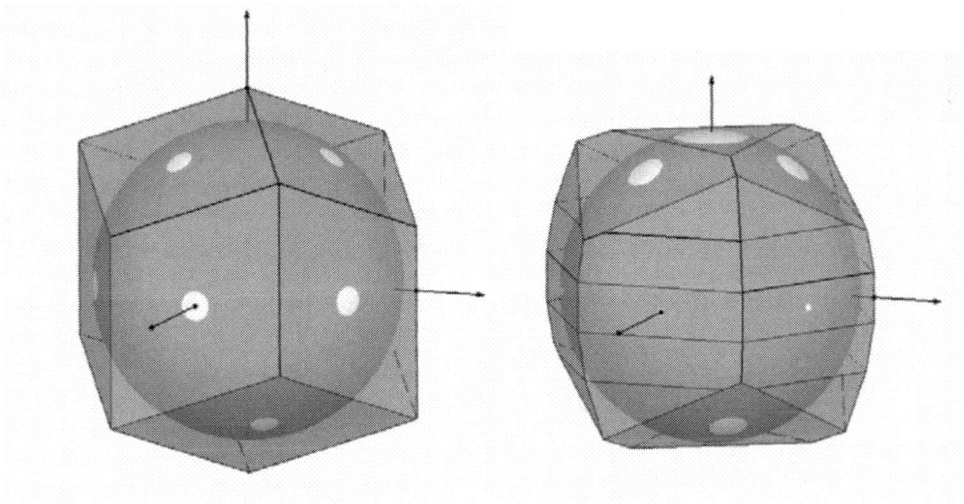

Figure 3. The configuration of the Brillouin zone and the Fermi sphere for the bcc structure (left) and the modulated trigonal structure (right).

4. Summary and conclusions.

We have reported a structural distortion of the bcc phase of CuZn at pressures above 40 GPa and proposed a tentative structure solution based on a modulated trigonal cell. The analysis of the interaction of the Brillouin-Jones zone and the Fermi sphere showed a pressure-induced enhancement of the Hume-Rothery effect as a mechanism for structure stabilization. These findings are relevant to the understanding of the formation and stability of the high-pressure phases of Li that have been suggested to stabilize due to the Hume-Rothery mechanism [2].

Acknowledgments. Authors thank Natalia I Novokhatskaya for alloy preparation and Maddury Somayazulu for help with experiments. V.F.D. is grateful to R.J. Hemley and H.-k. Mao for hosting her visit to the Geophysical Laboratory and for the support of this work. This work was performed in part at the Geophysical Laboratory, Carnegie Institution of Washington and at HPCAT (Sector 16), Advanced Photon Source (APS), Argonne National Laboratory. HPCAT is supported by DOE-BES, DOE-NNSA, NSF, and the W.M. Keck Foundation. APS is supported by DOE-BES, under Contract No. DE-AC02-06CH11357. Financial support by the Russian Foundation for Basic Research, grant No. 07-02-00901 is gratefully acknowledged. O.D. acknowledges support from the Royal Society.

References

[1] McMahon M I and Nelmes R J 2006 *Chem. Soc. Rev.* **35** 943
[2] Degtyareva V F 2006 *Phys.-Usp.* **49** 369
[3] Degtyareva O arXiv:0909.5356v1 [cond-mat.mtrl-sci]
[4] Maksimov E G, Magnitskaya M V, and Fortov V E 2005 *Phys.- Usp.* **48** 761
[5] Hume-Rothery W 1926 *J. Inst. Met.* **35** 319
[6] Degtyareva V F, Degtyareva O, Sakharov M K, Novokhatskaya N I, Dera P, Mao H-k and Hemley R J 2005 *J. Phys.: Condens. Matter* **17** 7955
[7] Degtyareva V F, Degtyareva O, Mao H-k and Hemley R J (to be published)
[8] Mao H-k, Xu J and Bell P M 1986 *J. Geophys. Res.* **91** 4673
[9] Hammersley A P, Svensson S O, Hanfland M, Fitch A N, and Häusermann D 1996 *High Pressure Res.* **14** 235
[10] Petricek V, Dusek M, and Palatinus L 2000 *JANA2000 - The Crystallographic Computing System*. Institute of Physics, Praha, Czech Republic.
[11] Evain M 1992 *U-FIT - A cell parameter refinement program*, Institut des Matriaux Jean Rouxel, Nantes.
[12] Gregoryanz E, Goncharov A F, Sanloup C, Somayazulu M, Mao H-k, and Hemley R J 2007 *J. Chem. Phys.* **126** 184505
[13] Ding Y, Ahuja R, Shu J, Chow P, Luo W, and H-k Mao 2007 *Phys. Rev. Lett.* **98**, 085502
[14] *Mathematical, Physical and Chemical Tables* 1999 2nd ed. edited by Prince E, Vol. C of *International Tables for Crystallography* (Kluwer Academic, Dordrecht), Sec. 9.8.
[15] Degtyareva V F and Smirnova I S 2007 *Z. Kristallogr.* **222** 718
[16] Sato H and Toth R S 1962 *Phys. Rev. Lett.* **8** 239
[17] Yamamoto A 1982 *Acta Crystallogr.* B**38** 1446

Journal of Physics: Conference Series **226** (2010) 012019 doi:10.1088/1742-6596/226/1/012019

Electronic origin of the incommensurate modulation in the structure of phosphorus IV

V F Degtyareva

Institute of Solid State Physics Russian Academy of Sciences, Chernogolovka, Russia

degtyar@issp.ac.ru

Abstract. An incommensurate modulated structure was found recently in a light group V element phosphorous in the phase P-IV stable in the pressure range 107 – 137 GPa. We consider configurations of the Brillouin zone and Fermi sphere within a nearly-free-electron model in order to analyze the importance of these configurations for the crystal structure energy. For the phase P-IV with the base-centered orthorhombic structure, *oC*2, we consider a commensurate approximant with an 11- fold supercell along the c-axis and a modulation wave vector equal 3/11 which is close to the experimentally observed value of 0.267. Atomic shifts due to the modulation result in appearance of satellite reflections and hence in a formation of additional Brillouin zone planes. The stability of this structure is attributed to the lowering of the electronic band structure energy due to Brillouin zone – Fermi surface interactions.

1. Introduction

Recent high-pressure x-ray diffraction studies revealed unusual complex and low symmetry structures in some simple elements [1], including incommensurate modulated (IM) structures. IM structures were found in the elements of group VII (I and Br) and group VI (S, Se and Te). Very recently, an IM structure was found in a light group V element phosphorous in the phase P-IV stable in the pressure range 107 – 137 GPa [2]. This phase is intermediate between simple cubic and simple hexagonal structures that have atomic coordination equal 6 and 8, respectively. All IM structures were observed when elements become metallic under pressure. This implies the importance of the two main contributions to the lattice energy: electrostatic (Ewald) and electronic (band structure) energies. The latter can be lowered due to a formation of a Brillouin zone plane and an opening of an energy gap at this plane. This effect is important for stability of so-called Hume-Rothery phases in Cu – Zn and related alloys [3]. Under pressure, the band structure energy part becomes more important leading to a formation of complex low-symmetry structures [4].

For a classical Hume-Rothery phase Cu_5Zn_8, the Brillouin zone filling by electron states is equal to 93%, and is around this number for many other phases stabilized by the Hume-Rothery mechanism [5]. Diffraction patterns of these phases have a group of strong reflections with their scattering vectors q_{hkl} lying near $2k_F$ and the Brillouin zone planes corresponding to these q_{hkl} form a polyhedron that is very close to the Fermi sphere. We consider configurations of the Brillouin zone and Fermi sphere within a nearly-free-electron model in order to analyze the importance of these configurations for the lattice energy. A computer program, BRIZ, has been developed for the visualization of Brillouin zone – Fermi sphere configurations to estimate Brillouin zone filling by electron states [5]. Thus, with the BRIZ program one can obtain a qualitative picture and some quantitative characteristics on how a structure matches the criteria of the Hume-Rothery mechanism.

2. Results and Discussion

The phase P-IV has an incommensurately modulated crystal structure with superspace group *Cmmm* (00γ) *s*00 [2,6], with the modulation wave vector $\gamma = 0.2673$. The basic lattice is base-centered orthorhombic with lattice parameters $a = 2.772$ Å, $b = 3.215$ Å, $c = 2.063$ Å at 125 GPa [2]. Diffraction pattern and suggested structural model are shown in Figure 1. We consider a commensurate approximant with an 11-fold supercell along the c-axis and a modulation wave vector equal to 3/11 = 0.273 which is close to the experimentally observed value of 0.267.

Taking the experimentally found atomic volume 9.19 Å ^{3} and the number of valence electrons for group V element P equal z = 5 we estimate the value of $2k_F$ equal to 5.05 Å^{-1}. Within a nearly free-electron model the Fermi sphere radius is defined as $k_F = (3\pi^2 z/V)^{1/3}$, where z is the number of valence electrons per atom and V is the atomic volume. The $2k_F$ position is indicated on Figure 1 (right inset). Atomic shifts due to the modulation result in appearance of satellite reflections and hence in a formation of additional Brillouin zone planes. Reflections have scattering vectors $H = ha^* + kb^* + lc^* + m\gamma c^* = ha^* + kb^* + (l+m\gamma)c^*$. For the commensurate approximation, $\gamma = 3/11$, an 11-fold supercell has been employed.

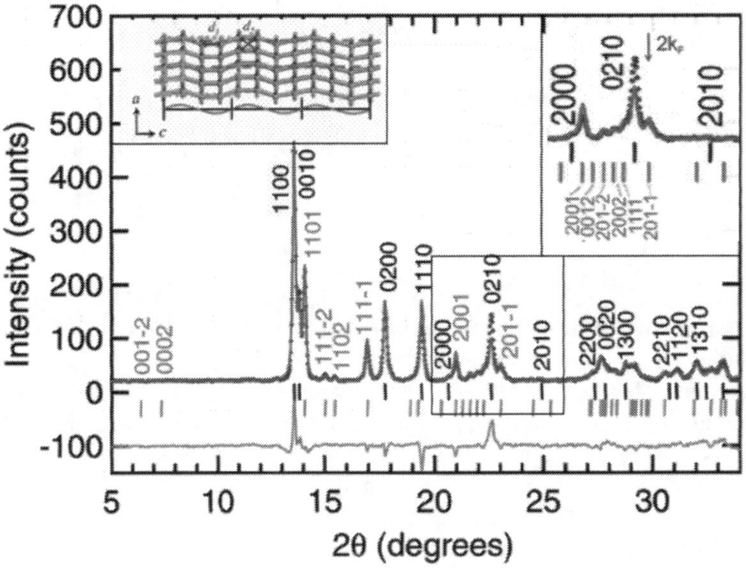

Figure 1. The experimental diffraction pattern of P-IV at 125 GPa (from Ref.2). The tick marks show the peak positions for main (upper) and modulated (lower) reflections. The left inset shows structural model of P-IV in *ac* plane. The basic cell is shown by solid (blue) line. The commensurate approximant with 11-fold supercell along the c-axis is shown by dashed (red) line. A modulation wave 3/11 is shown below. The right inset shows reflections near $2k_F$ indicated by arrow. Indices for reflections are given for the incommensurate cell (4D).

We selected a group of reflections near $2k_F$ for construction of the Brillouin-Jones zone. At first step we constructed planes of reflections (20*lm*) type and the (0012) reflection as shown in Figure 2.

On projection down b* one can see a set of planes lying close to the Fermi sphere. This configuration implies that some deformation of the sphere to an ellipsoid would be necessary. Within the free electron model it was suggested to consider a ratio of the Fermi radius to the ½*q* (a distance to a plane from the origin) called as "truncation" factor [7] which is usually 1.05 and may increase up to 1.10 as in the case of (2000) plane for phosphorous.

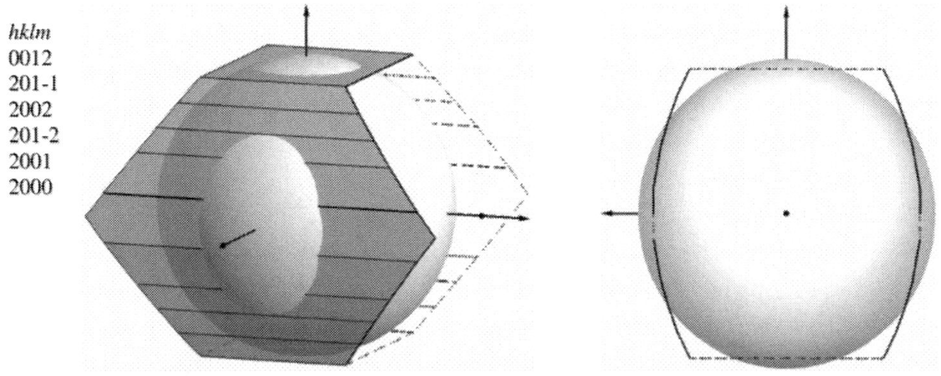

Figure 2. Configuration of the Brillouin planes (20*lm*) type and (0012) with the inscribed Fermi sphere for the valence electron number z = 5: (left) common view; (right) the view down b*.

Close to the Fermi sphere are planes (1111) and (0210) that are added to the Brillouin – Jones zone shown in Figure 3. This zone is filled to 86 % by electron states which satisfies the Hume-Rothery mechanism of phase stabilization [5].

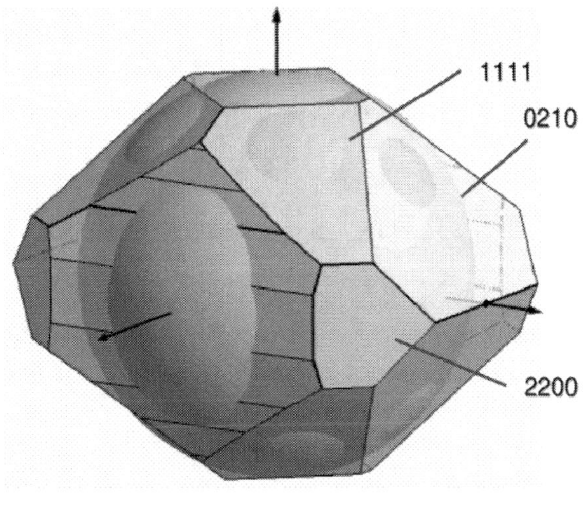

Figure 3. The Brillouin – Jones zone for P-IV constructed with planes close to $2k_F$, shown in the inset to Figure 1. Planes (1111), (0210) and (2200) are added to the planes shown in Figure 2.

Now we consider Fermi sphere – Brillouin zone configuration to find factors defining magnitude of the modulation wave vector. It was noted in Ref. [6] that the $2k_F$ vector is matching the (201-1) peak on the diffraction pattern. In Figure 4 we construct the Brillouin planes (001) and (200) in projection down b* and mark nodes (001), (200) and (201) of the basic cell. Constructing the $2k_F$ vector through the point where the (200) plane crosses the Fermi sphere we define the position of the (201-1) reflection. The shift of this reflection down the c* axis from (201) is equal to γc^* and the value of γ determined from this construction is 0.268 which is very close to the experimentally found value $\gamma = 0.2673$ [2].

Thus, the plane (201-1) corresponding to the modulation peak (201-1) appears just to cut an empty section where the (200) plane crosses the Fermi sphere. This mechanism is similar to the "nesting" effect and allowed the lowering of electron energy on a Brillouin plane close to the Fermi sphere. Formation of the modulation wave accounts for appearance of other modulation peaks and consequently a group of Brillouin planes near the Fermi sphere as shown in Figures 2 and 3.

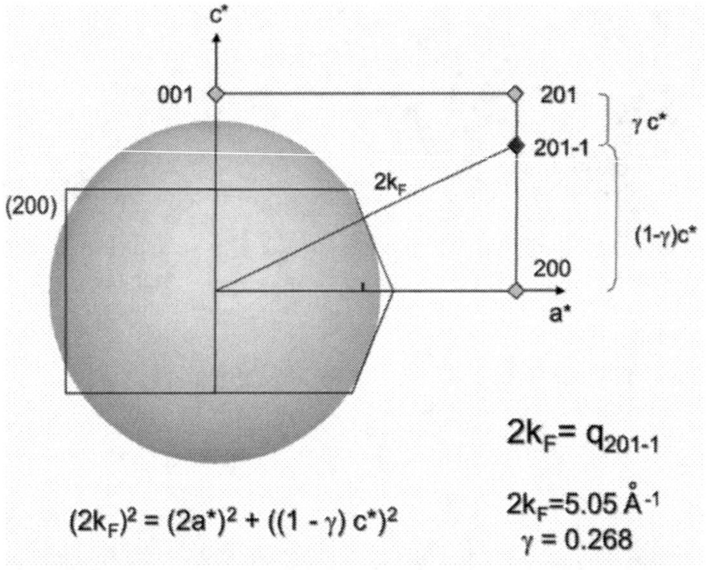

Figure 4. Configuration of the rreciprocal lattice for P-IV and the Fermi sphere down b*. Three-integer indices correspond to the basic unit cell. The position of the (201-1) reflection is defined by a reciprocal vector equal to $2k_F$ through the point of crossing of the (200) plane with the Fermi sphere. From this construction one can estimate the value of the modulation wave vector γ.

It is interesting to compare formation of the modulated structure P-IV to the supercell structure in the CuAu alloy [7]. A common feature is appearance of Brillouin planes close to the Fermi sphere. In the CuAu, an already existing reflection is adjusted to the Fermi sphere by the formation of the commensurately modulated structure with a shift by 1/2 of the period in each 5 cell blocks of a 10-fold supercell. In P-IV structure, the commensurate approximant is nearly as large as in CuAu - an 11-fold supercell. However, the plane lying on the Fermi sphere that defines the modulation is a newly formed one. The modulation wave for the positional displacement of atoms produces a group of Brillouin planes close to the Fermi sphere leading to a more significant effect on the electronic energy.

In conclusion, the stability of the IM structure in P-IV is attributed to the lowering of the electronic band structure energy due to Brillouin zone – Fermi surface interaction. Similar approach can also be applied to the IM structure of the group VI elements.

Acknowledgments
Author wish to acknowledge assistance from Dr. Olga Degtyareva and financial support from the Russian Foundation for Basic Research, grant No. 10-02-00552.

References

[1] McMahon M I and Nelmes R J 2006 *Chem. Soc. Rev.* **35** 943

[2] Fujihisa H, Akahama Y, Kawamura H, Ohishi Y, Gotoh Y, Yamawaki H, Sakashita M, Takeya S, and Honda K 2007 *Phys. Rev. Lett.* 98 175501

[3] Mott N F and Jones H 1936 *The Theory of the Properties of Metals and Alloys* (London: Oxford University Press)

[4] Degtyareva V F 2006 *Phys.-Usp.* **49** 369

[5] Degtyareva V F and Smirnova I S 2007 *Z. Kristallogr.* **222** 718

[6] Marqués M, Ackland G J, Lundegaard L F, Falconi S, Hejny C, and McMahon M I 2008 *Phys. Rev.* B **78** 054120

[7] Sato H and Toth R S 1962 *Phys. Rev. Lett.* **8** 239

Journal of Physics: Conference Series **226** (2010) 012020 doi:10.1088/1742-6596/226/1/012020

MLD Relations of Pisot Substitution Tilings

Franz Gähler

Faculty of Mathematics, University of Bielefeld, D-33615 Bielefeld, Germany

E-mail: `gaehler@math.uni-bielefeld.de`

Abstract. We consider 1-dimensional, unimodular Pisot substitution tilings with three intervals, and discuss conditions under which pairs of such tilings are locally isomorhphic (LI), or mutually locally derivable (MDL). For this purpose, we regard the substitutions as homomorphisms of the underlying free group with three generators. Then, if two substitutions are conjugated by an inner automorphism of the free group, the two tilings are LI, and a conjugating outer automorphism between two substitutions can often be used to prove that the two tilings are MLD. We present several examples illustrating the different phenomena that can occur in this context. In particular, we show how two substitution tilings can be MLD even if their substitution matrices are not equal, but only conjugate in $GL(n, \mathbb{Z})$. We also illustrate how the (in our case fractal) windows of MLD tilings can be reconstructed from each other, and discuss how the conjugating group automorphism affects the substitution generating the window boundaries.

1. Introduction

In this article, we consider substitutions σ on an alphabet of three letters, whose abelianisation matrix (substitution matrix) M is primitive and unimodular, has irreducible characteristic polynomial, and a leading (Perron-Frobenius, PF) eigenvalue which is a Pisot number. The words generated by such a substitution can be regarded as elements of a free group with three generators, and automorphisms of the free group give rise to transformations of words. Alternatively, we can work with a geometric realisation of the substitution, by letting it act on three intervals, whose lengths are chosen proportional to the components of the left eigenvector associated with the leading PF-eigenvalue λ of M. Each tile is then substituted with a sequence of tiles, whose total length is equal to λ times the original length. Such a geometric realisation generates a tiling of the line, instead of a sequence of symbols, or a word in a free group.

The matrix M represents a linear mapping A of $\mathbb{R}^3$, expressed with respect to some basis $\{b_i\}$. As M is unimodular, this mapping is an automorphism of the lattice L generated by this basis. We choose the geometry of L such that the expanding and contracting eigenspaces of A are perpendicular to each other, so that A commutes with the orthogonal projections on these eigenspaces. This can be realised as follows. After appropriate rescaling, the tile lengths, being components of the PF-eigenvector, are contained in the algebraic field $Q(\lambda)$, and so are all coordinates of lattice points in the expanding eigenspace of V of A. The corresponding coordinates in the contracting eigenspace W can be chosen as the $d-1$ Galois conjugates of the coordinate in V. We then have a cut-and-project scheme (CPS) defined by the lattice L, and the eigenspaces V and W of A:

$$\begin{array}{ccccc} V \cong \mathbb{R} & \stackrel{\pi_1}{\longleftarrow} & \mathbb{R}^3 & \stackrel{\pi_2}{\longrightarrow} & W \cong \mathbb{R}^2 \\ \cup & & \cup & & \cup \\ \Lambda & & L & & \Omega \end{array} \tag{1}$$

One of the formulations of the Pisot conjecture states that the vertex set Λ of a Pisot substitution tiling always is a model set, which means that there exists a window set $\Omega \subset W$ which is the closure of its interior, and which has boundary of measure zero, such that $\Lambda = \{\pi_1(x) \mid x \in L, \pi_2(x) \in \Omega\}$. Similarly, the subsets of the left end points of all tiles of a given type in a Pisot substitution tiling are model sets, too, with appropriate subwindows Ω_i. For all examples considered below, the Pisot conjecture can be shown to hold, even though a proof for the general case is still missing. For a more detailed description of Pisot substitution tilings and their associated CPS, we refer to [1]. In particular, we note that one can derive also a dual, contractive substitution acting on window sets in W, whose fixed points are the subwindows Ω_i. We remark that the CPS (1) also admits a canonical projection tiling, whose window Ω is the image under the projection π_2 of the parallelepiped spanned by the basis $\{b_i\}$ of L. The subwindow for tile i, whose length is equal to the length of $\pi_1 b_i$, is simply the parallelogram spanned by the vectors $\pi_2 b_j$ and $\pi_2 b_k$, where j and k are the two indices different from i. This canonical projection tiling is not substitutional in general, but its subwindows can serve as convenient starting points for the dual substitution determining the windows. Acting on the canonical windows as seeds, the dual substitution overlap free.

2. LI and MLD Relations

The CPS (1) does not specify the window Ω yet. Substitutions having the same abelianisation matrix M (but differ in the order of the letters within a substituted word) give rise to the same CPS, but will have different windows in general. As we shall see below, even substitutions with different abelianisation matrices may belong to a common CPS.

In the following, we shall study relations between certain substitution tilings belonging to a common CPS. For this, besides the geometric realisation of a substitution tiling it is also useful to consider the substitution action on the underlying free group with three generators. In our examples, the substitution acts with a group automorphism. If for two substitutions σ_1 and σ_2 there exists a fixed word w in the group, such that $\sigma_1(g) = w^{-1}\sigma_2(g)w$ for every generator g of the group, then the two substitutions produce tilings wich are locally isomorphic (LI), meaning that all their finite subpatterns are the same. This can be seen as follows. One first observes that there exists some power of σ_1, such that σ_1^k has a bi-infinite fixed point, and that σ_1^k and σ_2^k are still conjugate in the same way, with a (longer) word w'. In a second step, one can then show that the fixed point of σ_1^k is also a fixed point of σ_2^k, which implies that the two substitutions generate the same tilings.

A more delicate relation is mutual local derivability (MLD) [2]. Two tilings are MLD, if one can be reconstructed from the other in a *local* way, and vice versa. For this to work, the two tilings must first be brought to the appropriate relative scale and position. A good starting point is to consider two tilings belonging to a common CPS. In fact, two (model set) tilings are MLD if and only if the window of one can be constructed by finite unions and intersections of lattice translates of the window of the other, and vice versa. Looking at the windows can suggest an MLD relation, but for proving such a relation it is very helpful if one substitution can be written as a conjugate of the other, $\sigma_1 = \rho^{-1} \circ \sigma_2 \circ \rho$, where ρ is an outer automorphism of the free group (an inner automorphism would lead to an LI relation). Such an automorphism will make the transformation of one tiling into the other explicit.

If a substitution σ acts invertibly on the underlying free group, the boundaries of its windows are generated by a substitution, too. This boundary substitution is given by $\sigma_b = \tilde{\sigma}^{-1}$, which

Figure 1. Windows of the substitution $a \to cb$, $b \to c$, $c \to cab$.

Figure 2. Windows of the substitution $a \to bc$, $b \to c$, $c \to cba$.

is the inverse of σ, read backwards [4, 5]. As a seed for the iteration, it is again convenient to take the windows of the canonical projection tiling belonging to the same CPS. For instance, the canonical window for tile a is bounded by the closed path consisting of the four consecutive segments $\pi_2 b_2$, $\pi_2 b_3$,$-\pi_2 b_2$, and $-\pi_2 b_3$. This path is represented by the word $bcb^{-1}c^{-1}$ in the free group, on which the boundary substitution σ_b acts. In each step, the boundary path is transformed into one with more, but shorter segments, eventually converging to the fractal boundary of the final window.

If we now have two substitutions σ and σ' which are conjugated, $\sigma' = \rho^{-1} \circ \sigma \circ \rho$, their boundary substitutions satisfy $\sigma'_b = \tilde{\rho}^{-1} \circ \sigma_b \circ \tilde{\rho}$. Iterating this, we find $\sigma'^n_b = \tilde{\rho}^{-1} \circ \sigma^n_b \circ \tilde{\rho}$. In the limit of a fully fractalized window, the action of $\tilde{\rho}^{-1}$ can be neglected (it is local at the scale of the then infinitesimally small segments), and σ'_b can be understood as σ_b acting on the seed of σ'_b, transformed by $\tilde{\rho}$. As the dual substitution acting on the canonical windows and their iterates is overlap free, the transformation $\tilde{\rho}$ commutes with the fractalization induced by σ_b, and it becomes manifest that the windows of σ are obtained from those of σ' via the transformation induced by $\tilde{\rho}$. In particular, the windows of the two substitutions have the same fractal structure.

In the following, different phenomena arising in this context are illustrated with a number of examples. As a short-hand notation, we write the action of a substitution σ on a free group as the list of images of the generators, in our case a triple $[\sigma(a), \sigma(b), \sigma(c)]$.

3. Examples

As a first example, we consider the substitutions $\sigma_1 = [cb, c, cab]$ and $\sigma'_1 = [bc, c, cba]$, which have the same abelianisation matrix. These two substitutions are conjugate by the free group automorphism $\rho_1 = [bab^{-1}, b, c]$, with inverse $\rho_1^{-1} = [b^{-1}ab, b, c]$. It is easily checked that indeed we have $\sigma_1 = \rho_1^{-1} \circ \sigma'_1 \circ \rho_1$. In a word generated by σ_1, there is always a b to the right of an a. ρ_1 eats up that b, and adds a b to the left of the a instead, effectively replacing ab pairs by ba pairs. ρ_1^{-1} performs the opposite operation. This is obviously a local operation, no matter whether one works with words in a free group, with symbolic sequences, or with tilings. The LI classes of tilings generated by the two substitutions are MLD. The windows of the two substitutions σ_1 and σ'_1 are shown in Figures 1 and 2, respectively. The windows for the a, b, and c tiles are in red (medium gray), green (light gray), and blue (dark gray). When transforming from Figure 1 to Figure 2, part of the b tiles (green), namely those to the right of an a tile, move to where the

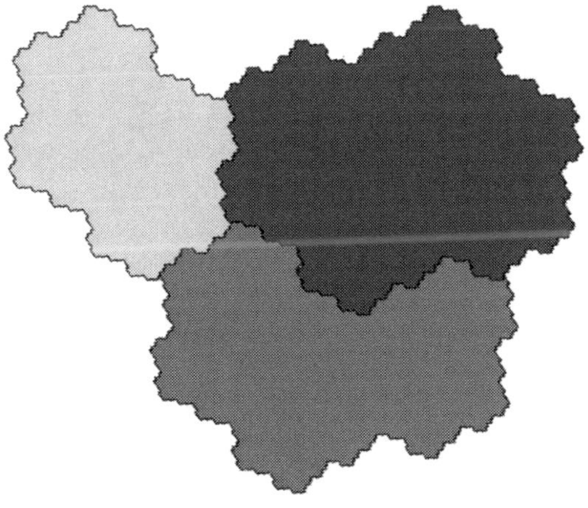

Figure 3. Windows of the substitution $a \to c$, $b \to a$, $c \to cab$.

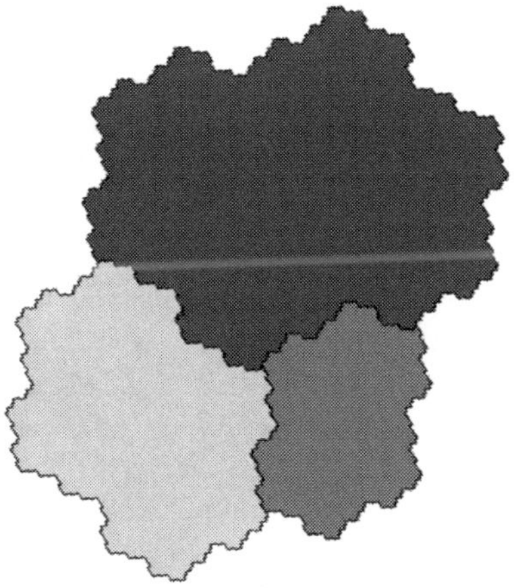

Figure 4. Windows of the substitution $a \to c$, $b \to ca$, $c \to cb$.

a tiles were before. The subwindow of the a tiles (red) thus becomes green, and a congruent copy of it is cut away from the original subwindow of the b tiles in green. The red subwindow of the a tiles instead moves to a different place, because the a tiles are now to the right of a b tile. As a and b tiles have different lengths, the left endpoint of the second tile of ab and ba pairs differs, and so the corresponding subwindows are at different places.

In the second example, we consider two substitutions with different abelianisation matrices, $\sigma_2 = [c, a, cab]$ and $\sigma_2' = [c, ca, cb]$. Again, there is a conjugating automorphism $\rho_2 = [a, a^{-1}b, c]$, with inverse $\rho_2^{-1} = [a, ab, c]$, so that $\sigma_2 = \rho_2^{-1} \circ \sigma_2' \circ \rho_2$. Here, in words produced by σ_2, all b tiles are to the right of an a tile. ρ_2 eats up the a tile to the left of a b tile, effectively replacing all ab pairs by just one b. Other a tiles (not to the left of a b) and all c tiles are left as they are. Conversely, ρ_2^{-1} splits all b in a σ_2'-word into ab pairs. On the tiling level, this operation is local if and only if the length of an ab pair of tiles in the σ_2-tiling is the same as the length of a b tile in the σ_2'-tiling, whereas a and c tiles have the same length for both tilings. With appropriate global scalings, this is indeed the case. $\sigma_2 = \rho_2^{-1} \circ \sigma_2' \circ \rho_2$ implies that the two abelianisation matrices are conjugate in $GL_3(\mathbb{Z})$. In fact, the two substitutions have the same CPS, with the same lattice L. The only difference is, that the linear mapping A is expressed with respect to two different lattice bases, yielding two different matrix representations of A, and different tile lengths (which are the lengths of the projected basis vectors). It is therefore not surprising, that the length of tile b in the σ_2'-tiling is the sum of the lengths of the two tiles a and b of the σ_2-tiling. The windows of the substitutions σ_2 and σ_2' are shown in Figures 3 and 4, respectively, using the same coloring as for the previous example. In Figure 3, part of the a tiles (in red), namely those to the left of a b tile, become the new b tiles in Figure 4 (green), whereas the old b tiles in Figure 3 (green) are discarded. MLD relations can therefore arise also if the two abelianisation matrices are not equal, but conjugate in $GL_3(\mathbb{Z})$, because the two substitutions then share a common CPS. We emphasise, however, that this relation is local only for the tilings with properly sized tiles. This pair of examples had been discussed in detail already in [3]. σ_2' is LI to the Rauzy or Tribonacci substitution.

Finally, as a third example, we consider a quartet of substitutions, all with the same abelianisation matrix M. These substitutions are $\sigma_A = [ca, ab, cab]$, $\sigma_B = [ac, ab, abc]$, $\sigma_C = [ca, ba, bac]$, and $\sigma_D = [ac, ab, bac]$. σ_A and σ_D are conjugate in a way already seen in the first example: $\sigma_D = \rho_3^{-1} \circ \sigma_A \circ \rho_3$, where $\rho_3 = [a, b, a^{-1}ca]$ simply replaces ac pairs by ca pairs. In order to discuss the relations to the other substitutions, we introduce the automorphisms

Figure 5. Windows of the substitution $a \to ca$, $b \to ab$, $c \to cab$.

Figure 6. Windows of the substitution $a \to ac$, $b \to ab$, $c \to abc$.

Figure 7. Windows of the substitution $a \to ca$, $b \to ba$, $c \to bac$.

Figure 8. Windows of the substitution $a \to ac$, $b \to ab$, $c \to bac$.

$u_1 = [c, a, ab]$ and $u_2 = [c, a, ba]$. We then have $\sigma_A = u_1 \circ u_1 \circ u_2$, $\sigma_B = u_1 \circ u_2 \circ u_1$, and $\sigma_C = u_2 \circ u_1 \circ u_1$, so that $\sigma_A = u_1 \circ \sigma_B \circ u_1^{-1}$ and $\sigma_C = u_1^{-1} \circ \sigma_B \circ u_1$. The conjugating automorphism u_1 has an abelianisation matrix U which commutes with the common abelianisation matrix M of the substitutions, even though U is non-trivial. This is possible because M is equal to the third power of U. Therefore, M is conjugate to itself by some non-trivial mapping, which acts non-trivially on the lattice L, changing the scale of the tiling by the cubic root of the inflation factor λ of the substitution (or its inverse). Consequently, in order to be MLD, the tilings produced by σ_A, σ_B and σ_C must be at relative scales $\lambda^{-\frac{1}{3}}$, 1, and $\lambda^{\frac{1}{3}}$, respectively. The situtation is in fact similar to the second example, where the substitutions share a common CPS, but different lattice bases of L are used. Here, these different lattice bases still lead to the same abelianisation matrix M, but produce tiles of different sizes. The windows of the substitutions σ_A, σ_B, σ_C, and σ_D are shown if Figures 5 to 8, respectively. We don't discuss the transformations between them in detail here, but it is quite obvious that each subwindow can be obtained as translate or union of translates of subwindows of the other substitutions.

4. Conclusions

The CPS of a Pisot substitution tiling can accommodate many other substitution tilings as well. Some of these are obtained by permuting the letters in the substituted words, but there may be others arising from the choice of a different basis for the lattice of the CPS, as we have seen in the second example. The more complicated a substitution is, the richer is the set of substitution tilings supported by its CPS. Some of the tilings sharing a common CPS can be related in interesting ways, however. In particular, there may local isomorphism or mutual local derivability relations, sometimes in surprising ways. We have discussed some of the phenomena that may arise in this context, and have illustrated them with a number of examples. The key tool was to formulate the substitution as an automorphism of an underlying free group, which allowed to find the relations with algebraic methods, and to make the transformations between the related tilings explicit.

Acknowledgments

The author would like to thank Pierre Arnoux, Dirk Frettlöh, and Edmund Harriss for fruitful discussions.

References

[1] Frettlöh D 2005, Duality of Model Sets Generated by Substitutions, *Romanian J. of Pure and Applied Math.* **50** 619–39.

[2] Baake M, Schlottmann M and Jarvis P D 1991, Quasiperiodic tilings with tenfold symmerty and equivalence with respect to local derivability, *J. Phys. A: Math. Gen.* **24** 4637–54.

[3] Arnoux P, Berthé V, Hilion A and Siegel A 2006, Fractal representation of the attractive lamination of an automorphism of the free group, *Ann. Inst. Fourier (Grenoble)* **56** 2161–212.

[4] Ei H 2003, Some properties of invertible substitutions of rank d, and higher dimensional substitutions, *Osaka Journal of Mathematics* **40** 543–62.

[5] Arnoux P, Furukado M, Harriss E and Ito A 2010, Algebraic numbers, free group automorphisms and substitutions of the plane, accepted for *Transactions of the AMS.*

A family of ternary decagonal tilings

Nobuhisa Fujita
Institute of Multidisciplinary Research for Advanced Materials, Tohoku University, Sendai 980-8577, Japan
E-mail: nobuhisa@tagen.tohoku.ac.jp

Abstract. A new family of decagonal quasiperiodic tilings are constructed by the use of generalized point substitution processes, which is a new substitution formalism developed by the author [N. Fujita, *Acta Cryst. A 65, 342 (2009)*]. These tilings are composed of three prototiles: an acute rhombus, a regular pentagon and a barrel shaped hexagon. In the perpendicular space, these tilings have windows with fractal boundaries, and the windows are analytically derived as the fixed sets of the conjugate maps associated with the relevant substitution rules. It is shown that the family contains an infinite number of local isomorphism classes which can be grouped into several symmetry classes (e.g., C_{10}, D_5, etc.). The member tilings are transformed into one another through collective simpleton flips, which are associated with the reorganization in the window boundaries.

1. Introduction

The standard projection formalism [1, 2, 3, 4] for constructing quasiperiodic point sets had evolved into the so called *dualization method*, which can be used to construct extensive *quasiperiodic tilings*; a complete account of the dualization method can be found in [5] and references cited therein. The generated tilings however are restricted to those whose windows are given as convex polytopes. Recall that some quasiperiodic tilings are known to have windows which have fractal boundaries, take for instance the dodecagonal square-triangle tilings [6, 7]. Such tilings, including many unknowns too, are unlikely to be constructed by way of projection because of the following difficulties. (i) Handling directly the broad range of fractal windows, which also fulfil the non-trivial constraints for tilings with a moderately small number of prototiles, is generally rather difficult. (ii) Even if the window is known, it is practically difficult to avoid minute numerical errors associated with the fine structures of the window when projecting the higher-dimensional structure down to the physical space.

These difficulties however are not relevant if instead a substitution (or an inflation) formalism is used for construction. Recall that various substitution rules for constructing quasiperiodic tilings with fractal windows can be found in the literature [8, 9, 10, 11, 12]. We describe a substitution rule for a tiling by the following two steps: (S1) expansive similarity transformation of the tiling and (S2) division of the expanded tiles into tiles of the original size. For a deterministic tilings, it is commonly assumed that the division of every expanded tile is determined uniquely by the local surroundings of that tile. Moreover, division rules for neighbouring expanded tiles must not compete with each other because tiles are not allowed to overlap. To search for a substitution rule of a new tiling, trial and error has been commonplace. This uneconomic situation has led me to propose a more systematic way for composing substitution rules for generating quasiperiodic tilings. The new method is called *generalized point substitution processes* (GPSPs) [13], in which Step S2 is divided into two sub-steps: (S2a) decoration of the expanded tiles with points which would be the candidate positions of the vertices of the next tiling and (S2b) choose an appropriate subset of the candidate positions so that the chosen points can be connected by unit edges to generate the next tiling. These sub-steps operate in a local manner so that they conform to the local uniqueness of the division Step S2.

In the subsequent two sections, the basic information about the decagonal Bravais module and the GPSP scheme is summarized. Then in Sec.4 the scheme is extensively applied to construct the new family of ternary decagonal quasiperiodic tilings. Note that Similar tilings have already been discussed in [13]. The aim is to somewhat extend the method used there so that it can handle a broader class of tilings. It turns out that the present method can generate tilings whose point groups are D_5, C_5, C_2, D_1 and C_1, in addition to C_{10} which has been the only point group for the ternary tilings discussed in [13]. In the final section, we introduce a randomization procedure into the GPSP scheme, where we obtain further variety of tilings with partial disorder. The effects of simpleton flips in the tilings are also considered.

2. Decagonal Bravais module

The five crystallographic axes for decagonal systems can be represented by the five unit vectors, $\boldsymbol{e}_j = (\cos(j\theta), \sin(j\theta))$ with $j = 0, 1, ..., 4$ and $\theta = 2\pi/5$. The first four members are taken as the basis set for indexing the lattice points in the system. The basis set generates a Z-module of rank four called the decagonal Bravais module [14], denoted by the symbol $\boldsymbol{\Lambda}_{10}$. $\boldsymbol{\Lambda}_{10}$ has the point symmetry of the dihedral group D_{10}, while it has an important property of scaling invariance $\tau\boldsymbol{\Lambda}_{10} = \boldsymbol{\Lambda}_{10}$, where $\tau = (1+\sqrt{5})/2$ is the golden mean.

The two-dimensional axis vectors in $\mathbb{E}^{\parallel}$ are lifted into the four-dimensional Euclidean hyperspace $\mathbb{E}_4$ as $\boldsymbol{\epsilon}_j = (\boldsymbol{e}_j, \boldsymbol{e}_j^{\perp})$, where the complementary components are defined by $\boldsymbol{e}_j^{\perp} := \boldsymbol{e}_{2j \pmod 5}$ where $j = 0, 1, ..., 4$. The first four members of $\{\boldsymbol{\epsilon}_j\}$ are the generators of the decagonal lattice $\boldsymbol{L}_{10}$. The orthogonal projection of $\boldsymbol{L}_{10}$ onto $\mathbb{E}^{\parallel}$ gives nothing but the decagonal Bravais module $\boldsymbol{\Lambda}_{10}$ [15]. The two-dimensional perpendicular space $\mathbb{E}^{\perp}$ is defined as the orthogonal complement to $\mathbb{E}^{\parallel}$ in $\mathbb{E}_4$; that is, $\mathbb{E}_4 = \mathbb{E}^{\parallel} \oplus \mathbb{E}^{\perp}$. Then $\boldsymbol{L}_{10}$ can be projected onto $\mathbb{E}^{\perp}$ as well, generating the conjugate module $\boldsymbol{\Lambda}_{10}^{\perp}$. Since the orthogonal projections from $\boldsymbol{L}_{10}$ to both $\boldsymbol{\Lambda}_{10}$ and $\boldsymbol{\Lambda}_{10}^{\perp}$ are bijections, one can introduce a natural bijection $\hat{\pi}$ between the modules; $\boldsymbol{\Lambda}_{10}^{\perp} = \hat{\pi}\boldsymbol{\Lambda}_{10}$.

3. Generalized point substitution processes

We make a restricted use of the term *planar tilings* for disjoint coverings of the plane by copies of a finite number of polygonal prototiles while fulfilling edge-to-edge condition. From such a tiling, a point set can always be defined by taking the set of vertices. In contrast, however, it is not always possible to define a tiling from a point set. Let us confine ourselves to the case when points from the set can be connected with uncrossed unit edges given by the vectors $\boldsymbol{e}_j$ $(j = 0, 1, ..., 4)$ and these edges define a tessellation of the plane by a finite kinds of polygons. Then the point set can be called *unit connective*. It is further assumed that the vertex set $\boldsymbol{\Sigma}_{\mathcal{T}}$ of a tiling $\mathcal{T}$ is a subset of the Bravais module $\boldsymbol{\Lambda}_{10}$.

The window $\boldsymbol{W}_{\mathcal{T}}(\subset \mathbb{E}^{\perp})$ is so defined that $\boldsymbol{\Sigma}_{\mathcal{T}}$ is the orthogonal projection of $(\mathbb{E}^{\parallel} + \boldsymbol{W}_{\mathcal{T}}) \cap \boldsymbol{L}_{10}$ onto $\mathbb{E}^{\parallel}$. [1] Then the image of $\boldsymbol{\Sigma}_{\mathcal{T}}$ in $\mathbb{E}^{\perp}$, denoted as $\boldsymbol{\Sigma}_{\mathcal{T}}^{\perp} := \hat{\pi}\boldsymbol{\Sigma}_{\mathcal{T}}$, is a dense subset of the window $\boldsymbol{W}_{\mathcal{T}}$, which on the other hand should be included in the closure of $\boldsymbol{\Sigma}_{\mathcal{T}}^{\perp}$, i.e. $\boldsymbol{\Sigma}_{\mathcal{T}}^{\perp} \subset \boldsymbol{W}_{\mathcal{T}} \subseteq \overline{\boldsymbol{\Sigma}_{\mathcal{T}}^{\perp}}$.

A GPSP [13] is described as a three-step process for composing a substitution rule: (G1) *expansive similarity transformation of the original tiling by the ratio* $\sigma = \tau^n$ *(*τ *: the Pisot unit, or the golden mean for the decagonal case)*, (G2) *replication of the basic motif* $\boldsymbol{S}$ *on every vertex*

[1] Here the $+$ symbol implies that $\boldsymbol{A} + \boldsymbol{B} \equiv \{a + b | ^{\forall} a \in \boldsymbol{A}, ^{\forall} b \in \boldsymbol{B}\}$.

and (G3) *elimination of excess points, so that the remaining points are unit connective.* Steps G1 and G2 together constitute a *point inflation rule*, which has been introduced for constructing a quasiperiodic point set [16], whereas Step G3 is an crucial step for constructing quasiperiodic tilings [13].

4. RPH tilings

RPH tilings have three prototiles, R (a rhombus with acute angles of 36°), P (a regular pentagon) and H (a barrel shaped hexagon). Every edge corresponds to one of the five unit vectors $\{\boldsymbol{e}_j\}$ defined above. In [13], two GPSPs have been identified as substitution rules for RPH tilings. These GPSPs are the mirror images of one another and are connected through simpleton flips involving three adjacent tiles, R, P and H. The two GPSPs can be applied in a mixed and arbitrary order so that an infinite number of tilings with the point group C_{10} can be obtained. The conjugate maps for these GPSPs are defined in the perpendicular space, and the windows with fractal boundaries for the relevant RPH tilings are derived analytically as the fixed sets of these conjugate maps.

Let us provide some information about the two GPSPs for RPH tilings. The scaling ratio for the expansion step G1 is taken as the square of the golden mean τ. The basic motif $\boldsymbol{S}$ for Step G2 is a centred decagon, with the representative indices being [0000] (centre) and [1100] (the decagonal shell). The first two steps, G1 and G2, would generate the candidate positions for the vertices of the next RPH tiling. These positions are superfluous for the vertices of an RPH tiling, so that an excess part should be eliminated at Step G3. To be specific, either of the two points that lie on every acute angle of expanded rhombi at the unit distance from the apex is to be eliminated. There are four ways to eliminate such points from an expanded rhombus, as shown in Fig.1; one can assign an arrow to each of the acute angles according to the choice of a point that is eliminated.

Since the two elimination rules labelled l and r (signifying the left- and right-handed chiralities, respectively) in Fig.1 are invariant under the two-fold rotation of the rhombus, either of these rules can be applied uniquely to all the expanded rhombi in a single iteration. It has been argued that the relevant two chiral GPSPs could be applied in an arbitrary order, resulting in a family of RPH tilings with the point group C_{10} [13]. For the left-handed GPSP, the arrows assigned to all the acute angles are symmetrically equivalent. Let us represent this situation as the wheel diagram shown in the upper left part of Fig.3.

It turns out that the window for an RPH tiling generated by the two GPSPs has a fractal boundary with the point symmetry C_{10}. The window can be analytically represented as the limit figure obtained by successively applying the conjugate maps associated with the GPSPs. A conjugate map operates in $\mathbb{E}^{\perp}$ and comprises the following three steps: (D1) *contractive similarity transformation by the ratio* $\bar{\sigma}$, *which is the algebraic conjugate to* σ *and* $|\bar{\sigma}| = 1/|\sigma| < 1$, (D2) *replication of the contracted figure onto every point of the conjugate motif* $\boldsymbol{S}^{\perp} = \hat{\pi}\boldsymbol{S}$ and (D3) *subtraction of an excess part of the resulting figure corresponding to the elimination step G3 in the physical space.* Note that, for decagonal RPH tilings, the algebraic conjugate to $\sigma = \tau^2$ is given by $\bar{\sigma} = 1/\tau^2$ for Step D1. It has been shown [13] that Step D3 for RPH tilings can be described as a carving process of a small portion near the boundary of the figure obtained after Step D2. Note that ten strips are indicated by the broken lines in the window diagrams in Fig.3. Corresponding to the arrows for the ten acute angles in the wheel diagram, arrows can be associated with the ten strips. Then the aft-end of each strip (tail of the arrow) is so carved that it would coincide with the fore-end of the strip through the τ-translation.

In Fig.2, a 1/20 sector of the boundary of the window is shown for first several iterations of the conjugate map for the left-handed GPSP. At each iteration, each line segment is replaced with an array of three segments, which are $1/\tau^2$ times shorter than the original one. As the conjugate map is repeated infinitely, the boundary would be a generalized von Koch curve whose fractal dimension is simply calculated as $dim(\partial \boldsymbol{W}_T) = \ln 3/\ln(\tau^2) \doteqdot 1.1415$.

It is also tempting to allow different elimination rules (i.e., arrow designations) for each expanded rhombus. Let us confine ourselves to the case when the choice of an elimination rule for each expanded rhombus only depends on the orientation of the rhombus. The choice for all the orientations can be specified by an wheel diagram. There are $2^{10}(= 1024)$ distinct arrow

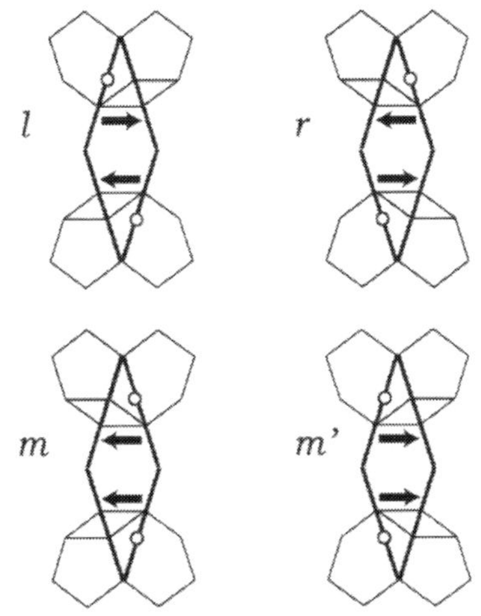

Figure 1. Points can be eliminated in four distinct ways from an expanded rhombus. The eliminated points are represented as open circles. The choice of a side for each acute angle can be represented by an arrow.

Figure 2. A 1/20 sector of the boundary of the window is constructed in a recursive manner. The corresponding window (Fig.3, upper left) is for the case when only the l-type elimination rule is applied.

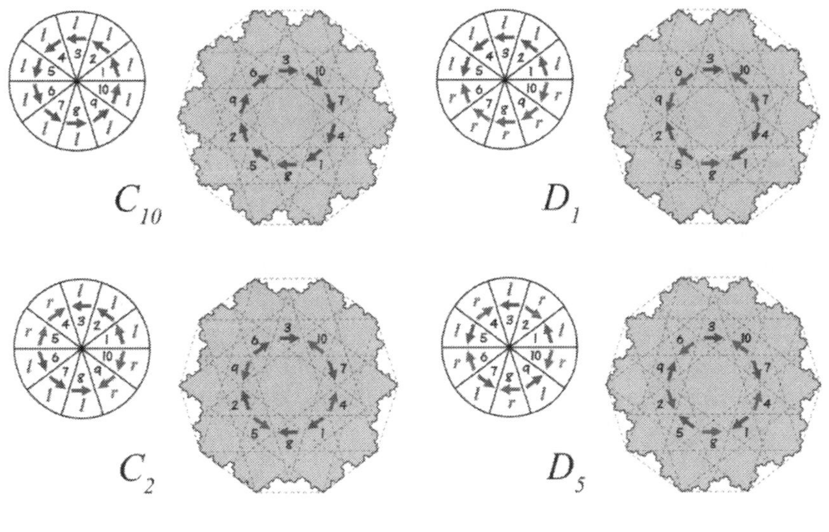

Figure 3. Four different ways to assign arrows to acute angles depending on their directions are represented by the wheel diagrams. Their point group symmetries are as indicated. The windows for the RPH tilings generated by successively applying the relevant GPSPs are depicted.

designations for wheel diagrams. In Fig.3, four wheel diagrams as well as the corresponding windows are shown. Different shapes of the windows imply that the relevant tilings belong to distinct local isomorphism classes. The global point symmetry of a tiling (which is the same as that of the window) can be inferred from the point symmetry of the relevant wheel diagram.

It is readily seen that a tiling with a mirror symmetry could be generated only if the elimination rule m or m' (see Fig.1) would takes place for at least one orientation of expanded rhombi. Since these elimination rules do not allow a two-fold axis, a two-fold rotational symmetry is incompatible with a mirror symmetry. Therefore, a global mirror symmetry exists only when a two-fold rotational symmetry is absent; the only allowed point groups are D_5 and D_1. In addition, C_2 and C_1 may also be the point group of a wheel diagram with arrows. Moreover, if different GPSPs are applied in a mixed and arbitrary order, the resulting point group would be a common subgroup to the point groups of the relevant GPSPs; C_5 is allowed in this way. Obviously, most of the 1024 possible wheel diagrams have the point group C_1. In Fig.4, an RPH tiling generated by four-cycles of GPSPs is demonstrated along with the corresponding window.

In [13], statistics of the local tile arrangements in RPH tilings have been analysed by two substitution matrices associated with the GPSPs of the types l and r. It has been shown that the relative frequencies of the three prototiles are invariably given by $n(R) : n(P) : n(H) = 1 : 2 : 1/\tau$. An alternative way to prove the statistics is to take the advantage of the four-dimensional description of the structures.

Let us denote by v the density of the vertices of an RPH tiling. Then it is given by the formula: $v = w/\Omega_4$, where w is the area of the window and Ω_4 the four-dimensional volume of the primitive unit cell of the decagonal lattice $\boldsymbol{L}_{10}$. For the family of RPH tilings, $w = 2\sqrt{5}a$ and $\Omega_4 = 5\sqrt{5}/4$, leading to $v = (8/5)a$, where $a = \sin(\pi/5) = \tau^{-1/2}5^{1/4}/2$ is the area of an R tile. The point density is related to the densities of R, P and H tiles in the following way:

$$n(R) + \frac{3}{2}n(P) + 2n(H) = \frac{8}{5}a. \tag{1}$$

Recall also that an RPH tiling is a projection of a 'puckered' net which is embedded in $\boldsymbol{L}_{10}$ along

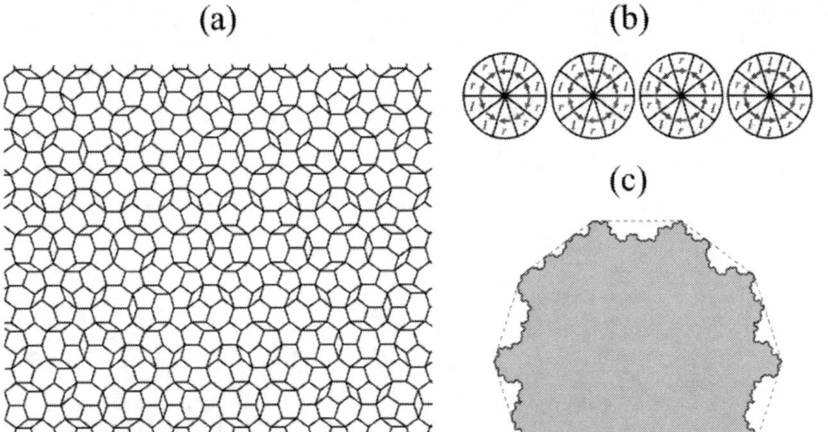

Figure 4. Part (a) shows a square patch of the RPH tiling generated by successively applying the four-cycle of GPSPs that are represented in Part (b). Part (c) depicts the window for the tiling. There is no point group symmetry except for the trivial one.

the physical space. The puckered net can be projected onto a lattice plane spanned by two of the five axis vectors $\boldsymbol{\epsilon}_i$ and $\boldsymbol{\epsilon}_j$ $(i \neq j)$. A specific shape (R, P or H) and orientation of tiles in the physical space corresponds to a specific contribution to the area of the projected image. The total area of the projected image is proportional to that of the tiling in $\mathbb{E}^{\parallel}$. There are ten lattice planes onto which the tiling can be mapped, and accordingly there are ten equations for the densities of the relevant tiles. By appropriately adding the left and right hand sides of these equations, one derives the following two equations for the density of the three prototiles:

$$n(R) + \frac{1}{2}n(P) + n(H) = \frac{4}{5}a, \tag{2}$$

$$\frac{3}{2}n(P) + 2n(H) = \frac{4}{5}\tau a. \tag{3}$$

From Eqs.(1-3), it turns out that the densities for the three prototiles are

$$n(R) = \frac{4}{5\tau^2}a, \ n(P) = \frac{8}{5\tau^2}a, \ n(H) = \frac{4}{5\tau^3}a. \tag{4}$$

The right hand sides of Eqs.(1-3) may be modified if there exists a linear phason strain, which however is assumed to be absent in the present report.

5. Random RPH tilings

The 1024 GPSPs as described above can be applied in an arbitrary order, resulting in an unlimited number of RPH tilings or, more precisely, local isomorphism classes; these tilings still maintain perfect quasiperiodicity. Let us now lift the constraint on the uniqueness of the elimination rule for all the expanded rhombi in the same orientation. If a new GPSP, in which the elimination rule for each expanded rhombus is chosen at random, is used at each iteration, a member of a stochastic ensemble of RPH tilings will be obtained. It will still fulfil the edge-to-edge condition. The number of possible patterns within a patch of diameter R will diverge exponentially as R is increased. Such a randomized GPSP can be thought of as a realization of the concept of *random substitution rules* proposed by Gummelt [17], who gave the example of randomized Penrose/Tübingen-triangle tilings.

In every GPSP described so far, irrespective of it being randomized or not, the elimination step G3 takes a subset of a larger point set, which is generated through Steps G1 and G2. It follows that by repeating the latter two steps only, a unique super-set can be defined from which the vertices of any RPH tiling originate. The super-set is the set of vertices of the *para-Penrose tiling* [16], whose window is a moth-eaten version of the regular decagon; see Fig.4-no.1 in [16]. Then for any of the randomized RPH tilings described above, points close to the boundary of this window is partially removed. The partial occupancies in certain triangular regions close to the boundary of the window are the same as the deterministic case as given in the end of the caption of Fig.8 in [13]; the only difference lies in that in the present case points are removed in a stochastic manner and that there is no clear boundary between occupied and unoccupied parts. In all cases the main part of the window is fully occupied except the partially occupied boundary regions, hence the quasiperiodicity is almost maintained. The relatively small amount of randomness will cause weak diffuse components superposed in the diffraction spectra.

The elimination rules for expanded rhombus as shown in Fig.1 can be mutually connected through simpleton flips involving three tiles, R, P and H. Every flip is caused by a hop of a vertex by a distance of $1/\tau$ in the parallel direction to the relevant arrow. This corresponds to a transport of a point in the perpendicular space by a distance of τ along the associated strip; see Fig.3. Moreover, all the RPH tilings we have introduced so far are mutually connected through collective simpleton flips of the same kind, because such flips can be repeated indefinitely so that the boundary regions of the window can be reorganized.

We have also performed a preliminary Monte Carlo study in which simpleton flips are excited stochastically when there is the combination of three tiles R, P and H. It has turned out that the resulting tilings are genuinely random, where the quasiperiodic order in the initial tiling has been completely dissolved while only the orientational order has been maintained.

Finally, it may be helpful to redirect the reader to some physical counterparts. The same kind of simpleton flips as discussed above has been observed in situ in a *d*-Al-Cu-Co quasicrystal at a temperature of 1123 K [18]. Furthermore, the centres of atomic clusters in a high-temperature phase of *d*-Al-Ni-Co quasicrystal can be described rather nicely by the vertices of an RPH tiling [19, 20]. Therefore, the family of RPH tilings may serve as templates for physical models of decagonal quasicrystals.

References

[1] de Bruijn N G 1981 Algebraic-theory of Penrose non-periodic tilings of the plane *Ned. Akad. Wetensh. Proc. Ser. A* **84** 38
[2] Mackay A L 1982 Crystallography and the Penrose pattern *Physica A* **114** 609
[3] Kramer P and Neri R 1984 On periodic and non-periodic space fillings of Em obtained by projection *Acta Cryst. A* **40** 580
[4] Duneau M and Katz A 1985 *Phys. Rev. Lett.* Quasiperiodic patterns **54** 2688
[5] Schlottmann M 1993 Periodic and quasi-periodic Laguerre tilings *Int. J. Mod. Phys. B* **7** 1351
[6] Baake M, Klitzing R and Schlottmann M 1992 Fractally shaped acceptance domains of quasi-periodic square triangle tilings with dodecagonal symmetry *Physica A* **191** 554
[7] Smith A P 1993 The sphere packing problem in quasi-crystals *J. Non-Cryst. Solids* **153-154** 258
[8] Stampfli P 1986 A dodecagonal quasiperiodic lattice in two dimensions *Helv. Phys. Acta* **59** 1260
[9] Zobetz E 1992 A pentagonal quasi-periodic tiling with fractal acceptance domain *Acta Cryst. A* **48** 328
[10] Luck J M, Godrèche C, Janner A and Janssen T 1993 The nature of the atomic surfaces of quasi-periodic self-similar structures *J. Phys. A: Math. Gen.* **26** 1951
[11] Godreche C, Luck J M, Janner A and Janssen T 1993 Fractal atomic surfaces of self-similar quasi-periodic tilings of the plane *J. Phys. I France* **3** 1921
[12] Hermisson J, Richard C and Baake M 1997 A guide to the symmetry structure of quasiperiodic tiling classes *J. Phys. I France* **7** 1003
[13] Fujita N 2009 Point substitution processes for decagonal quasiperiodic tilings *Acta Cryst. A* **65** 342
[14] Rokhsar D S, Mermin N D and Wright D C 1987 Rudimentary quasicrystallography: The icosahedral and decagonal reciprocal lattices *Phys. Rev. B* **35** 5487
[15] Niizeki K 1989 Self-similarity of quasilattices in 2 dimensions: I. the n-gonal quasilattice *J. Phys. A: Math. Gen.* **22** 193
[16] Niizeki K 2008 Self-similar quasilattices with windows having fractal boundaries *J. Phys. A: Math. Theor.* **41** 175208
[17] Gummelt P 2008 Generalized substitution rules: concepts and examples *Z. Kristallogr.* **223** 805
[18] Edagawa K, Suzuki K and Takeuchi S 2000 High resolution transmission electron microscopy observation of thermally fluctuating phasons in decagonal Al-Cu-Co *Phys. Rev. Lett.* **85** 1674
[19] Hiraga K, Lincoln F J and Sun W 1991 Structure and structural change of Al-Ni-Co decagonal quasicrystal by high-resolution electron microscopy *Mater. Trans. JIM* **32** 308
[20] Niizeki K, Sugihara H and Matsumura A 1994 Real-space analysis of phason disorders in decagonal quasi-crystals *Mater. Sci. Eng. A* **181-182** 742

Crystallization by stochastic flips

Olivier Bodini[1], **Thomas Fernique**[2] **and Damien Regnault**[2]
[1]LIP6, Univ. Paris 6, 104 av. du Président Kennedy, 75016 Paris, FRANCE
[2]LIF, Univ. Aix-Marseille & CNRS, 39 rue Joliot-Curie, 13453 Marseille, FRANCE
E-mail: [1]`olivier.bodini@lip6.fr`, [2]`{thomas.fernique,damien.regnault}@lif.univ-mrs.fr`

Abstract. Tilings are often used as a toy model for quasicrystals, with the ground states corresponding to the tilings satisfying some local properties (matching rules). In this context, a challenging problem is to provide a theory for quasicrystals growth. One of the proposed theories is the relaxation process. One assumes that the entropy of a tiling increases with the number of tilings which can be formed with the same tiles, while its energy is proportional to the ratio of satisfied matching rules. Then, by starting from an entropically stabilized tiling at high temperature and by decreasing the temperature, the phason flips which decrease (resp. increase) the energy would become more and more favoured (resp. inhibited). Ideally, the tiling eventually satisfies all the matching rules, and thus shows a quasicrystalline structure. This paper describes a stochastic process inspired by this and discusses its convergence rate.

1. Introduction

The discovery of non-periodic metallic alloys whose diffraction patterns show sharp peaks – soon called *quasicrystals* – produced a paradigm shift in the fields of crystallography. Known mathematical notions, in particular *aperiodic tilings*, quickly turned out to provide a suitable model of such structures. On the one hand, so-called *matching rules*, which specify how tiles can be assembled, can model the *energetic stabilization* at low temperature. But this fails to explain *growth* of quasicrystals because of the bad dynamics of tiles assembly. On the other hand, the fact that quasicrystals correspond to the tilings allowing the most numerous rearrangements of tiles could explain their growth, at high temperature, by *entropic stabilization*. But this fails to explain the experimentally observed stability of quasicrystals at low temperature.

We are here interested in a *relaxation process*, as suggested *e.g.* by Janot [3]: can we obtain an energetically stabilized quasicrystal by cooling an entropically stabilized quasicrystal? This raises the questions of the cooling process and its rate, that we discuss here. Other important questions, *e.g.*, the existence of a *phase transition* during the cooling or the role of *boundary conditions* (which are assumed to be fixed) are however not discussed. This paper just aims to state the problem and describe first results.

The rest of the paper is organized as follows. We formally state in Sec. 2 the notions we need to describe in Sec. 3 the process we are interested in. Sec. 4 then provides examples of the way this process runs, namely on *dimer tilings* and on the *Penrose tiling*. Last, we consider in Sec. 5 a simplified process and we provide conjectures and some (partial) results about its convergence rate. This is a work in progress.

2. General settings

2.1. Canonical tilings

Let $\vec{v}_1, \ldots, \vec{v}_n$ be n non-colinear vectors of $\mathbb{R}^d$, $n > d \geq 1$. For each $\{i_1, \ldots, i_d\} \subset \{1, \ldots, n\}$, the *proto-tile of type* $\{i_1, \ldots, i_d\}$ is the compact set of $\mathbb{R}^d$ denoted by $T_{i_1,\ldots,i_d}$ and defined by:

$$T_{i_1,\ldots,i_d} = \{\lambda_1 \vec{v}_{i_1} + \ldots + \lambda_d \vec{v}_{i_d} \mid 0 \leq \lambda_1, \ldots, \lambda_d \leq 1\}.$$

Translated proto-tiles are called *tiles*. Coverings of $\mathbb{R}^d$ by tiles with non-intersecting interiors are called $n \to d$ *canonical tilings*, or simply $n \to d$ *tilings*. One also speaks about *partial* (resp. *finite*) tilings when only subsets (resp. finite subsets) of $\mathbb{R}^d$ are covered.

A $n \to d$ tiling is said to be k*-periodic*, $0 \leq k \leq d$, if it is invariant by translation along k linearly independant vectors. One simply says *periodic* if $k = d$, *non-periodic* otherwise.

An interesting property of $n \to d$ tilings is that they can be seen as d-dimensional hypersurfaces living in a n-dimensional space. Indeed, let $(\vec{e}_1, \ldots, \vec{e}_n)$ denote the canonical basis of $\mathbb{R}^n$. The *lift* of a $n \to d$ tiling is the image by a map ϕ which is linear over tiles and defined for two connected vertices $\vec{x}$ and $\vec{x} + \vec{v}_k$ by $\phi(\vec{x} + \vec{v}_k) = \phi(\vec{x}) + \vec{e}_k$. Tiles are thus mapped onto d-dimensional hyperfaces of unit hypercubes of $\mathbb{Z}^n$ and tilings are mapped onto d-dimensional "stepped" hypersurfaces of $\mathbb{R}^n$.

Figure 1. Three coloring of the same patch of a $3 \to 2$ canonical tiling: white tiles (left), shaded tiles making easy to see the lift in $\mathbb{R}^3$ (middle) and tiles colored according to the distance (modulo 3) of their lift to the V-cut defined by the hyperplane of normal vector $\vec{e}_1 + \vec{e}_2 + \vec{e}_3$ (right).

A $n \to d$ tiling is said to have *thickness* at most $k \geq 0$ if its lift lies into a "slice" $V + [0, k]^d$, where V is a d-dimensional affine space of $\mathbb{R}^n$. Tilings of thickness at most 1 are called V*-cuts*. They are the thinnest tilings (one indeed checks that thinner slices yield holes in the lift). Not every tiling has a well-defined thickness; those which have are said to be *perp-bounded*. Intuitively, a thick tiling corresponds (under the lift) to a bumby stepped surface, while a thin one corresponds to a straight stepped surface. In particular, the V-cuts correspond to the straightest stepped surfaces, and are therefore sometimes used as digitizations of affine spaces.

Physically, a tile model a stable local configuration of atoms, while tilings model the possible global configurations. Although this is only a toy-model, it is widely used and provides a good insight in some properties of crystals and quasicrystals, which are respectively modeled by periodic and non-periodic V-cuts. For example, with $\vec{v}_k = e^{\frac{2ik\pi}{n}}$ and V generated by the vectors

$$(\cos(2k\pi/n))_{0 \leq k < n} \quad \text{and} \quad (\sin(2k\pi/n))_{0 \leq k < n},$$

the obtained V-cut turns out to be the lift of a $n \to 2$ with having n-fold rotational symmetry. The *crystallographic restriction* ensures that such a tiling can be periodic only for $n \in \{2, 3, 4, 6\}$.

2.2. Matching rules

We call *decoration* of a tile a real function defined over the boundary of this tile. Two decorated tiles are said to *match* if their decorations take the same value in each intersecting point (if any). Intuitively, decorations can be seen as colors on facets of tiles, with two matching tiles sharing the same color where they meet. Note that decorations are not necessarily constant over facets[1].

Given a set τ of decorated proto-tiles, we denote by X_τ the set of tilings by the corresponding tiles in which any two tiles match. The tilings of X_τ are said to be *enforced by matching rules*. The case of a finite set τ is of special interest, since it provides a finite local description of X_τ. In this case, X_τ is said to have *finite type*.

It is easily seen that a periodic tiling has finite type. The first example of a finite type set of tilings which does not contain any periodic tiling has been constructed in 1966 [4]. Although rather complicated, this construction discarded the idea that a non-local property (non-periodicity) cannot be enforced by local constraints. Simpler examples have been since obtained (*e.g.*, [5, 2, 6, 7]), as well as some generic constructions [8, 9]. This is an active domain of research.

Physically, matching rules model the short-range interactions between atoms of adjacent tiles, with matching tiles corresponding to low-energy local configuration (some refinement considering that tiles can "more or less" match can be introduced). Hence, in this model, the internal energy of a tiling varies as the proportion of matching tiles, with minimal energy tilings (ground states) being thus enforced by matching rules. In particular, quasicrystals can be seen as ground states of matching rules enforcing non-periodic tilings.

Non-periodic finite type tilings can thus model the *structure* of quasicrystals, but they however fail to model their *growth*. Indeed, whenever a set of decorated tiles can form only non-periodic tilings, aggregating one decorated tile at time such that it matches with the previous ones often yields a *deception*, *i.e.*, a finite configuration which cannot be extended to a tiling of the whole space without introducing mismatches [10]. Some refinements seem to avoid this problem [11], but a non-local inspection at each time is needed (what could be seen as physically unrealistic).

2.3. Flips

Whenever a vertex x of a $n \to d$ tiling belongs to exactly $d+1$ tiles, one gets a new tiling of the same subset by translating each of these tiles by the vector which belongs to the d other tiles. This local rearrangement of tiles is called a *flip* or *phason-flip*, said to be performed *around* x.

Figure 2. Flips on $n \to 2$ and $n \to 3$ tilings respectively involve three parallelogram tiles (left, in $\mathbb{R}^2$) and four parallelepiped tiles (right, in $\mathbb{R}^3$), which still fill the same space in both cases.

A tiling is said to be (finitely) flip-accessible from another tiling if the former can be obtained by performing a (finite) sequence of flips on the latter[2]. It is proven [12] that two finite $n \to 2$

[1] One checks that this allows, for example, to model the celebrated arrow-matching for Penrose tilings.
[2] Flip-accessibility is also sometimes called *ergodicity*.

tilings covering the same subset are mutually finitely flip-accessible. A characterization of flip-accessibility among infinite $n \to 2$ tilings is also known [13]. Flip-accessibility turns out to not always hold for $n \to 3$ tilings [14], but a complete characterization still remains to be obtained.

Physically, flips model local rearrangements of atoms that have been experimentally observed [15, 16]. They provide a mechanism for moving atoms (of controversial importance, however), allowing to explore different structures of the matter, *i.e.*, different tilings. Let us stress that, although theoretically defined as translations of tiles, flips do not necessarily need many moves of atoms: small variations of atoms positions can correspond to such rearrangements of tiles.

3. Crystallization by stochastic flips

The stochastic process we are interested in is the Markov chain $(x_t)_{t=0,1,\dots}$ defined as follows.

(i) We introduce a non-negative parameter T - the temperature. It starts from a very high initial value T_0 and then varies in a specified way (*temperature schedule*) up to reach zero.

(ii) We start from a finite decorated $n \to d$ tiling x_0 whose tiles do not necessarily all match, but can do it up to some rearrangement (which do not modify the covered subset of $\mathbb{R}^d$).

(iii) Given $x = x_t$ and $T > 0$, we choose uniformly at random a flippable vertex of x and we denote by y the tiling obtained by performing this flip. We then choose uniformly at random $\alpha \in [0,1]$, and we set $x_{t+1} := y$ or $x_{t+1} := x$, depending whether or not α is lower than

$$\frac{\exp(-E(y)/T) \times F(x)}{\exp(-E(x)/T) \times F(y)},$$

where $E(z)$ and $F(z)$ respectively denotes the number of mismatches and the number of flippable vertices of a tiling z.

Roughly, the flips decreasing the number of mismatches are all the more favoured since the temperature is low. This naturally raises the "(quasi)crystallization" question: does this process converge towards a mismatch-free tiling when the temperature goes to zero, and at which rate? Note that the hypothesis on the initial tiling x_0 is a necessary condition for such a convergence. This convergence (and its rate) would moreover depends on the temperature schedule and on x_0.

Let us give a brief "physical justification" for the study of such a process. Recall that the stability of a material at temperature T is governed by the minimization of its *free energy* $F = U - TS$, where U and S respectively denote its *internal energy* and its *entropy*. In terms of tilings, internal energy and entropy are usually defined as follows: the internal energy of a finite decorated tiling is proportional to its number of mismatches, while its entropy is proportional to the logarithm of the number of different tilings which can be obtain by rearranging tiles (without modifying the covered subset of $\mathbb{R}^d$). In the light of this, our process turns out to be, when T is fixed, a Metropolis-Hasting algorithm drawing samples from the Boltzmann distribution. Moreover, according to the so-called *random tiling hypothesis* [17], the initial tiling x_0 of our process has, under some conditions, maximal entropy among tilings with the same number of tiles. It can thus be seen as rather stable at the initial temperature of our process since, at high temperature, minimizing the free energy roughly means maximizing the entropy.

4. Two examples

Let us first consider the $3 \to 2$ tiling with 6-fold rotational symmetry. There is only one tile up to rotation, usually called a *dimer* because it covers exactly two cells of the triangular grid. This tiling can model only crystals since it is periodic.

This tiling is easily enforced by matching rules: decorations shall simply make that two translation-equivalent tiles which are adjacent along an edge do not match. One checks that the number of mismatches can vary from −6 to 6 by performing a flip. Fig. 3 provides an example.

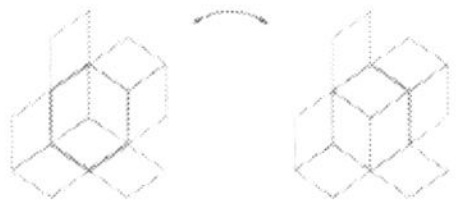

Figure 3. Here, performing a flip on the central vertex decreases (from left to right) or increases (from right to left) by 2 the number of mismatches (emphasized by the thick lines).

Fig. 4 shows the evolution of a random initial tiling x_0 when the temperature linearly decreases as a function of the number of performed flips. Tiles are colored as in Fig. 1, left. Mismatch-free regions thus correspond to unicolor regions.

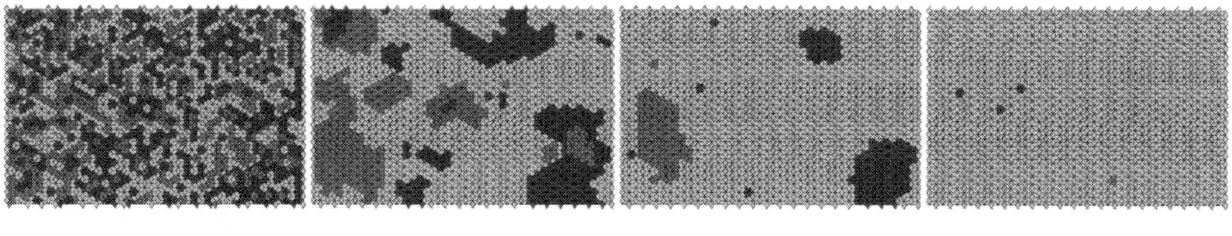

Figure 4. Snapshots (every 3000 flips) of the cooling of a dimer tiling.

Let us now consider the $5 \to 2$ tiling with 5-fold rotational symmetry. This is a non-periodic tiling of finite type, called the *Penrose tiling*. Unfortunately, mismatches turn out to be sparse, instead of forming level sets as in the $3 \to 2$ case (what allowed, by choosing suitable colors, a nice visualization of mismatch-free regions). This is because the lift of this tiling is a 2-dim. surface living in a 5-dim. space. We hence just assign different colors to the thin and fat rhombs. Fig. 5 shows the evolution of a random initial tiling x_0 when the temperature linearly decreases.

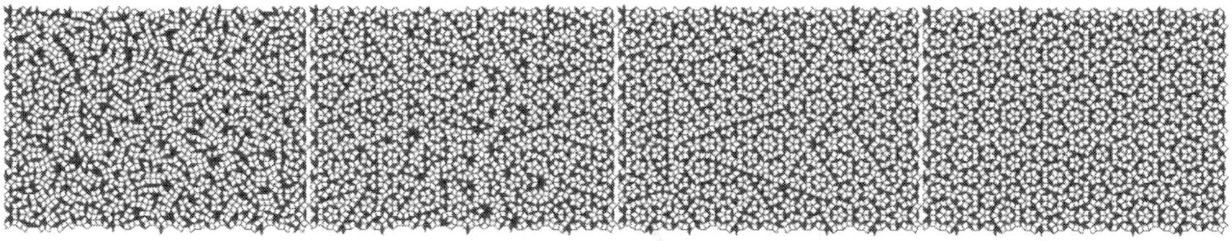

Figure 5. Snapshots (every 10000 flips) of the cooling of a Penrose tiling.

5. Towards theoretical results

Obtaining bounds on the number of steps of the process described in Sec. 3 and characterizing the corresponding optimal temperature schedules would be theoretical results of great interest. This however seems very hard. We thus chose to first consider the following simplified process. There is no more temperature parameter. We still start from a tiling x_0 satisfying the same hypothesis and we proceed step by step. Given the tiling x_t, we choose uniformly at random a flip which does not increase the number of mismatches. If there is no such flip, the process stops. Otherwise, x_{t+1} is the tiling obtained by performing this flip.

This new Markov chain experimentally runs similarly as the previous one, only faster and without the "noise" due to sparse flips creating isolated mismatches. Ergodicity is however no more ensured by the flip-accessibility results recalled in Sec. 2.3 because some flips are now forbidden. The notion of *constrained* flip-accessibility shall be investigated.

Recall from the introduction that we are interested in the rate of our cooling process. Indeed, quasicrystals can be experimentally grown, *e.g.* by the Bridgman-Stockbarger technique, with low cooling rates, but not *too* low. In particular, a cooling time exponential in the tiling's size would make our model highly unrealistic. We thus focus on the *convergence time* of our process.

The *convergence time* is the random variable $C(x_0) := \min\{t \mid E(x_t) = 0\}$, where $E(z)$ denotes the number of mismatches of z. Let also $\langle z\rangle$ denotes the set of tilings which can be obtained by rearranging tiles of a tiling z without modifying the covered subset of $\mathbb{R}^d$. The *worst convergence time* $\widehat{C}$ and, given a distribution μ_{x_0} over $\langle x_0\rangle$, the *mean convergence time* $\overline{C}$ are the random variables defined by

$$\widehat{C}(x_0) := \max_{x\in\langle x_0\rangle} C(x) \quad \text{and} \quad \overline{C}(\mu_{x_0}) := \sum_{x\in\langle x_0\rangle} C(x)\mathrm{d}\mu_{x_0}(x).$$

We are interested in asymptotics of their moments (at least their expected value $\mathbb{E}$) when the size $|x_0|$ of x_0 goes to infinity, in particular when μ_{x_0} is the uniform distribution U_{x_0} over $\langle x_0\rangle$.

- For $n \to 1$ tilings, numerous simulations led us to conjecture $\mathbb{E}(\widehat{C}(x_0)) = O(|x_0|^3)$ and $\mathbb{E}(\overline{C}(U_{x_0})) = O(|x_0|^{5/2})$. We have so far proven both bounds for $2 \to 1$ tilings in [18].
- For $n \to 2$ tilings, numerous simulations led us to conjecture $\mathbb{E}(\widehat{C}(x_0)) = O(|x_0|^2)$ and $\mathbb{E}(\overline{C}(U_{x_0})) = O(|x_0|^{3/2})$. We have so far proven $\mathbb{E}(\widehat{C}(x_0)) = O(|x_0|^3)$ for $3 \to 2$ tilings.

Note that bounds for $n \to 2$ tilings are smaller that for $n \to 1$ tilings. We do not have yet considered $n \to 3$ tilings, but we expect lower bounds.

Acknowledgments

The second author thanks Michael Baake, Leticia Cugliandolo, Franz Gähler, Christian Mercat, Éric Rémila and Mathieu Sablik for fruitful discussions. We also thank the anonymous referee.

References

[1] Shechtman D, Blech I, Gratias D and Cahn J W 1984 *Phys. Rev. Lett.* **53** 1951–54
[2] Penrose R 1974 *Bull. Inst. Maths. Appl.* **10** 266
[3] Janot C 1992 *Quasicrystals, a Primer* (Oxford: Oxford University Press)
[4] Berger R 1966 *Memoirs of the American Mathematical Society* **66** 1–72
[5] Robinson R M 1971 *Inv. Math* **12** 177-90
[6] Kari J 1996 *Discrete Mathematics* **160** 259–64
[7] Socolar J E S 1990 *Comm. Math. Phys.* **129** 599–619
[8] Mozes S (1989) *J. D'Analyse Math.* **53** 139-86
[9] Goodman-Strauss C (1998) *Annals of Mathematics* **147** 181-223
[10] Dworkin S and Shieh J I 1995 *Comm. Math. Phys.* **168** 337–52
[11] Onoda G Y, P J Steinhardt, DiVincenzo D P and Socolar J E S 1988 *Phys. rev. lett.* **60** 2653–56
[12] Kenyon R 1993 *Algorithmica* **9** 382–97
[13] Bodini O, Fernique Th and Rémila É 2008 *Information and Computation* **206** 1065–73
[14] Desoutter V and Destainville N 2005 *J. Phys. A: Math. Gen.* **38** 17–45
[15] Lyonnard S, Coddens G, Calvayrac Y and Gratias D 1996 *Phys. Rev. Lett B* **53** 3150–60
[16] Adagawa K, Mandal P, Hosono K,3, Suzuki K, and Takeuchi S 2004 *Phys. Rev. Lett B* **70**
[17] Henley C 1990 *Quasicrystals: The State of the Art* ed D. P. DiVincenzo et P. J. Steinhardt (Singapore: World Scientific)
[18] Bodini O, Fernique Th and Regnault D 2010 *Proc. of ANALCO'10*

Journal of Physics: Conference Series **226** (2010) 012023 doi:10.1088/1742-6596/226/1/012023

Surprises in aperiodic diffraction

Michael Baake[1] **and Uwe Grimm**[2]

[1] Fakultät für Mathematik, Universität Bielefeld, Postfach 100131, 33501 Bielefeld, Germany
[2] Department of Mathematics and Statistics, The Open University, Walton Hall, Milton Keynes MK7 6AA, UK

E-mail: `mbaake@math.uni-bielefeld.de` and `u.g.grimm@open.ac.uk`

Abstract. Mathematical diffraction theory is concerned with the diffraction image of a given structure and the corresponding inverse problem of structure determination. In recent years, the understanding of systems with continuous and mixed spectra has improved considerably. Moreover, the phenomenon of homometry shows various unexpected new facets. Here, we report on some of the recent results in an exemplary and informal fashion.

1. Introduction

The diffraction measure is a characteristic quantity of a translation bounded measure ω on Euclidean space (or on any locally compact Abelian group). It emerges as the Fourier transform $\widehat{\gamma}$ of the autocorrelation measure γ of ω, and has important applications in crystallography, because it describes the outcome of kinematic diffraction (from X-rays or neutron scattering, say). In recent years, initiated by the discovery of quasicrystals (which are non-periodic but nevertheless show pure Bragg diffraction), a systematic study by many people has produced a reasonably satisfactory understanding of the class of measures with a pure point diffraction measure, meaning that $\widehat{\gamma}$ is a pure point measure, without any continuous component.

Clearly, reality is more complicated than that, in the sense that real world structures will (and do) show lots of continuous components as well. Unfortunately, the methods around dynamical systems that are used to establish pure point spectra do not seem to extend to the treatment of systems with mixed spectrum, at least not in sufficient generality. Nevertheless, systems with continuous spectral components have recently been investigated also from a rigorous mathematical point of view, with a number of unexpected results. In particular, the phenomenon of homometry becomes more subtle, as we will see below.

In this informal exposition, we summarise some classic results that have recently resurfaced in the context of diffraction, with special emphasis on singular continuous and absolutely continuous spectra. For proofs (and the formal details) we refer to the original papers or to work in progress.

2. Mathematical setting and pure point spectra

Below, we mainly consider various (weighted) Dirac combs on the real line, with support on $\mathbb{Z}$ (the set of integers), and their diffraction. Even in this simple and seemingly restricted setting, unexpected phenomena show up. If $\omega = \sum_{n\in\mathbb{Z}} w(n)\delta_n$, with δ_n the normalised point measure at n, the *natural autocorrelation* (if it exists) of ω is

$$\gamma = \omega \circledast \widetilde{\omega} := \lim_{N\to\infty} \frac{\omega_N * \widetilde{\omega_N}}{2N}, \tag{1}$$

which we view as a measure on $\mathbb{R}$. Here, $\omega_N = \omega|_{[-N,N]}$ is the restriction of ω to the closed interval $[-N, N]$, and $\widetilde{\mu}$ is the 'flipped over' measure, defined by $\widetilde{\mu}(g) = \overline{\mu(\widetilde{g})}$ with g a continuous test function of compact support and $\widetilde{g}(x) = \overline{g(-x)}$. Here, and in analogous situations below, the bar denotes complex conjugation. We will only consider situations where the limit in (1) exists, either always (in the deterministic cases) or almost surely (in the probabilistic cases). An explicit calculation shows that $\mathrm{supp}(\omega) \subset \mathbb{Z}$ implies that γ must be of the form

$$\gamma = \sum_{m\in\mathbb{Z}} \eta(m)\, \delta_m \tag{2}$$

with the autocorrelation coefficients

$$\eta(m) = \lim_{N\to\infty} \frac{1}{2N+1} \sum_{n=-N}^{N} w(n)\overline{w(n-m)}. \tag{3}$$

The existence of the limit in (1) is equivalent to the existence of the $\eta(m)$ for all $m \in \mathbb{Z}$.

By construction, γ is a positive definite measure, and hence Fourier transformable. The result is $\widehat{\gamma}$, the *diffraction measure*, which is a positive, translation bounded measure on $\mathbb{R}$. It describes (kinematic) diffraction from the measure ω; compare [12] for background and [18, 9] for the development of this approach to diffraction theory. One of the benefits of this approach is the unique decomposition

$$\widehat{\gamma} = (\widehat{\gamma})_{\mathsf{pp}} + (\widehat{\gamma})_{\mathsf{sc}} + (\widehat{\gamma})_{\mathsf{ac}} \tag{4}$$

of the diffraction measure into its pure point, singular continuous and absolutely continuous parts, the latter splitting relative to Lebesgue measure λ, which is the natural reference measure for volume in Euclidean space.

The best known example is the lattice Dirac comb $\omega = \delta_{\mathbb{Z}} = \sum_{n\in\mathbb{Z}} \delta_n$, with autocorrelation $\gamma = \delta_{\mathbb{Z}}$ and diffraction $\widehat{\gamma} = \delta_{\mathbb{Z}}$. Here, γ follows from an elementary calculation, while the formula for $\widehat{\gamma}$ is a consequence of the *Poisson summation formula*

$$\widehat{\delta_{\Gamma}} = \mathrm{dens}(\Gamma)\, \delta_{\Gamma^*} \tag{5}$$

for an arbitrary lattice Γ (in $\mathbb{R}^d$) of density $\mathrm{dens}(\Gamma)$, with dual lattice Γ^*, the latter defined by $\Gamma^* := \{x \in \mathbb{R}^d \mid xy \in \mathbb{Z} \text{ for all } y \in \Gamma\}$; see [9] and references therein. The integer lattice is perhaps the simplest example for a system with pure point diffraction. More generally, if ω is any $\mathbb{Z}$-periodic measure, it can be written as $\omega = \mu * \delta_{\mathbb{Z}}$, with μ a finite measure. This leads to $\gamma = (\mu * \widetilde{\mu}) * \delta_{\mathbb{Z}}$ and $\widehat{\gamma} = |\widehat{\mu}|^2\, \delta_{\mathbb{Z}}$, where $|\widehat{\mu}|^2$ is a continuous positive function on $\mathbb{R}$. Its values at integer points are the *intensities* of the Bragg peaks.

A similarly nice formula formula holds for regular model sets. Within a given cut and project scheme (CPS) $(\mathbb{R}^d, \mathbb{R}^m, \mathcal{L})$ with lattice $\mathcal{L} \subset \mathbb{R}^{d+m}$, consider $\Lambda = \{x \in L \mid x^\star \in W\}$, where $L = \pi(\mathcal{L})$ is the projection of $\mathcal{L}$ in $\mathbb{R}^d$, W is the (compact) window in $\mathbb{R}^m$ (with boundary of measure 0), and $\star$ denotes the star map of the CPS. By the general model set theorem [18, 27, 9], the diffraction $\widehat{\gamma}$ of the Dirac comb δ_Λ is then the pure point measure

$$\widehat{\gamma} = \sum_{k\in L^{\circledast}} |A(k)|^2\, \delta_k, \tag{6}$$

where $L^{\circledast} = \pi(\mathcal{L}^*)$ is the projection of the dual lattice $\mathcal{L}^*$ and the amplitude $A(k)$ is given by

$$A(k) = \frac{\mathrm{dens}(\Lambda)}{\mathrm{vol}(W)} \int_W e^{2\pi i k^\star y}\, \mathrm{d}y = \frac{\mathrm{dens}(\Lambda)}{\mathrm{vol}(W)}\, \widehat{1_W}(-k^\star).$$

Table 1. Weights for the homometric pair from [17].

j	0	1	2	3	4	5
c_j	11	25	42	45	31	14
c_j	10	21	39	46	35	17

This formula has various generalisations, for instance to model sets with other internal spaces or to weighted model sets; see [27, 9] and references therein for more.

As an example, we consider the *period doubling sequence*, as defined by the substitution

$$\varrho = \varrho_{\text{pd}} \colon a \mapsto ab\,,\; b \mapsto aa. \tag{7}$$

A two-sided sequence can be obtained from a fixed point of ϱ^2 via the iteration

$$a|a \xrightarrow{\varrho^2} abaa|abaa \xrightarrow{\varrho^2} \ldots \longrightarrow w = \varrho^2(w), \tag{8}$$

with convergence in the (obvious) product topology. Here and below, we write two-sided sequences as $w = \ldots w_{-2}w_{-1}|w_0w_1\ldots$ and use $|$ to mark the origin. Note that $a|a$ in (8) is a legal seed, so that w is a fixed point of ϱ^2 in the strict sense.

We attach a Dirac comb to w by $\omega = \sum_{n\in\mathbb{Z}} h(w_n)\,\delta_n$, with $h(a) = h_+$ and $h(b) = h_-$. This turns out to define a (weighted) regular model set [10, 9], so that the diffraction is indeed a pure point measure. It is given by

$$\widehat{\gamma_{\text{pd}}} = \sum_{k\in L^{\circledast}} \bigl|h_+A(k) + h_-B(k)\bigr|^2\,\delta_k,$$

where $L^{\circledast} = \bigcup_{\ell\geq 1}\mathbb{Z}/2^{\ell} = \bigl\{\frac{m}{2^r} \mid (r=0,\, m\in\mathbb{Z}) \text{ or } (r\geq 1,\, m \text{ odd})\bigr\}$ is the Fourier module of the period doubling sequence. The amplitudes read

$$A(k) = \frac{2}{3\cdot(-2)^r}\,e^{2\pi i k} \quad\text{and}\quad B(k) = \delta_{r,0} - A(k),$$

where we implicitly refer to the parametrisation of $L^{\circledast}$. We state this formula here without further details; see [10, 9] for a proof.

Let us briefly mention the *homometry* problem. It refers to the possibility that distinct measures can still possess the same autocorrelation. An interesting example is constructed in [17], based on 6-periodic Dirac combs of the form $\delta_{6\mathbb{Z}} * \sum_{j=0}^{5} c_j\delta_j$. In particular, the two choices of Table 1 lead to the same autocorrelation – and, in fact, even to identical correlation functions up to order 5. It is only the order 6 correlation that tells the two combs apart.

For model sets, there are further possibilities, because different windows can have the same covariogram. The latter (for a compact set K) is defined as $\mathrm{cvg}_K(x) = \mathrm{vol}\bigl(K\cap(x+K)\bigr) = \bigl(1_K * \widetilde{1_K}\bigr)(x)$. This phenomenon results in identical autocorrelations and hence in homometric model sets. A simple planar example was constructed in [4]; see also [16, 14, 15]. We skip further details here and shift our attention to continuous spectra now.

3. Singular continuous spectra

The paradigm of a singular continuous measure is the distribution function for the classic middle-thirds Cantor measure, which is also called the Devil's staircase. Here, we show a class of structures that lead to somewhat similar functions, yet with significant differences.

The classic Thue-Morse (or Prouhet-Thue-Morse) sequence [1] can be defined via a fixed point of the substitution

$$\varrho = \varrho_{\mathrm{TM}} \colon \; 1 \mapsto 1\bar{1}\,, \; \bar{1} \mapsto \bar{1}1 \tag{9}$$

on the binary alphabet $\{1, \bar{1}\}$. The one-sided fixed point starting with 1 reads

$$v = v_0 v_1 v_2 \ldots = 1\bar{1}\bar{1}1\bar{1}11\bar{1}\ldots, \tag{10}$$

while $\bar{v}$ is the fixed point starting with $\bar{1}$. One can now define a two-sided sequence w by

$$w_i = \begin{cases} v_i, & i \geq 0, \\ v_{-i-1}, & i < 0. \end{cases}$$

It is easy to check that w defines a 2-cycle under ϱ, and hence a fixed point for ϱ^2. Since the central seed $1|1$ is legal, an iteration of ϱ^2 applied to it converges to w in the product topology.

The sequence w defines a dynamical system (under the action of the group $\mathbb{Z}$). Its compact space is the (discrete) *hull*, obtained as the closure of the $\mathbb{Z}$-orbit of w,

$$\mathbb{X}_{\mathrm{TM}} = \overline{\{S^i w \mid i \in \mathbb{Z}\}},$$

where S denotes the shift operator and the closure is taken in the local (or product) topology, where two sequences are close when they agree on a large segment around the origin. Now, $(\mathbb{X}_{\mathrm{TM}}, \mathbb{Z})$ is a strictly ergodic dynamical system (hence uniquely ergodic and minimal [25, 29]). Its unique invariant probability measure is given via the (absolute) frequencies of finite words (or patches) as the measures of the corresponding cylinder sets, which then generate the σ-algebra.

Here, we are interested in the diffraction of the (signed) Dirac comb $\omega_{\mathrm{TM}} = \sum_{n\in\mathbb{Z}} w_n \delta_n$, where we interpret 1 and $\bar{1}$ as weights 1 and -1. Its autocorrelation exists, as a consequence of unique ergodicity, and is actually the same measure for *all* sequences of $\mathbb{X}_{\mathrm{TM}}$. We can now employ the special structure of our fixed point w to analyse it. The autocorrelation is of the general form (2), with coefficients $\eta(0) = 1$, $\eta(-m) = \eta(m)$ for all $m \in \mathbb{Z}$ and

$$\eta(m) = \lim_{N\to\infty} \frac{1}{N} \sum_{n=0}^{N-1} v_n v_{n+m} \tag{11}$$

for all $m \geq 0$. Here, the structure of w and its relation to v was used to derive (11) from (3). Observing that v satisfies $v_{2n} = v_n$ and $v_{2n+1} = \bar{v}_n$ for all $n \geq 0$, one can derive from (11) the recursions

$$\eta(2m) = \eta(m) \quad \text{and} \quad \eta(2m+1) = -\frac{1}{2}\big(\eta(m) + \eta(m+1)\big), \tag{12}$$

which actually hold for all $m \in \mathbb{Z}$. One finds $\eta(\pm 1) = -1/3$ from solving the recursion for $m = 0$ and $m = -1$ with $\eta(0) = 1$, while all other values are then recursively determined.

To analyse the diffraction measure $\widehat{\gamma}$ of the TM sequence (following [23, 20]), one can start with its pure point part. Defining $\Sigma(N) = \sum_{n=-N}^{N} \big(\eta(n)\big)^2$, one derives via the recursion (12) that $\Sigma(4N) \leq \frac{3}{2}\Sigma(2N)$, which implies $\frac{1}{N}\Sigma(N) \xrightarrow{N\to\infty} 0$. By Wiener's criterion [30], this means $\big(\widehat{\gamma}\big)_{\mathsf{pp}} = 0$, so that $\widehat{\gamma}$ is a continuous measure.

Defining the (continuous) distribution function F via $F(x) = \widehat{\gamma}([0,x])$, another consequence of (12) is the pair of functional relations

$$\mathrm{d}F(\tfrac{x}{2}) \pm \mathrm{d}F(\tfrac{x+1}{2}) = \begin{Bmatrix} 1 \\ -\cos(\pi x) \end{Bmatrix} \mathrm{d}F(x).$$

Splitting F into its sc and ac parts (which is unique) now implies backwards that the recursion (12) holds separately for the two sets of autocorrelation coefficients, η_{sc} and η_{ac}, with yet unknown initial conditions at 0. An application of the Riemann-Lebesgue lemma, however, forces $\eta_{\mathsf{ac}}(0) = 0$, and hence $\eta_{\mathsf{ac}}(m) = 0$ for all $m \in \mathbb{Z}$, so that also $(\widehat{\gamma})_{\mathsf{ac}} = 0$; compare [20]. This shows that $\widehat{\gamma}$, which is not the zero measure, is purely singular continuous. Figure 1 shows an image, where we have used the Volterra iteration

$$F_{n+1}(x) = \frac{1}{2}\int_0^{2x} \bigl(1-\cos(\pi y)\bigr) F_n'(y)\,\mathrm{d}y \qquad \text{with} \qquad F_0(x) = x$$

to calculate F with sufficient precision (note that $F(x+1) = F(x) + 1$, so that a display on $[0,1]$ suffices). In contrast to the Devil's staircase, the TM function is *strictly* increasing, which means that there is no plateau (which would indicate a gap in the support of $\widehat{\gamma}$); see [5] and references therein for details.

Despite the above result, the TM sequence is closely related to the period doubling sequence, via the (continuous) block map

$$\varphi\colon \quad 1\bar{1}, \bar{1}1 \mapsto a\,, \quad 11, \bar{1}\bar{1} \mapsto b\,, \tag{13}$$

which defines an exact 2-to-1 surjection from the hull $\mathbb{X}_{\mathrm{TM}}$ to $\mathbb{X}_{\mathrm{pd}}$.

The TM sequence is often considered as a rare and special example, which is misleading. To demonstrate the point (following [21]), consider the generalised Morse sequences defined by

$$\varrho = \varrho_{k,\ell}\colon\ 1 \mapsto 1^k \bar{1}^\ell\,,\ \bar{1} \mapsto \bar{1}^k 1^\ell \tag{14}$$

for arbitrary $k, \ell \in \mathbb{N}$. Here, the one-sided fixed point starting with $v_0 = 1$ satisfies

$$v_{m(k+\ell)+r} = \begin{cases} v_m, & \text{if } 0 \le r < k, \\ \bar{v}_m, & \text{if } k \le r < k+\ell, \end{cases} \tag{15}$$

for $m \ge 0$. A two-sided sequence can be constructed as above. Since each choice of k, ℓ leads to a strictly ergodic dynamical system, we know that all autocorrelation coefficients exist. Clearly (again with $\bar{1} \,\widehat{=}\, -1$), we have $\eta(0) = 1$, while several possibilities exist to calculate $\eta(\pm 1) = \frac{k+\ell-3}{k+\ell+1}$. In general, we obtain the recursion

$$\eta\bigl((k+\ell)m + r\bigr) = \frac{1}{k+\ell}\bigl(\alpha_{k,\ell,r}\,\eta(m) + \alpha_{k,\ell,k+\ell-r}\,\eta(m+1)\bigr), \tag{16}$$

with $\alpha_{k,\ell,r} = k+\ell-r-2\min(k,\ell,r,k+\ell-r)$, which holds for all $m \in \mathbb{Z}$ and $0 \le r \le k+\ell-1$.

The recursion can once again be used to show the absence of pure point components (by Wiener's criterion) as well as that of absolutely continuous components (by the Riemann-Lebesgue lemma), thus establishing that each sequence in this family leads to a signed Dirac comb with purely singular continuous diffraction. The distribution function satisfies

$$F(x) = \widehat{\gamma}([0,x]) = x + \sum_{m\ge 1} \frac{\eta(m)}{m\pi}\sin(2\pi m x),$$

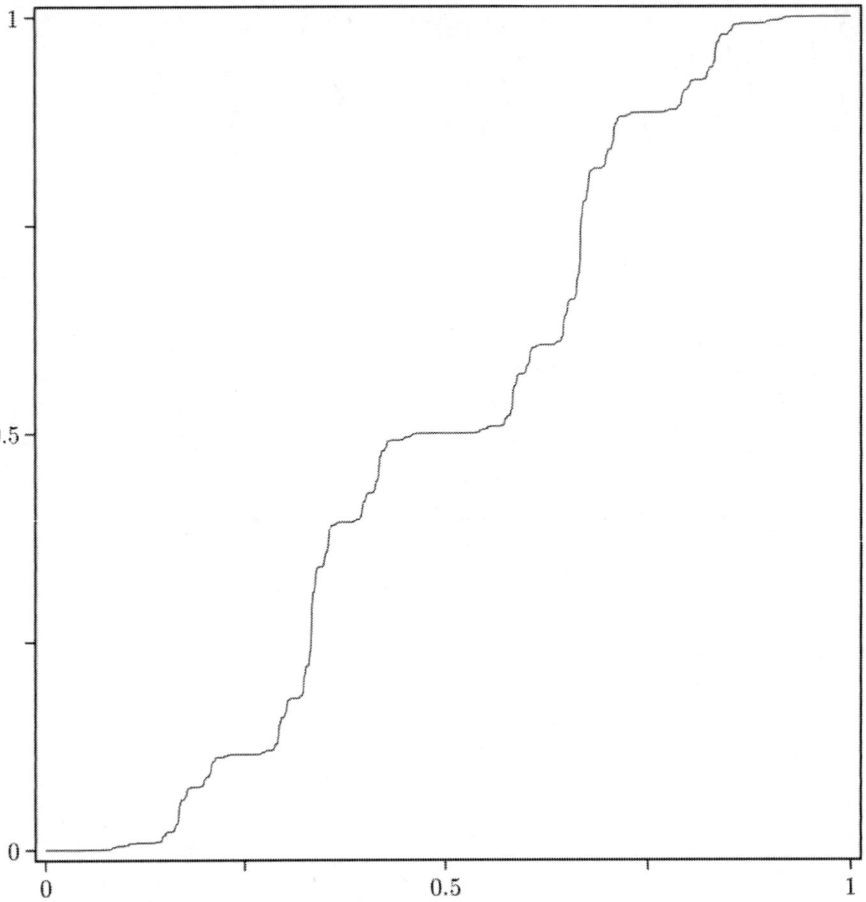

Figure 1. The distribution function of the TM measure on $[0, 1]$.

which is a uniformly converging series. Moreover, the measure $\widehat{\gamma}$ has a (vaguely convergent) representation as an infinite *Riesz product*: With $\theta(x) := 1 + \frac{2}{k+\ell} \sum_{r=1}^{k+\ell-1} \alpha_{k,\ell,r} \cos(2\pi r x)$, it is given by $\prod_{n\geq 0} \theta\big((k+\ell)^n x\big)$. The entire analysis is thus completely analogous to that of the original TM sequence and shows that the latter is a typical example in an infinite family. Further details (and proofs) will be given in [3].

Let us finally mention that the block map (13) applies to any member of this family, and always gives a 2-to-1 cover of the hull that belongs to the generalised period doubling substitution

$$\varrho' \colon\; a \mapsto b^{k-1} a b^{\ell-1} b,\; b \mapsto b^{k-1} a b^{\ell-1} a.$$

Since we always have a coincidence in the sense of Dekking [13], they all define systems with pure point spectrum – another analogy to the classic case $k = \ell = 1$.

4. Absolutely continuous spectra

The simplest example in this category is the Bernoulli (or coin tossing) comb. Let $(W_n)_{n\in\mathbb{Z}}$ be a family of independent and identically distributed (i.i.d.) random variables. We assume that W represents this family and takes values 1 and -1 with probabilities p and $1-p$. We then consider the random measure

$$\omega_{\mathrm{B}} = \sum_{n\in\mathbb{Z}} W_n \, \delta_n \,, \tag{17}$$

which (almost surely in the probabilistic sense) has a natural autocorrelation of the form (2), with coefficients $\eta_{\mathrm{B}}(0) = 1$ and (a.s.) $\eta_{\mathrm{B}}(m) = (2p-1)^2$ for all $0 \neq m \in \mathbb{Z}$, so that we have

$$\gamma_{\mathrm{B}} = (2p-1)^2 \delta_{\mathbb{Z}} + 4p(1-p)\delta_0 \qquad \text{(a.s.)}. \tag{18}$$

In particular, when $p = 1/2$, it gives $\gamma_{\mathrm{B}} = \delta_0$. Eq. (18) can be proved either by an application of Birkhoff's ergodic theorem for $\mathbb{Z}$-action or by the strong law of large numbers (SLLN). We refer to [6] for details and further references. Let us also mention that (18) can be rewritten as $\gamma_{\mathrm{B}} = \beta^2 \delta_{\mathbb{Z}} + (1-\beta^2)\delta_0$, where β has the meaning of the drift velocity of a one-dimensional random walk on the line; compare [11, 8].

As a consequence, we obtain a diffraction measure of mixed type,

$$\widehat{\gamma_{\mathrm{B}}} = (2p-1)^2 \delta_{\mathbb{Z}} + 4p(1-p)\lambda \qquad \text{(a.s.)}, \tag{19}$$

where λ denotes Lebesgue measure. The special case of a fair coin ($p = 1/2$) leads to $\widehat{\gamma_{\mathrm{B}}} = \lambda$, which is purely absolutely continuous. More generally, when W takes the (possibly complex) values h_+ and h_-, each with probability $1/2$, one obtains

$$\widehat{\gamma} = \left|\tfrac{h_+ + h_-}{2}\right|^2 \delta_{\mathbb{Z}} + \left|\tfrac{h_+ - h_-}{2}\right|^2 \lambda,$$

which is another simple example of a mixed spectrum with pure point and absolutely continuous components.

An interesting deterministic counterpart is the *Rudin-Shapiro sequence* [26, 28, 25, 1]. Its quaternary version is usually defined by the substitution

$$\varrho = \varrho_{\mathrm{RS}} \colon \; a \mapsto ac \,, \; b \mapsto dc \,, c \mapsto ab \,, \; d \mapsto db.$$

A fixed point $u = \varrho^2(u)$ can be constructed from the legal seed $b|a$ via iteration of ϱ^2. The *binary* version follows then as the reduction $w = \varphi(u)$ with the mapping φ given by $\varphi(a) = \varphi(c) = 1$ and $\varphi(b) = \varphi(d) = -1$. The two-sided sequence w satisfies $w_{-1} = -1$ and $w_0 = 1$ together with the recursion

$$w(4n+\ell) = \begin{cases} w(n), & \text{for } \ell \in \{0,1\}, \\ (-1)^{n+\ell} w(n), & \text{for } \ell \in \{2,3\}, \end{cases} \tag{20}$$

for all $n \in \mathbb{Z}$. The sequence w and its hull once again define a strictly ergodic dynamical system, so that the natural autocorrelation γ_{RS} exists and is the same for all members of the hull. It is of the general form (2). To calculate it, one needs a little trick. Define the coefficients

$$\left.\begin{matrix} \eta(m) \\ \vartheta(m) \end{matrix}\right\} := \lim_{N\to\infty} \frac{1}{2N+1} \sum_{n=-N}^{N} w(n) w(n+m) \begin{cases} 1 \\ (-1)^n \end{cases} \tag{21}$$

for $m \in \mathbb{Z}$. Note that all these limits exist, by Birkhoff's ergodic theorem (this is immediately clear for $\eta(m)$, but also follows for $\vartheta(m)$ on the basis of the quarternary sequence u, which also gives rise to a strictly ergodic system); compare [24] for an alternative approach.

It is easy to check that $\eta(0) = 1$ and $\vartheta(0) = 0$, while a slightly more involved calculation reveals the following coupled system of recursions,

$$\begin{aligned} \eta(4m) &= \frac{1+(-1)^m}{2}\, \eta(m), \qquad \eta(4m{+}2) \, = \, 0, \\ \eta(4m{+}1) &= \frac{1-(-1)^m}{4}\, \eta(m) + \frac{(-1)^m}{4}\, \vartheta(m) - \frac{1}{4}\, \vartheta(m{+}1), \\ \eta(4m{+}3) &= \frac{1+(-1)^m}{4}\, \eta(m{+}1) - \frac{(-1)^m}{4}\, \vartheta(m) + \frac{1}{4}\vartheta(m{+}1), \end{aligned}$$

together with

$$\begin{aligned}
\vartheta(4m) &= 0, \qquad \vartheta(4m+2) = \frac{(-1)^m}{2}\,\vartheta(m) + \frac{1}{2}\,\vartheta(m+1),\\
\vartheta(4m+1) &= \frac{1-(-1)^m}{4}\,\eta(m) - \frac{(-1)^m}{4}\,\vartheta(m) + \frac{1}{4}\,\vartheta(m+1),\\
\vartheta(4m+3) &= -\frac{1+(-1)^m}{4}\,\eta(m+1) - \frac{(-1)^m}{4}\,\vartheta(m) + \frac{1}{4}\,\vartheta(m+1).
\end{aligned}$$

These equations are valid for all $m \in \mathbb{Z}$. Extracting $\vartheta(1) = \vartheta(-1) = 0$ from the last two equations (with $m = 0$ resp. $m = -1$), one can then recursively conclude that $\vartheta(m) = 0$ for all $m \in \mathbb{Z}$ and $\eta(m) = 0$ for all $m \neq 0$. This shows

$$\gamma_{\mathrm{RS}} = \delta_0 \quad \text{and} \quad \widehat{\gamma_{\mathrm{RS}}} = \lambda. \tag{22}$$

In particular, the (deterministic) diffraction (22) of the binary RS sequence is the same as the (almost sure) diffraction of the coin tossing sequence (19) for $p = 1/2$, despite the fact that their entropies are different (0 versus $\log(2)$); see [19, 6] for further details.

The situation is actually 'worse' in the following sense. Let $S \in \{\pm 1\}^{\mathbb{Z}}$ be a two-sided sequence, assumed ergodic, with associated Dirac comb $\omega_S = \sum_{n\in\mathbb{Z}} S_n \delta_n$ and autocorrelation γ_S. The latter exists due to the ergodicity assumption. Let $(W_n)_{n\in\mathbb{Z}}$ be the above family of i.i.d. random variables with values in $\{\pm 1\}$. Consider now the new random Dirac comb

$$\omega = \sum_{n\in\mathbb{Z}} S_n W_n \,\delta_n, \tag{23}$$

which we call the *Bernoullisation* of the original sequence. It can be considered as a 'randomisation via second thoughts', since the action of the W_n is to either keep S_n (with probability p) or to change its sign (with probability $1 - p$). By another (slightly more complicated) application of the SLLN, one finds [6]

$$\gamma = (2p-1)^2 \gamma_S + 4p(1-p)\delta_0 \qquad \text{(a.s.)} \tag{24}$$

together with the corresponding almost sure diffraction $\widehat{\gamma} = (2p-1)^2\,\widehat{\gamma_S} + 4p(1-p)\lambda$. If S is the binary RS sequence from above, one finds $\gamma = \delta_0$ and $\widehat{\gamma} = \lambda$, *independently* of the parameter p. This proves that the entire family defined by (23) for the RS sequence is homometric, with the entropy varying continuously between 0 and $\log(2)$; compare [7] for the connection between diffraction and entropy in the pure point case.

5. Outlook

The diffraction measure is a useful tool, both for the understanding of experiments and for various theoretical questions. Beyond the well-studied case of pure point spectra, also continuous spectra are explicitly accessible and sometimes (as above) even computable. However, as was to be expected, various aspects are more involved, and this includes the inverse problem. The latter is increasing in complexity also with growing dimension. For instance, the Ledrappier system [22] in the plane has the same autocorrelation as the Bernoulli comb (both living on $\mathbb{Z}^2$), but has rank-1 entropy despite being genuinely two-dimensional.

Various generalisations exist to non-lattice systems, which are best formulated via the theory of point processes [2]. Here, the classic Poisson process (of mean point density 1) is another example (with random positions) that gives diffraction measure λ when the points are randomly weighted with 1 and -1. It is clear that higher order correlations can tell these systems apart, but many questions are still open.

Finally, more realistic models of randomness have to include interactions. The theory of Gibbs measures is a necessary and useful tool here, but only first steps in this direction have been taken; see [2] and references therein. Our present day understanding is still rather limited.

Acknowledgments
It is a pleasure to thank Shelomo I. Ben-Abraham, Daniel Lenz, Robert V. Moody and Tom Ward for various discussions. This work was supported by the German Research Council (DFG), within the CRC 701, by EPSRC, via Grant EP/D058465, and by a Leverhulme Trust Visiting Professorship Grant.

References
[1] Allouche J-P and Shallit J 2003 *Automatic Sequences: Theory, Applications, Generalizations* (Cambridge: Cambridge University Press)
[2] Baake M, Birkner M and Moody R V 2008 Diffraction of stochastic point sets: Explicitly computable examples *Preprint* arXiv:0803.1266
[3] Baake M, Blomenkamp T, Gähler F and Grimm U 2009 A family of generalised Morse sequences with purely singular continuous diffraction (*in preparation*)
[4] Baake M and Grimm U 2007 Homometric model sets and window covariograms *Z. Krist.* **222** 54–58
[5] Baake M and Grimm U 2008 The singular continuous diffraction measure of the Thue-Morse chain *J. Phys. A: Math. Theor.* **41** 422001
[6] Baake M and Grimm U 2009 Kinematic diffraction is insufficient to distinguish order from disorder *Phys. Rev. B* **79** 020203(R) and **80** 029903(E)
[7] Baake M, Lenz D and Richard C 2007 Pure point diffraction implies zero entropy for Delone sets with uniform cluster frequencies *Lett. Math. Phys.* **82** 61–77
[8] Baake M and Löwe M 2004 Comment on: Curious properties of simple random walks *J. Stat. Phys.* **116** 1449–1451
[9] Baake M and Moody R V 2004 Weighted Dirac combs with pure point diffraction *J. reine angew. Math. (Crelle)* **573** 61–94
[10] Baake M, Moody R V and Schlottmann M 1998 Limit-(quasi)periodic point sets as quasicrystals with p-adic internal spaces *J. Phys. A: Math. Gen.* **31** 5755–5765
[11] Ben-Abraham S I 1993 Curious properties of simple random walks *J. Stat. Phys.* **73** 441–445
[12] Cowley J M 1995 *Diffraction Physics* 3rd edition (Amsterdam: North-Holland)
[13] Dekking F M 1978 The spectrum of dynamical systems arising from substitutions of constant length *Z. Wahrscheinlichkeitsth. verw. Geb.* **41** 221–239
[14] Deng X and Moody R V 2009 How model sets can be determined by their two-point and three-point correlations *J. Stat. Phys.* (in press) (*Preprint* arXiv:0901.4381)
[15] Deng X and Moody R V 2009 Weighted model sets and their higher point-correlations *Preprint* arXiv:0904.4552
[16] Grimm U and Baake M 2008 Homometric point sets and inverse problems *Z. Krist.* **223** 777–781
[17] Grünbaum F A and Moore C C 1995 The use of higher-order invariants in the determination of generalized Patterson cyclotomic sets *Acta Cryst. A* **51** 310–323
[18] Hof A 1995 On diffraction by aperiodic structures *Commun. Math. Phys.* **169** 25–43
[19] Höffe M and Baake M 2000 Surprises in diffuse scattering *Z. Krist.* **215** 441–444
[20] Kakutani S 1972 Strictly ergodic symbolic dynamical systems *Proc. 6th Berkeley Symposium on Math. Statistics and Probability* ed L M LeCam L M, J Neyman and E L Scott (Berkeley: Univ. of California Press) pp 319–326
[21] Keane M 1968 Generalized Morse sequences *Z. Wahrscheinlichkeitsth. Verw. Geb.* **10** 335–353
[22] Ledrappier F 1978 Un champ markovien peut être d'entropie nulle et mélangeant *C. R. Acad. Sci. Paris Sér. A-B* **287** A561–A563
[23] Mahler K 1927 The spectrum of an array and its application to the study of the translation properties of a simple class of arithmetical functions. Part II: On the translation properties of a simple class of arithmetical functions *J. Math. Massachusetts* **6** 158–163
[24] Pytheas Fogg N 2002 *Substitutions in Dynamics, Arithmetics and Combinatorics* LNM 1794 (Berlin: Springer)
[25] Queffélec M 1987 *Substitution Dynamical Systems – Spectral Analysis* LNM 1294 (Berlin: Springer)
[26] Rudin W 1959 Some theorems on Fourier coefficients *Proc. AMS* **10** 855–859
[27] Schlottmann M 2000 Generalised model sets and dynamical systems *Directions in Mathematical Quasicrystals* CRM Monograph Series vol. 13, ed M Baake and R V Moody (Rhode Island: AMS) pp 143–159
[28] Shapiro H S 1951 *Extremal Problems for Polynomials and Power Series* Masters Thesis (Boston: MIT)
[29] Walters P 2000 *An Introduction to Ergodic Theory* reprint (New York: Springer)
[30] Wiener N 1927 The spectrum of an array and its application to the study of the translation properties of a simple class of arithmetical functions. Part I: The spectrum of an array *J. Math. Massachusetts* **6** 145–157

CSLs of the root lattice A_4

M Heuer[1] and P Zeiner[2]

[1] Department of Mathematics and Statistics, The Open University, Milton Keynes, UK
[2] Fakultät für Mathematik, Universität Bielefeld, Germany

E-mail: m.heuer@open.ac.uk, pzeiner@math.uni-bielefeld.de

Abstract. Recently, the group of coincidence isometries of the root lattice A_4 has been determined providing a classification of these isometries with respect to their coincidence indices. A more difficult task is the classification of all CSLs, since different coincidence isometries may generate the same CSL. In contrast to the typical examples in dimensions $d \leq 3$, where coincidence isometries generating the same CSL can only differ by a symmetry operation, the situation is more involved in 4 dimensions. Here, we find coincidence isometries that are not related by a symmetry operation but nevertheless give rise to the same CSL. We indicate how the classification of CSLs can be obtained by making use of the icosian ring and provide explicit criteria for two isometries to generate the same CSL. Moreover, we determine the number of CSLs of a given index and encapsulate the result in a Dirichlet series generating function.

1. Introduction

In crystallography, CSLs are a well established tool for the description and classification of grain boundaries, see [1] and references given there. CSLs have been studied for several decades in dimensions $d \leq 3$. In higher dimensions they became interesting after the discovery of quasicrystals. Here, we discuss a 4-dimensional lattice, namely the root lattice A_4, which is intimately related to quasicrystals with fivefold symmetry and the Penrose tiling in particular.

Let us recall some basic notions first. A *lattice* Γ in the Euclidean space $\mathbb{R}^d$, can be described as the set of all integer linear combinations of a basis $\{b_1, \ldots, b_d\}$ of $\mathbb{R}^d$, which is denoted by $\Gamma = \langle b_1, \ldots, b_d \rangle_{\mathbb{Z}}$. A *coincidence site lattice* (CSL) of a lattice Γ is defined as $\Gamma \cap R\Gamma$, where R is a linear isometry and the corresponding *coincidence index* $\Sigma(R) = [\Gamma : (\Gamma \cap R\Gamma)]$ is finite, i.e. R and $R\Gamma$ share a common sublattice. The orthogonal group, i.e. the group of all linear isometries of $\mathbb{R}^d$, is denoted by $\mathrm{O}(d, \mathbb{R})$. We define the set of all *coincidence isometries* as

$$\mathrm{OC}(\Gamma) := \{R \in \mathrm{O}(d, \mathbb{R}) \mid \Sigma(R) < \infty\}$$

and its restriction to rotations as $\mathrm{SOC}(\Gamma)$. According to [1–3], $\mathrm{OC}(\Gamma)$ and $\mathrm{SOC}(\Gamma)$ are not only subgroups of $\mathrm{O}(d, \mathbb{R})$, but subgroups of the group of all similarity isometries which is defined as $\mathrm{OS}(\Gamma) := \{R \in \mathrm{O}(d, \mathbb{R}) \mid \alpha R\Gamma \subset \Gamma \text{ for some } \alpha \in \mathbb{R}^+\}$. A sublattice of Γ of the form $\alpha R\Gamma$ is called a *similar sublattice* (SSL) of Γ. If we denote the *denominator* of a matrix $R \in \mathrm{OS}(\Gamma)$ relative to the lattice Γ as $\mathrm{den}_\Gamma(R) := \min\{\alpha \in \mathbb{R}^+ \mid \alpha R\Gamma \subset \Gamma\}$, we have the following characterisation, see [2] for details:

$$\mathrm{OC}(\Gamma) = \{R \in \mathrm{OS}(\Gamma) \mid \mathrm{den}_\Gamma(R) \in \mathbb{N}\} \tag{1}$$

In the following sections we first describe the root lattice A_4 in a suitable setting. Then, we briefly introduce the key objects required for the analysis of its CSLs, as well as their relations to each other. We continue by reviewing some parts of the classification of the coincidence rotations of the root lattice A_4, see [2; 4] for details, and extend this analysis by providing explicit criteria for two coincidence isometries to generate the same CSL. Finally, we classify the CSLs of the root lattice A_4 with respect to their coincidence indices and encapsulate the result in a Dirichlet series generating function which gives the number of CSLs for every index.

2. The root lattice A_4

The root lattice A_4 is usually defined in a 4-dimensional hyperplane of $\mathbb{R}^5$ as follows:

$$A_4 := \{(x_1, \ldots, x_5) \in \mathbb{Z}^5 \mid x_1 + \ldots + x_5 = 0\}$$

However, this realisation of A_4 in $\mathbb{R}^5$ is not suitable for our purposes. We prefer the following realisation in $\mathbb{R}^4$

$$L := \left\langle (1,0,0,0), \tfrac{1}{2}(-1,1,1,1), (0,-1,0,0), \tfrac{1}{2}(0,1,\tau-1,-\tau) \right\rangle_{\mathbb{Z}}$$

where $\tau = (1+\sqrt{5})/2$ is the golden ratio; see [2; 5; 6] for details. Note that L is just rescaled by a factor $\frac{1}{\sqrt{2}}$ in comparison to A_4. This particular description L is very convenient for our problem, as it enables us to use the arithmetic of the quaternion algebra $\mathbb{H}(\mathbb{Q}(\sqrt{5}\,))$; see [7] for a detailed introduction to Hamilton's quaternions. To simplify the notation we define

$$K := \mathbb{Q}(\sqrt{5}\,) = \{r + s\sqrt{5} \mid r, s \in \mathbb{Q}\},$$

which is a quadratic number field. The algebra $\mathbb{H}(K)$ is explicitly given as

$$\mathbb{H}(K) = K \oplus \mathrm{i}K \oplus \mathrm{j}K \oplus \mathrm{k}K,$$

where the generating elements satisfy Hamilton's relations $\mathrm{i}^2 = \mathrm{j}^2 = \mathrm{k}^2 = \mathrm{ijk} = -1$. It is equipped with a *conjugation* $\bar{.}$ which is the unique mapping that fixes the elements of the centre of the algebra K and reverses the sign on its complement. If we write

$$q = (a,b,c,d) = a + \mathrm{i}b + \mathrm{j}c + \mathrm{k}d, \quad \text{this means} \quad \bar{q} = (a,-b,-c,-d).$$

The reduced norm and trace in $\mathbb{H}(K)$ are defined by

$$\mathrm{nr}(q) := q\bar{q} = |q|^2 \quad \text{and} \quad \mathrm{tr}(q) := q + \bar{q},$$

where we canonically identify an element $\alpha \in K$ with the quaternion $(\alpha, 0, 0, 0)$. For any $q \in \mathbb{H}(K)$, $|q|$ is its Euclidean length, which need not be an element of K. Nevertheless, one has $|rs| = |r||s|$ for arbitrary $r, s \in \mathbb{H}(K)$. Due to the geometric meaning, we use the notations $|q|^2$ and $\mathrm{nr}(q)$ in parallel. An element $q \in \mathbb{H}(K)$ is called *integral* when both $\mathrm{nr}(q)$ and $\mathrm{tr}(q)$ are elements of

$$\mathcal{O} := \mathbb{Z}[\tau] := \{m + n\tau \mid m, n \in \mathbb{Z}\},$$

which is the ring of integers of the quadratic number field K.

The *icosian ring* $\mathbb{I}$ consists of all linear combinations of the vectors

$$(1,0,0,0),(0,1,0,0),\tfrac{1}{2}(1,1,1,1),\tfrac{1}{2}(1-\tau,\tau,0,1)$$

with coefficients in $\mathcal{O}$. The elements of $\mathbb{I}$ are called *icosians*. For more information on this remarkable object, see for example [6; 8–10] and references given there.

An element $p \in \mathbb{I}$ is called $\mathbb{I}$*-primitive* when $\alpha p \in \mathbb{I}$, with $\alpha \in K$, is only possible with $\alpha \in \mathcal{O}$. Similarly, a sublattice Λ of L is called L*-primitive* when $\alpha\Lambda \subset L$, with $\alpha \in \mathbb{Q}$, implies $\alpha \in \mathbb{Z}$. Whenever the context is clear, we simply use the term "primitive" in both cases.

The detailed arithmetic structure of $\mathbb{I}$ plays a key role in the characterisation of the coincidence rotations for L; see [2]. Another important object for this characterisation is the following map, called *twist map*. If $q = (a,b,c,d)$, it is defined by the mapping

$$q \mapsto \widetilde{q} := (a', b', d', c'),$$

where $'$ denotes the algebraic conjugation in K, as defined by the mapping $\sqrt{5} \mapsto -\sqrt{5}$. The algebraic conjugation in K is also needed to define the absolute norm on K, via $\mathrm{N}(\alpha) := |\alpha\alpha'|$. For the various properties of the twist map, e.g. $\widetilde{\widetilde{q}} = q$ and $\widetilde{pq} = \widetilde{q}\widetilde{p}$ where $p, q \in \mathbb{I}$, we refer the reader to [6] and references therein. The most important properties in our present context are summarised in the following Lemma, see [6, Proposition 1] and [2, Lemma 4], which describes the relations between L and $\mathbb{I}$.

Lemma 1. *Within* $\mathbb{H}(K)$, *one has* $\widetilde{\mathbb{I}} = \mathbb{I}$ *and*

$$L = \{x \in \mathbb{I} \mid \widetilde{x} = x\} = \{x + \widetilde{x} \mid x \in \mathbb{I}\} = \phi_{+}(\mathbb{I}),$$

where the $\mathbb{Q}$*-linear mapping* $\phi_{+}\colon \mathbb{H}(K) \longrightarrow \mathbb{H}(K)$, *is defined by* $\phi_{+}(x) = x + \widetilde{x}$.

The dual A_4^* of the the root lattice A_4, here in form of the dual of the lattice L, is given by

$$L^* := \{x \in \mathbb{R}^4 \mid \langle x|y\rangle \in \mathbb{Z} \text{ for all } y \in L\}.$$

3. CSLs

The investigation of CSLs of L can be restricted to rotations only, since $\overline{L} = L$, which means that any orientation reversing operation can be obtained from an orientation preserving one after applying conjugation first. Let $\mathrm{SOS}(L)$ be the group of all similarity rotations of the lattice L. Considering Eq. (1), it is obvious how $\mathrm{SOC}(L)$ and $\mathrm{SOS}(L)$ are related in general. By [6, Corollary 1] we know that any similarity rotation of L is of the form $x \mapsto \frac{1}{|q\widetilde{q}|} qx\widetilde{q}$ with $q \in \mathbb{I}$, which we refer to as $R(q)$, or more precisely we define $R(q)x := \frac{1}{|q\widetilde{q}|} qx\widetilde{q}$. We denote the denominator of $R(q)$ by $\mathrm{den}(q)$. Among the elements of $\mathrm{SOS}(L)$, we can identify the elements of $\mathrm{SOC}(L)$ as follows; see [2] for the proof.

Proposition 2. *Let* $0 \neq q \in \mathbb{I}$ *be an arbitrary icosian. Then, the lattice* $\frac{1}{|q\widetilde{q}|} qL\widetilde{q} \cap L$ *is a CSL of* L *if and only if* $|q\widetilde{q}| \in \mathbb{N}$. *If* q *is primitive, then* $\mathrm{den}(q) = |q\widetilde{q}|$.

We call an icosian $q \in \mathbb{I}$ *admissible* when $|q\widetilde{q}| \in \mathbb{N}$. As $\mathrm{nr}(\widetilde{q}) = \mathrm{nr}(q)'$, the admissibility of q implies that $\mathrm{N}\big(\mathrm{nr}(q)\big) = |q\widetilde{q}|^2$ is a square in $\mathbb{N}$. An immediate consequence of Proposition 2 is that

$$\mathrm{SOC}(L) = \{R(p) \mid p \in \mathbb{I} \text{ is primitive and admissible }\}, \tag{2}$$

which delivers the following classification.

Theorem 3. *The CSLs of L are precisely the lattices of the form $L \cap \frac{1}{|q\tilde{q}|} qL\tilde{q}$ with $q \in \mathbb{I}$ primitive and admissible.*

This is the first step to connect certain primitive right ideals $p\mathbb{I}$ of the icosian ring with the CSLs of L. Before we continue in this direction, let us consider the relation with the coincidence rotations, see [2, Lemma 5].

Lemma 4. *Let $r, s \in \mathbb{I}$ be primitive and admissible quaternions, with $r\mathbb{I} = s\mathbb{I}$. Then, one has $L \cap \frac{rL\tilde{r}}{|r\tilde{r}|} = L \cap \frac{sL\tilde{s}}{|s\tilde{s}|}$.*

The converse statement to Lemma 4 is not true, as the equality of two CSLs does *not* imply the corresponding ideals to be equal. An example is provided by $r = (\tau, 2\tau, 0, 0)$ and $s = (\tau^2, \tau, \tau, 1)$, which define the same CSL, though $s^{-1}r$ is not a unit in $\mathbb{I}$. The CSL is spanned by the basis $\{(1, 2, 0, 0), (2, -1, 0, 0), (\frac{3}{2}, \frac{1}{2}, \frac{1}{2}, \frac{1}{2}), (-1, \frac{1}{2}, \frac{\tau-1}{2}, -\frac{\tau}{2})\}$. Note that, as a direct consequence of [6, Lemma 5], two primitive quaternions $r, s \in \mathbb{I}$ are related by a rotation symmetry of L if and only if $r\mathbb{I} = s\mathbb{I}$.

For our further discussion we need to replace the primitive and admissible icosian p, and with it $\tilde{p}$, by certain $\mathcal{O}$-multiples, such that their norms have the same prime divisors in $\mathcal{O}$. In view of the form of the rotation $x \mapsto \frac{1}{|p\tilde{p}|} px\tilde{p}$, this is actually rather natural because it restores some kind of symmetry of the expressions in relation to the two quaternions involved. For a primitive and admissible icosian $p \in \mathbb{I}$ we choose explicitly

$$\alpha_q = \sqrt{\frac{\mathrm{lcm}(\mathrm{nr}(q), \mathrm{nr}(\tilde{q}))}{\mathrm{nr}(q)}} \in \mathcal{O}, \tag{3}$$

where we assume a suitable standardisation for the lcm (least common multiple) of two elements of $\mathcal{O}$; see [2] for details. Moreover, we have the relation $\alpha_{\tilde{q}} = \widetilde{\alpha_q} = \alpha_q'$. The icosian $\alpha_q q$ is called the *extension* of the primitive admissible element $q \in \mathbb{I}$, and $(\alpha_q q, \alpha_q' \tilde{q})$ the corresponding *extension pair*. Since α_q and α_q' are central, the extension does not change the rotation, i.e.

$$\frac{qx\tilde{q}}{|q\tilde{q}|} = \frac{q_\alpha x \widetilde{q_\alpha}}{|q_\alpha \widetilde{q_\alpha}|}$$

holds for all quaternions x. Note that the definition of the extension pair is unique up to units of $\mathcal{O}$, and that one has the relation

$$\mathrm{nr}(q_\alpha) = \mathrm{lcm}\big(\mathrm{nr}(q), \mathrm{nr}(\tilde{q})\big) = \mathrm{nr}(\widetilde{q_\alpha}) = |q_\alpha \widetilde{q_\alpha}| \in \mathbb{N}.$$

Following [2] we define the set

$$L(q) = \{qx + \widetilde{x}\widetilde{q} \mid x \in \mathbb{I}\} = \phi_+(q\mathbb{I}).$$

Note that, due to $\widetilde{\mathbb{I}} = \mathbb{I}$, one has $L(q) = \widetilde{L(q)}$. This leads to the following parametrisation of the CSLs of L, compare [2, Theorem 2].

Theorem 5. *Let $q \in \mathbb{I}$ be admissible and primitive, and let $q_\alpha = \alpha_q q$ be its extension. Then, the CSL defined by q is given by*

$$L \cap \frac{1}{|q\tilde{q}|} qL\tilde{q} = L(q_\alpha) = (q_\alpha \mathbb{I} + \mathbb{I}\widetilde{q_\alpha}) \cap L.$$

and its index is $\Sigma_L(q) = \mathrm{nr}(q_\alpha) = \mathrm{lcm}(\mathrm{nr}(q), \mathrm{nr}(q)')$.

4. The Number of CSLs of each Index

Until now we have followed the argumentation in [2], which lead to the classification of all coincidence rotations of the lattice L. In order to obtain a classification of the actual CSLs, we make use of the corresponding classification for the icosian ring $\mathbb{I}$ and its submodul $L[\tau]$, which can either be seen as $\mathcal{O}$-modules of rank 4 or as lattices in $\mathbb{R}^8$, so all our definitions of the introductory section apply. Since any coincidence rotation of L is a coincidence rotation of $L[\tau]$ and $\mathbb{I}$, we can start our analysis from the classification of the CSLs of $\mathbb{I}$, which directly gives a first sufficient condition when two CSLs of L are equal, see [11] for details.

Lemma 6. *Let $p_1, p_2 \in \mathbb{I}$ be primitive and admissible, such that*

$$|p_1|^2 = |p_2|^2 \quad \textit{and} \quad \mathrm{glcd}(p_1, |p_1\widetilde{p_1}|) = \mathrm{glcd}(p_2, |p_2\widetilde{p_2}|).$$

Then $L \cap \frac{1}{|p_1\widetilde{p_1}|} p_1 L \widetilde{p_1} = L \cap \frac{1}{|p_2\widetilde{p_2}|} p_2 L \widetilde{p_2}$.

In fact, the converse holds only if $|p_1|^2 = |p_2|^2$ is not divisible by the splitting prime $p = 5$. For the general situation we have the following

Theorem 7. *Let $p_1, p_2 \in \mathbb{I}$ be primitive and admissible. Then*

$$L \cap \frac{1}{|p_1\widetilde{p_1}|} p_1 L \widetilde{p_1} = L \cap \frac{1}{|p_2\widetilde{p_2}|} p_2 L \widetilde{p_2}$$

if and only if

$$|p_1|^2 = |p_2|^2 \quad \textit{and} \quad \mathrm{glcd}(p_1, \tfrac{|p_1\widetilde{p_1}|}{c}) = \mathrm{glcd}(p_2, \tfrac{|p_2\widetilde{p_2}|}{c}), \tag{4}$$

where $c = \sqrt{5}$, if $|p_1|^2 = |p_2|^2$ is divisible by 5 and $c = 1$ otherwise.

Note that $|p_1|^2 = |p_2|^2$ directly implies that the coincidence indices and denominators coincide, which is clearly necessary for the CSLs to be equal. The second condition of (4) tells us how much p_1 and p_2 are allowed to differ in terms of their decomposition in irreducible elements of $\mathbb{I}$. If $|p_1|^2 = |p_2|^2$ is not divisible by 5 and $\mathrm{nr}(p_1) = \mathrm{nr}(p_1)'$, then (and only then) the second condition is equivalent to $p_1 = p_2\epsilon$ where ϵ is a unit of $\mathbb{I}$, which means that R_1 and R_2 are symmetry related. Hence, in this special case, and only then, we can deduce from two equal CSLs that they are generated by coincidence rotations which are symmetry related, whereas in all other cases there are at least two rotations that are not symmetry related but give the same CSL.

Now, we can conclude with the derivation of the Dirichlet series for the number of CSLs of L of a given index, see [12] for a introduction to Dirichlet series. Note, that by [1, Lemma 2.5, Theorem 2.2] the number of CSLs of a given index n is the same for the lattices A_4 and L as well as their dual lattice L^*. Let $f(n)$ be the number of CSLs of index n. Then, $f(n)$ is a multiplicative arithmetic function, given at prime powers p^r with $r \geq 1$ by

$$f(p^r) = \begin{cases} 6 \cdot 5^{2r-2} & \text{for } p = 5 \\ (p^2+1)p^{2r-2} & \text{for } p \equiv \pm 2 \pmod 5 \\ \frac{(p+1)^2}{p^3-1}\left(p^{2r+1} + p^{2r-2} - 2p^{\frac{(r-1)}{2}}\right) & \text{for } p \equiv \pm 1 \pmod 5, r \text{ odd}, \\ \frac{(p+1)^2}{p^3-1}\left(p^{2r+1} + p^{2r-2} - 2\frac{(p^2+1)}{p+1}p^{\frac{(r-2)}{2}}\right) & \text{for } p \equiv \pm 1 \pmod 5, r \text{ even}. \end{cases}$$

Finally elementary calculations involving geometric series yield its Dirichlet series generating function compare [11; 13]:

$$D(s) = \sum_{n=1}^{\infty} \frac{f(n)}{n^s} = \Bigl(1 + 6\tfrac{5^{-s}}{1-5^{2-s}}\Bigr) \prod_{p \equiv \pm 1\,(5)} \tfrac{1+p^{-s}+2p^{1-s}+2p^{-2s}+p^{1-2s}+p^{1-3s}}{(1-p^{2-s})\,(1-p^{1-2s})} \prod_{p \equiv \pm 2\,(5)} \tfrac{1+p^{-s}}{1-p^{2-s}}$$
$$= 1 + \tfrac{5}{2^s} + \tfrac{10}{3^s} + \tfrac{20}{4^s} + \tfrac{6}{5^s} + \tfrac{50}{6^s} + \tfrac{50}{7^s} + \tfrac{80}{8^s} + \tfrac{90}{9^s} + \tfrac{30}{10^s} + \tfrac{144}{11^s} + \dots$$

Acknowledgments

Acknowledgements It is a pleasure to thank M. Baake and U. Grimm for their suggestions and helpful discussions. This work was supported by EPSRC grant EP/D058465/1 and the German Research Council (DFG), within the CRC 701.

References

[1] Baake M 1997 *The mathematics of long-range aperiodic order (Waterloo, ON, 1995)* (*NATO Adv. Sci. Inst. Ser. C Math. Phys. Sci.* vol 489) (Dordrecht: Kluwer) pp 9–44 `arXiv:math/0605222v1 [math.MG]`

[2] Baake M, Grimm U, Heuer M and Zeiner P 2008 *European J. Combin.* **29** 1808–1819 `arXiv:0709.1341v1 [math.MG]`

[3] Glied S and Baake M 2008 *Z. Krist.* **223** 770–772 `arXiv:0808.0109v1 [math.MG]`

[4] Heuer M 2008 *Z. Krist.* **223** 817–821 `arXiv:0808.1228v1 [math.MG]`

[5] Chen L, Moody R V and Patera J 1998 *Quasicrystals and discrete geometry (Toronto, ON, 1995)* (*Fields Inst. Monogr.* vol 10) (Providence, RI: Amer. Math. Soc.) pp 135–178

[6] Baake M, Heuer M and Moody R V 2008 *J. Algebra* **320** 1391–1408 `arXiv:math/0702448v2 [math.MG]`

[7] Koecher M and Remmert R 1991 *H.-D. Ebbinghaus, et al (Eds.), Numbers* (New York: Springer) pp 189–220

[8] Reiner I 2003 *Maximal orders* (*London Mathematical Society Monographs. New Series* vol 28) (Oxford: The Clarendon Press Oxford University Press) corrected reprint of the 1975 original

[9] Moody R V and Patera J 1993 *J. Phys. A* **26** 2829–2853

[10] Moody R V and Weiss A 1994 *J. Number Theory* **47** 405–412

[11] Heuer M 2010 Combinatorial aspects of root lattices and words Ph.D. Thesis in preparation

[12] Apostol T M 1998 *Introduction to analytic number theory* (New York: Springer-Verlag) 5th corr. printing

[13] Baake M and Zeiner P 2008 *Phil. Mag.* **88** 2025–2032 `arXiv:0712.0363v1 [math.MG]`

Multiplicativity in the theory of coincidence site lattices

Peter Zeiner

Fakultät für Mathematik, Universität Bielefeld, Bielefeld, Germany

E-mail: pzeiner@math.uni-bielefeld.de

Abstract. Coincidence Site Lattices (CSLs) are a well established tool in the theory of grain boundaries. For several lattices up to dimension $d = 4$, the CSLs are known explicitly as well as their indices and multiplicity functions. Many of them share a particular property: their multiplicity functions are multiplicative. We show how multiplicativity is connected to certain decompositions of CSLs and the corresponding coincidence rotations and present some criteria for multiplicativity. In general, however, multiplicativity is violated, while supermultiplicativity still holds.

1. Introduction

In crystallography CSLs have been used for several decades to classify and describe grain boundaries, see e.g. [1–4] and references therein. In the beginning research concentrated on lattices in dimensions $d \leq 3$, but since the discovery of quasicrystals also lattices in higher dimensions and $\mathbb{Z}$-modules have been studied, see e.g. [4–6]. Here we discuss a special aspect, namely the multiplicativity of certain combinatorial functions associated with CSLs.

Let us recall some basic concepts. Let $\Gamma \subseteq \mathbb{R}^d$ be a d-dimensional lattice and $R \in \mathrm{O}(d)$ a linear isometry. Then, R is called a (linear) *coincidence isometry* of Γ if $\Gamma(R) := \Gamma \cap R\Gamma$ is a lattice of finite index in Γ, and $\Gamma(R)$ is called an (ordinary or simple) *coincidence site lattice (CSL)* (for an introduction, see [4]). The group of all coincidence isometries of Γ is denoted by $OC(\Gamma)$, whereas the subgroup of all orientation preserving isometries is called the group of coincidence rotations and referred to as $SOC(\Gamma)$. The *coincidence index* $\Sigma(R) := [\Gamma : \Gamma(R)]$ is defined as the (group theoretical) index of $\Gamma(R)$ in Γ and is just the ratio of the volume of the corresponding unit cells.

These concepts can be generalized to include the possibility of multiple CSLs, see [7–9]. In particular, the lattice

$$\Gamma(R_1, \ldots, R_n) := \Gamma \cap R_1\Gamma \cap \ldots \cap R_n\Gamma = \Gamma(R_1) \cap \ldots \cap \Gamma(R_n) \tag{1}$$

is called a *multiple CSL (MCSL)* of order n, where R_i, $i \in \{1, \ldots n\}$, are coincidence isometries of Γ. Its index in Γ is denoted by $\Sigma(R_1, \ldots, R_n)$.

There arise several interesting combinatorial questions in this context. In particular, one is interested in the number of coincidence isometries, coincidence rotations and the number of CSLs of a given index m, which we denote by $|\mathcal{P}|f^{iso}(m)$, $|\mathcal{P}'|f^{rot}(m)$ and $f(m)$, respectively. Here $|\mathcal{P}|$ is the order of the point group $\mathcal{P}$ of Γ, and $\mathcal{P}' \subseteq \mathcal{P}$ is the (normal) subgroup of orientation preserving symmetry operations. These factors have been chosen to guarantee that $f^{iso}(m)$ and $f^{rot}(m)$ are normalized such that $f^{iso}(1) = 1$ and $f^{rot}(1) = 1$, respectively. In general we have $f^{iso}(m) \geq f(m)$, although $f^{iso}(m) = f^{rot}(m) = f(m)$ holds for the most important examples in $d = 2, 3$ like the square lattice and the cubic lattices. In $d \geq 4$ there exist several examples where $f^{iso}(m) = f^{rot}(m) > f(m)$ for infinitely many indices m, see e.g. [10, 11] for examples.

In many cases the multiplicity functions $f^{iso}(m)$, $f^{rot}(m)$ and $f(m)$ turn out to be multiplicative functions. Recall that a function $f : \mathbb{N} \to \mathbb{R}$ is called multiplicative if $f(mn) = f(m)f(n)$ holds whenever m and n are coprime. We call f supermultiplicative if the inequality $f(mn) \geq f(m)f(n)$ holds for m and n coprime. As an example we mention the square lattice [4], where

$$f^{iso}(m) = f^{rot}(m) = f(m) = \begin{cases} 1 & \text{for } m = 1 \\ 2^r & \text{if all prime factors } p \text{ of } m \text{ satisfy } p \equiv 1 \pmod 4 \\ & \text{and } r \text{ is the number of distinct prime factors} \\ 0 & \text{otherwise.} \end{cases} \tag{2}$$

Multiplicativity suggests to use a Dirichlet series as generating function, which reads in this case [4]

$$\begin{aligned} \Phi(s) &= \sum_{m=1}^{\infty} \frac{f(m)}{m^s} = \prod_{p \equiv 1(4)} \frac{1+p^{-s}}{1-p^{-s}} \\ &= 1 + \frac{2}{5^s} + \frac{2}{13^s} + \frac{2}{17^s} + \frac{2}{25^s} + \frac{2}{29^s} + \frac{2}{37^s} + \frac{2}{41^s} + \frac{2}{53^s} + \frac{2}{61^s} + \frac{4}{65^s} + \frac{2}{73^s} + \dots \end{aligned} \tag{3}$$

Note that it is the multiplicativity of $f(m)$ that guarantees that the product expansion exists. In case of the square lattice, multiplicativity is a consequence of the fact that $\mathbb{Z}[i]$ is a principal ideal domain. But multiplicativity also holds for the cubic lattices in 3 dimensions as well as for the A_4 root lattice and the hypercubic lattices in 4 dimensions [4, 10, 12]. In the latter cases multiplicativity is due to the unique prime factorization in certain quaternion algebras.

All these cases are quite special in the sense that they are related to algebras that allow a unique prime factorization. In fact, there are examples where $f(m)$ and $f^{iso}(m)$ are not multiplicative, e.g. for $\Gamma = 2\mathbb{Z} \times 3\mathbb{Z}$. This raises several questions: when are the multiplicity functions multiplicative, are there criteria for multiplicativity? Does the multiplicativity of $f(m)$ imply the multiplicativity of $f^{iso}(m)$ or $f^{rot}(m)$ or vice versa? Is there a connection between the multiplicativity of the multiplicity functions for ordinary CSLs and multiple CSLs? What can be said if $f(m)$ is not multiplicative?

We answer some of these questions below and show that there are some connections between multiplicativity and certain decompositions of CSLs into multiple CSLs. We explain and motivate our results and sketch some proofs, whereas detailed proofs will be published elsewhere.

For simplification we will consider only $f^{iso}(m)$ and $f(m)$ in the following. In fact, this is no restriction since all properties of $f^{iso}(m)$ have a direct counterpart for $f^{rot}(m)$. Moreover $f^{rot}(m) = f^{iso}(m)$ for all lattices Γ that contain an orientation reversing symmetry operation in their point group, since there exists an index preserving bijection between the symmetry preserving and symmetry reversing coincidence isometries — compare the remark at the beginning of Sec. 3 of [12].

2. Multiplicativity and supermultiplicativity of the coincidence index

Before looking at the multiplicity functions $f(m)$, it makes sense to have a closer look on the coincidence index $\Sigma(R)$. In particular we are interested in $\Sigma(R_1R_2)$, if R_1 and R_2 are coincidence isometries. We cannot expect to calculate $\Sigma(R_1R_2)$ from $\Sigma(R_1)$ and $\Sigma(R_2)$, but we get at least the following upper bound:

Lemma 1. *$\Sigma(R_1R_2)$ divides $\Sigma(R_1)\Sigma(R_2)$.*

The proof makes use of the second homomorphism theorem and can be deduced from the diagram in Fig. 1, which shows the relation of several CSLs and double CSLs. There we have set $m := \Sigma(R_1)$ and $n := \Sigma(R_2)$.

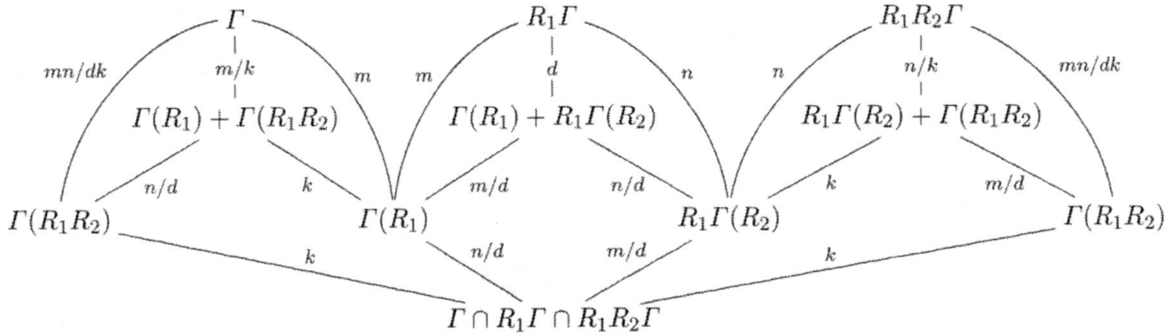

Figure 1. Relations between CSLs and their indices

We can even prove more:

Theorem 2. *If $\Sigma(R_1)$ and $\Sigma(R_2)$ are coprime, then*

$$\Sigma(R_1R_2) = \Sigma(R_1)\Sigma(R_2). \tag{4}$$

Note that the condition that $\Sigma(R_1)$ and $\Sigma(R_2)$ are coprime is essential. In general we cannot expect equality. A simple counter example is given by $R_2 = R_1^{-1}$, if $\Sigma(R_1) > 1$, since $\Sigma(R_1) = \Sigma(R_1^{-1})$ holds in general [4] and $\Sigma(E) = 1$.

For m and n coprime the diagram in Fig. 1 simplifies considerably and the result is shown in Fig. 2.

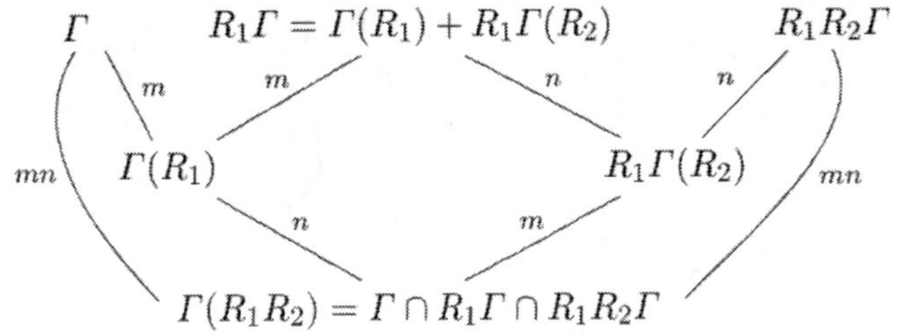

Figure 2. Relations between CSLs with $\Sigma(R_1)$ and $\Sigma(R_2)$ coprime

Moreover we can readily read off

Corollary 3. *If $\Sigma(R_1)$ and $\Sigma(R_2)$ are coprime, then*

$$\Gamma(R_1R_2) = \Gamma \cap R_1\Gamma \cap R_1R_2\Gamma = \Gamma(R_1) \cap R_1\Gamma(R_2) \tag{5}$$

This result is rather technical but plays an important role in the following, since it relates $\Gamma(R_1R_2)$ with some kind of multiple CSLs and provides the basis for something like a "prime decomposition" of CSLs.

Before we continue we want to point out that analogous results can be also obtained for similar sublattices, see e.g. [13]. There $\Sigma(R)$ has to be replaced by the corresponding index of the primitive similar sublattice, which is given by $\text{den}(R)^d$, where $\text{den}(R)$ is the denominator of the similarity rotation and d is the dimension. In fact, there is a close relationship between similar sublattices and CSLs, see [12, 14].

3. Multiplicity functions

We have already mentioned that the functions $f^{iso}(m)$ and $f(m)$ are in general not multiplicative. Nevertheless $f^{iso}(mn)$ (and likewise $f(mn)$) is not completely independent of $f^{iso}(m)$ and $f^{iso}(n)$. In fact we can prove [15]

Theorem 4. *$f^{iso}(m)$ is supermultiplicative, i.e. $f^{iso}(mn) \geq f^{iso}(m)f^{iso}(n)$ if m and n are coprime.*

The proof is mainly combinatorial and relies heavily on the multiplicativity of the index Σ as given in theorem 2. It is this theorem that guarantees that there are enough coincidence isometries of index mn.

Recall that $|\mathcal{P}|f^{iso}(m)$ counts the number of coincidence isometries of index m. Correspondingly $f^{iso}(m)$ counts the number of distinct symmetry classes $R\mathcal{P}$ of coincidence isometries of index m, i.e. there exist exactly $f^{iso}(m)$ different cosets $R_i\mathcal{P}$ and every R with $\Sigma(R) = m$ is contained in exactly one of these cosets. Furthermore $f^{iso}(m)$ is an upper bound for the number $f(m)$ of CSLs of index m. In fact, all $R \in R_i\mathcal{P}$ generate the same CSL, and hence the set of all R that generate a given CSL is the union of finitely many cosets $R_i\mathcal{P}$. Let us denote the set of all coincidence isometries S that generate the CSL $\Gamma(R)$ by $\mathcal{S}(R)$. Clearly there are exactly $f(m)$ different $\mathcal{S}(R_j)$ with index m. The fact that $\mathcal{S}(R)$ consists of more than one symmetry class $R\mathcal{P}$ in general makes the determination of $f(m)$ more difficult than the computation of $f^{iso}(m)$ and in general the expressions for $f(m)$ are more complicated than those for $f^{iso}(m)$, see [10, 11] for examples. Nevertheless we can show

Theorem 5. *$f(m)$ is supermultiplicative, i.e. $f(mn) \geq f(m)f(n)$ if m and n are coprime.*

The proof is again combinatorial, though slightly more difficult. In fact one needs the following lemma, which is also interesting on its own:

Lemma 6. *Assume that $\Sigma(R) =: m$ and $\Sigma(S) =: n$ are coprime. Then*

$$n\Gamma \cap \Gamma(RS) = n\Gamma(R) \qquad \text{and} \qquad mR\Gamma \cap \Gamma(RS) = nR\Gamma(S). \tag{6}$$

This lemma does not only tell us that we can recover $\Gamma(R)$ and $\Gamma(S)$ from $\Gamma(RS)$ alone but it also tells us how to do so: just by taking the intersection of $\Gamma(RS)$ with a suitable similar sublattice of Γ.

Multiplicativity can be destroyed for several reasons, and we can read them off from this lemma. First, there may be isometries Q of index $\Sigma(RS) = mn$ that cannot be written as a product $Q = RS$ with $\Sigma(R) = m$ and $\Sigma(S) = n$. As an example we mention $\Gamma = 2\mathbb{Z} \times 3\mathbb{Z}$. Here $f^{iso}(6) = f(6) = 1$, but $f^{iso}(2) = f(2) = 0 = f^{iso}(3) = f(3)$. Further examples can be found in [16].

Secondly, two isometries R,R' that generate the same CSL $\Gamma(R) = \Gamma(R')$ might give rise to different CSLs $\Gamma(RS)$ and $\Gamma(R'S)$. This is no problem as long as R and R' are symmetry related. In this case the set $\{\Gamma(R'S_k)\}_{k=1}^{f(n)}$ is just a permutation of $\{\Gamma(RS_k)\}_{k=1}^{f(n)}$, where S_k runs over a complete set of not symmetry related S_k of index $\Sigma(S_k) = n$. However, if R and R' are not symmetry related additional CSLs might occur.

Analogous results hold for similar sublattices. Let $g(m)$ be the number of similar sublattices of index m. Then the function $g(m)$ is in general only supermultiplicative. An example for a lattice with non–multiplicative $g(m)$ is again the lattice $\Gamma = 2\mathbb{Z} \times 3\mathbb{Z}$. But note that similar sublattices seem to be more sensitive to violation of multiplicativity than CSLs. E.g., for $\Gamma = \mathbb{Z} \times 5\mathbb{Z}$ multiplicativity is violated for $g(m)$ while $f(m)$ is still multiplicative [16, 17].

4. A criterion for multiplicativity

We have seen that $f(m)$ and $f^{iso}(m)$ are in general only supermultiplicative. Now the interesting question is whether there exist some criteria for multiplicativity and the answer is positive. A first hint is given by known examples in $d \leq 4$. For root lattices in $d \leq 4$ the multiplicity functions $f(m)$ and $f^{iso}(m)$ are usually multiplicative. The reason is that these lattices are related to principal ideal domains (and thus unique factorization domains) of algebraic integers or quaternions. So we expect that some kind of unique factorization property is essential. In fact we can prove

Theorem 7. *The following statements are equivalent:*

(i) *$f(m)$ is multiplicative.*

(ii) *Every (ordinary) CSL $\Gamma(R)$ can be written (uniquely) as $\Gamma(R) = \Gamma(R_1) \cap \ldots \cap \Gamma(R_n)$, where the indices $\Sigma(R_i)$ are powers of distinct primes.*

(iii) *Every MCSL $\Gamma(R_1, \ldots, R_n)$ of order n can be written (uniquely) as $\Gamma(R_1, \ldots, R_n) = \Gamma_1 \cap \ldots \cap \Gamma_k$, where the Γ_k are MCSLs of order at most n and whose indices Σ_k are powers of distinct primes.*

Note that Lemma 6 guarantees the uniqueness of the decomposition $\Gamma(R) = \Gamma(R_1) \cap \ldots \cap \Gamma(R_n)$, if it exists. Lemma 6 is also the main ingredient for the proof, which again also involves combinatorial arguments. But be careful. The decomposition $\Gamma(R) = \Gamma(R_1) \cap \ldots \cap \Gamma(R_n)$ does not imply the decomposition $R = R_1 \cdots R_n$, in general R is not even symmetry related to $R_1 \cdots R_n$. The situation is even worse. If $R = R_1 \cdots R_n$ is a decomposition of R then in general $\Gamma(R) \neq \Gamma(R_1) \cap \ldots \cap \Gamma(R_n)$, which is due to the fact that $O(n)$ is not Abelian for $n \geq 2$.

A corresponding criterion for the coincidence isometries exists as well. The formulation of it is a bit more intricate, since isometries usually do not commute. For CSLs the decomposition into its prime power constituents is unique (up to permutation), for isometries a decomposition will depend strongly on how the factors are ordered.

First notice that if the coincidence isometry R with $\Sigma(R) = mn$ can be factored as $R = R_1 R_2$ with $\Sigma(R_1) = m$ and $\Sigma(R_2) = n$ coprime, then R_1 and R_2 are uniquely determined up to elements of the point group $\mathcal{P}$, i.e. all other decompositions are of the form $R = (R_1 Q)(Q^{-1} R_2)$ with $Q \in \mathcal{P}$. Note that R_2 and $Q^{-1} R_2$ are usually not symmetry related, whereas R_1 and $R_1 Q$ are.

At this point it is not clear whether the existence of a decomposition $R = R_1 R_2$ implies a decomposition $R = R_2' R_1'$, where $\Sigma(R_1) = \Sigma(R_1') = m$ and $\Sigma(R_2) = \Sigma(R_2') = n$. This motivates the following definitions: We call a bijection $\pi = \{p_1, p_2 \ldots\}$ from the positive integers onto the prime numbers an ordering of the prime numbers. We call a decomposition of a coincidence isometry $R = R_1 \cdots R_n$ a π-decomposition of R if $\Sigma(R_i)$ is a power of p_i for any i (we allow $\Sigma(R_i) = p_i^0 = 1$). It is clear that any π-decomposition can be unique only up to point group elements.

We can now formulate the analogue of theorem 7 for $f^{iso}(m)$

Theorem 8. *The following statements are equivalent:*

(i) *$f^{iso}(m)$ is multiplicative.*

(ii) *There exists an ordering π of the prime numbers such that any coincidence isometry R has a (unique) π-decomposition.*

(iii) *For any ordering π of the prime numbers there exists a π-decomposition of every coincidence isometry R.*

Given these two quite similar criteria we may expect that there is some connection between the multiplicativity of $f(m)$ and $f^{iso}(m)$. In fact we can prove

Theorem 9. *$f(m)$ is multiplicative if $f^{iso}(m)$ is.*

Here it is worth to comment on the various decompositions that occur here. For simplicity we assume that only two prime powers are involved, say $\Sigma(R) = p_1^{r_1} p_2^{r_2}$. Then the multiplicativity of $f^{iso}(m)$ guarantees the existence of two decompositions $R = R_1 S_1$ and $R = R_2 S_2$ with $\Sigma(R_1) = p_1^{r_1} = \Sigma(S_2)$ and $\Sigma(R_2) = p_2^{r_2} = \Sigma(S_1)$. In this case the unique decomposition of $\Gamma(R)$ reads $\Gamma(R) = \Gamma(R_1) \cap \Gamma(R_2)$. So given the decompositions of R we get immediately the decomposition of $\Gamma(R)$. However, it does not work the other way round. So given a decomposition of $\Gamma(R)$ we do not get any information on the decompositions of R, even if we would know that they exist.

Now what about the converse of theorem 9? Does it exist? We do not know the answer so far. So we conclude with an open question: Does the multiplicativity of $f(m)$ imply the multiplicativity of $f^{iso}(m)$? Or are there lattices with multiplicative $f(m)$ but non–multiplicative $f^{iso}(m)$?

Acknowledgements

This work was supported by the German Research Council (DFG), within the CRC 701.

References

[1] Bollmann W 1970 *Crystal Defects and Crystalline Interfaces* (Berlin: Springer)
[2] Grimmer H, Bollmann W and Warrington D H 1974 *Acta Cryst.* **A 30** 197–207
[3] Grimmer H 1984 *Acta Cryst.* **A 40** 108–112
[4] Baake M 1997 *The Mathematics of Long-Range Aperiodic Order* ed Moody R V (Dordrecht: Kluwer) pp 9–44 rev. version: arXiv:math.MG/0605222
[5] Pleasants P A B, Baake M and Roth J 1996 *J. Math. Phys.* **37** 1029–1058 rev. version: arXiv:math/0511147
[6] Huck C 2009 *Z. Krist.* **224** 341–344
[7] Baake M and Grimm U 2006 *Z. Kristallogr.* **221** 571–581 arXiv:math.MG/0511306
[8] Zeiner P Multiple CSLs for cubic lattices in preparation
[9] Baake M and Zeiner P 2007 *Phil. Mag.* **87** 2869–2876
[10] Baake M and Zeiner P 2008 *Phil. Mag.* **88** 2025–2032 arXiv:0712.0363v1[math.MG]
[11] Heuer M and Zeiner P CSLs of the root lattice A_4 in this volume
[12] Baake M, Heuer M, Grimm U and Zeiner P 2008 *Europ. J. Combinatorics* **29** 1808–1819 arXiv:0709.1341
[13] Baake M, Heuer M and Moody R V 2008 *J. Algebra* **320** 1391–1408 arXiv:math/0702448v2[math.MG]
[14] Glied S and Baake M 2008 *Z. Krist* **223** 770–772 arXiv:0808.0109
[15] Zeiner P in preparation
[16] Freiberger J 2008 Koinzidenzgitter von Rechteckgittern diploma thesis (in German)
[17] Baake M, Scharlau R and Zeiner P Similar sublattices of planar lattices arXiv:0908.2558v1 [math.MG]

Coincidence isometries of a shifted square lattice

Manuel Joseph C. Loquias[1,2] **and Peter Zeiner**[1]

[1] Fakultät für Mathematik, Universität Bielefeld, Postfach 100131, 33501 Bielefeld, Germany
[2] Institute of Mathematics, University of the Philippines, Diliman, C.P. Garcia St., 1101 Diliman, Quezon City, Philippines

E-mail: {mloquias,pzeiner}@math.uni-bielefeld.de

Abstract. We consider the coincidence problem for the square lattice that is translated by an arbitrary vector. General results are obtained about the set of coincidence isometries and the coincidence site lattices of a shifted square lattice by identifying the square lattice with the ring of Gaussian integers. To illustrate them, we calculate the set of coincidence isometries, as well as generating functions for the number of coincidence site lattices and coincidence isometries, for specific examples.

1. Introduction

The sublattice of finite index formed by the points of intersection of a lattice and a rotated copy of the same lattice is called a coincidence site lattice or CSL. It was Friedel in 1911 who first recognized the use of CSLs in describing and classifying grain boundaries in crystals [1]. Since then, CSLs have proven to be an indispensable tool in the study of grain boundaries and interfaces [2, 3, 4, 5]. This prompted various authors to examine the CSLs of several lattices including cubic and hexagonal ones [6, 7, 8].

The discovery of quasicrystals triggered a renewed interest in CSLs. This led to the analysis of CSLs from a more mathematical point of view. Known results for lattices were again considered and reformulated so that they may be readily extended to aperiodic situations. This was necessary since the first stage in solving the coincidence problem for quasicrystals involved calculating the coincidence site modules (CSMs) of the underlying translation modules, such as modules with 5, 8, 10, and 12-fold symmetry (see [9, 10] and references therein, see also [11, 12, 13, 14]). Hence, coincidences of lattices and modules in dimensions $d \leq 4$ were investigated in [10, 15, 16, 17, 18]. Recent results include the decomposition of coincidence isometries of lattices and modules in Euclidean n-space as a product of at most n coincidence reflections [19, 20] and the relationship between the sets of coincidence and similarity isometries of lattices and modules [21, 22].

The mathematical treatment of the coincidence problem is very often restricted to linear coincidence isometries, that is, rotations and improper rotations, whereas isometries containing a translational part are ignored – as we did in the first two paragraphs above. Nevertheless, general (affine) isometries are important in crystallography. Indeed, the situation where one shifts the two component crystals against each other has been investigated in [23, 24] and references therein. It was shown that these shifts are needed to minimize the grain boundary energy, thus they are often referred to as "rigid relaxations". However, some authors claim that minimizing the energy may require shifts that destroy all coincidence sites.

Even though the idea of introducing a shift after applying a linear coincidence isometry has already been dealt with in the physical literature, not much can be found in the mathematical literature where a systematic treatment of the subject is still missing. Thus, we now aim to generalize the notion of a CSL and CSM, respectively, that is, we investigate the possible intersections of two lattices (modules) that are related by an isometry. To simplify the discussion, we restrict our attention here to the lattice case though most of the results also work in the module case. In fact, some steps in the general direction have been made in [9]. There, the authors have considered coincidence rotations around certain points which are not lattice (module) points. For instance, they determined the set of coincidence rotations about the center of a Delauney cell of the square lattice and calculated the corresponding indices.

In this paper we discuss a related and special case: the coincidence problem for shifted lattices. That is, after translating the lattice by some vector and upon rotation of this shifted lattice (with respect to the origin), we consider its intersection with the shifted lattice. This should be useful in the context of bicrystallography [25, 26]. Similar to the approach in [9, 10], we start our investigation with the square lattice and provide solutions that, when modified appropriately, also apply to planar modules.

The purpose of this paper is to shed further light on the geometry of CSLs. It is beyond the scope of this paper to discuss the actual grain boundary energy, which would require considering the actual Hamiltonians. In particular, the paper is not intended to determine which translations are the most favourable ones in terms of energy.

2. The coincidence problem for lattices

Let $\Gamma \subseteq \mathbb{R}^d$ be a d-dimensional lattice and $R \in O(d)$. We say that R is a *(linear) coincidence isometry* of Γ if $\Gamma(R) := \Gamma \cap R\Gamma$ is a sublattice of finite index in Γ and we call $\Gamma(R)$ a *coincidence site lattice* (CSL) of Γ. The *coincidence index* of a coincidence isometry R, denoted by $\Sigma(R)$, is given by $\Sigma(R) = [\Gamma : \Gamma(R)] = \frac{\mathrm{vol}(\Gamma(R))}{\mathrm{vol}(\Gamma)}$. Geometrically, $\Sigma(R)$ is equal to the ratio of the volume of a fundamental domain of Γ with the volume of a fundamental domain of $\Gamma(R)$.

We denote the set of (linear) coincidence isometries of Γ by $OC(\Gamma)$ and the set of coincidence rotations of Γ, that is, $OC(\Gamma) \cap SO(d)$, by $SOC(\Gamma)$. The set $OC(\Gamma)$ forms a group having $SOC(\Gamma)$ as a subgroup [10].

We summarize here the known results for the square lattice $\mathbb{Z}^2$ (see [9] for details). The group of coincidence rotations of $\mathbb{Z}^2$ is $SOC(\mathbb{Z}^2) = SO(2, \mathbb{Q})$, that is, the coincidence rotations of $\mathbb{Z}^2$ are the special orthogonal matrices having rational entries. In determining the structure of this group, the square lattice is identified with the ring of Gaussian integers $\Gamma = \mathbb{Z}[i]$, where $i = \sqrt{-1}$, embedded in the set of complex numbers $\mathbb{R}[i] = \mathbb{C}$. In this setting, a coincidence rotation R by an angle of θ corresponds to multipication by the complex number $e^{i\theta}$, where

$$e^{i\theta} = \varepsilon \cdot \prod_{p\equiv 1(4)} \left(\frac{\omega_p}{\overline{\omega_p}}\right)^{n_p} \tag{1}$$

with $n_p \in \mathbb{Z}$ and only a finite number of $n_p \neq 0$, ε is a unit in $\mathbb{Z}[i]$, p runs over the rational primes $p \equiv 1 \pmod 4$, and ω_p, and its complex conjugate $\overline{\omega_p}$ are the Gaussian prime factors of $p = \omega_p \cdot \overline{\omega_p}$. If we denote by z the numerator of $e^{i\theta}$, that is,

$$z = \prod_{\substack{p\equiv 1(4)\\ n_p>0}} {\omega_p}^{n_p} \cdot \prod_{\substack{p\equiv 1(4)\\ n_p<0}} (\overline{\omega_p})^{-n_p}, \tag{2}$$

then the coincidence index of R is the number theoretic norm of z, that is, $\Sigma(R) = N(z) = z \cdot \overline{z}$, and the CSL obtained from R, $\Gamma(R)$, is the principal ideal $(z) := z\mathbb{Z}[i]$. Consequently, the group of coincidence rotations of the square lattice is given by $SOC(\mathbb{Z}^2) = SOC(\Gamma) \cong C_4 \times \mathbb{Z}^{(\aleph_0)}$, where C_4 is the cyclic group of order 4 with generator i, and $\mathbb{Z}^{(\aleph_0)}$ is the direct sum of countably many infinite cyclic groups each of which is generated by $\omega_p/\overline{\omega_p}$ with $p \equiv 1 \pmod 4$.

Every coincidence isometry $T \in OC(\Gamma) \setminus SOC(\Gamma)$ can be written as $T = R \cdot T_r$, where $R \in SOC(\Gamma)$ and T_r is the reflection along the real axis (complex conjugation). Here, $\Sigma(T) = \Sigma(R)$ and $\Gamma(T) = \Gamma(R)$. Finally, $OC(\mathbb{Z}^2) = OC(\Gamma) \cong SOC(\Gamma) \rtimes C_2$ (semi-direct product), where C_2 is the cyclic group of order 2 generated by T_r.

The possible coincidence indices and the number of CSLs for a given index m may be described by means of a generating function. Let $\hat{f}(m)$ be the number of coincidence rotations of Γ and $f(m)$ be the number of CSLs of Γ, for a given index m. Then $\hat{f}(m) = 4f(m)$, where the factor 4 stems from the fact that there are four symmetry rotations. The function $f(m)$ is multiplicative (that is, $f(1) = 1$ and $f(mn) = f(m)f(n)$ when m, n are relatively prime), and $f(p^r) = 2$ for primes $p \equiv 1 \pmod 4$ whereas $f(p^r) = 0$ for primes $p \equiv 2, 3 \pmod 4$, where $r \in \mathbb{N}$. We write the generating function for $f(m)$ as a Dirichlet series $\Phi(s)$ given by

$$\begin{aligned}\Phi(s) &= \sum_{m=1}^{\infty} \frac{f(m)}{m^s} = \prod_{p\equiv 1(4)} \frac{1+p^{-s}}{1-p^{-s}} \\ &= 1 + \tfrac{2}{5^s} + \tfrac{2}{13^s} + \tfrac{2}{17^s} + \tfrac{2}{25^s} + \tfrac{2}{29^s} + \tfrac{2}{37^s} + \tfrac{2}{41^s} + \tfrac{2}{53^s} + \tfrac{2}{61^s} + \tfrac{4}{65^s} + \tfrac{2}{73^s} + \dots.\end{aligned}$$

3. Coincidences of shifted lattices

We now turn our attention to lattices Γ in $\mathbb{R}^d$ that are shifted by some vector $x \in \mathbb{R}^d$ and we look at intersections of the form $(x+\Gamma)\cap R(x+\Gamma)$, where $R \in O(d)$. We remark that we actually only need to consider values of x in a fundamental domain of Γ.

An $R \in O(d)$ is said to be a *(linear) coincidence isometry of the shifted lattice* $x + \Gamma$ if $(x + \Gamma) \cap R(x + \Gamma)$ is a sublattice of $x + \Gamma$ of finite index and we also call $(x + \Gamma) \cap R(x + \Gamma)$ a CSL of the shifted lattice $x + \Gamma$. We denote the set of all coincidence isometries of $x + \Gamma$ by $OC(x + \Gamma)$. The following theorem characterizes the set $OC(x + \Gamma)$ and relates the CSLs of $x + \Gamma$ with the CSLs of Γ [27].

Theorem 1. *Let* Γ *be a lattice in* $\mathbb{R}^d$ *and* $x \in \mathbb{R}^d$.

(i) $OC(x + \Gamma) = \{R \in OC(\Gamma) : Rx - x \in \Gamma + R\Gamma\}$

(ii) If $R \in OC(x + \Gamma)$ *with* $Rx - x = t + Rs$ *for some* $t, s \in \Gamma$, *then*

$$(x + \Gamma) \cap R(x + \Gamma) = x + (t + \Gamma(R)).$$

Theorem 1 tells us that a CSL of $x + \Gamma$ obtained from $R \in OC(x + \Gamma)$ is just a translate of the CSL $\Gamma(R)$ in Γ. Consequently, $[x + \Gamma : (x + \Gamma) \cap R(x + \Gamma)] = \Sigma(R)$ which means that shifting the lattice does not give rise to new values of coincidence indices. In addition, we see that $OC(x + \Gamma)$ is a subset of $OC(\Gamma)$. The set $OC(x + \Gamma)$ is non-empty because the identity $1 \in OC(x + \Gamma)$. Also, $OC(x + \Gamma)$ is closed under inverses, that is, $R^{-1} \in OC(x + \Gamma)$ whenever $R \in OC(x + \Gamma)$. However, given $R_1, R_2 \in OC(x + \Gamma)$, R_2R_1 is not necessarily in $OC(x + \Gamma)$. In fact, $OC(x + \Gamma)$ is not a group in general [27].

4. The coincidence problem for the shifted square lattice

From this point onwards, we take $\Gamma = \mathbb{Z}[i]$, the square lattice viewed as the ring of Gaussian integers, and $x \in \mathbb{C}$. From (1) and (2), we see that we can associate each $R \in SOC(\Gamma)$ to (z, ε), and we will write this as $R(z, \varepsilon)$. That is, $R(z, \varepsilon) \in SOC(\Gamma)$ stands for multiplication by the complex number $\varepsilon\frac{z}{\overline{z}}$. We may assume that $\frac{z}{\overline{z}}$ is reduced, that is, z and $\overline{z}$ have no factors in common. In addition, we shall simply set $z = 1$ whenever $R(z, \varepsilon) \in P(\Gamma)$, where $P(\Gamma)$ denotes the point group of Γ.

Let $SOC(x + \Gamma) := OC(x + \Gamma) \cap SO(d)$. We start with the following lemma.

Lemma 2. *Let* $\Gamma = \mathbb{Z}[i]$, $x \in \mathbb{C}$, $R = R(z, \varepsilon) \in SOC(\Gamma)$, *and* $T = RT_r$.

(i) $R \in SOC(x + \Gamma)$ *if and only if* $(\varepsilon z - \overline{z})x \in \mathbb{Z}[i]$

(ii) $T \in OC(x + \Gamma)$ *if and only if* $\varepsilon z\overline{x} - \overline{z}x \in \mathbb{Z}[i]$

Proof. Recall that Γ is a principal ideal domain. Since ε is a unit in Γ and z, $\overline{z}$ are relatively prime,

$$\Gamma + R\Gamma = \Gamma + \varepsilon\frac{z}{\overline{z}}\Gamma = \frac{1}{\overline{z}}(\overline{z}\Gamma + z\Gamma) = \frac{1}{\overline{z}}\gcd(z, \overline{z})\Gamma = \frac{1}{\overline{z}}\Gamma.$$

By Theorem 1, $R \in SOC(x + \Gamma) \Leftrightarrow Rx - x \in \Gamma + R\Gamma \Leftrightarrow \varepsilon\frac{z}{\overline{z}}x - x \in \frac{1}{\overline{z}}\Gamma \Leftrightarrow (\varepsilon z - \overline{z})x \in \mathbb{Z}[i]$. Similarly, $\Gamma + T\Gamma = \frac{1}{\overline{z}}\Gamma$. Applying again Theorem 1, we obtain the second statement. ∎

We now obtain the following results about $SOC(x + \Gamma)$ and $OC(x + \Gamma)$.

Theorem 3. *If* $\Gamma = \mathbb{Z}[i]$ *and* $x \in \mathbb{C}$ *then* $SOC(x + \Gamma)$ *is a subgroup of* $SOC(\Gamma)$.

Proof. We have already mentioned that $1 \in SOC(x+\Gamma)$ and $SOC(x+\Gamma)$ is closed under inverses. Let $R_j(z_j, \varepsilon_j) \in SOC(x + \Gamma)$ for $j = 1, 2$ and $g = \gcd(z_1, \overline{z_2})$. By Lemma 2, $(\varepsilon_j z_j - \overline{z_j})x \in \mathbb{Z}[i]$ for $j = 1, 2$. Write $z_1 = h_1 g$, $\overline{z_2} = \overline{h_2} g$, and hence, h_1 and $\overline{h_2}$ are relatively prime. This means that R_1R_2 corresponds to $(h_1h_2, \varepsilon_1\varepsilon_2)$ so that $R_1R_2 \in SOC(x+\Gamma)$ if $\left(\varepsilon_1\varepsilon_2 h_1h_2 - \overline{h_1h_2}\right)x \in \mathbb{Z}[i]$ from Lemma 2. Now,

$$\begin{aligned}
(\varepsilon_1\varepsilon_2 h_1h_2 - \overline{h_1h_2})x &= \frac{1}{g}(\varepsilon_1\varepsilon_2 z_1h_2 - \overline{h_1 z_2})x \\
&= \frac{1}{g}[(\varepsilon_1\varepsilon_2 z_1h_2 - \varepsilon_2h_2\overline{z_1}) + (\varepsilon_2\overline{h_1}z_2 - \overline{h_1 z_2})]x \\
&= \frac{1}{g}[\varepsilon_2h_2\underbrace{(\varepsilon_1z_1 - \overline{z_1})x}_{\in \mathbb{Z}[i]} + \overline{h_1}\underbrace{(\varepsilon_2z_2 - \overline{z_2})x}_{\in \mathbb{Z}[i]}] \in \frac{1}{g}\Gamma.
\end{aligned}$$

Similarly, we also obtain that $(\varepsilon_1\varepsilon_2h_1h_2 - \overline{h_1h_2})x \in \frac{1}{\overline{g}}\Gamma$. Hence,

$$(\varepsilon_1\varepsilon_2h_1h_2 - \overline{h_1h_2})x \in \frac{1}{g}\Gamma \cap \frac{1}{\overline{g}}\Gamma = \frac{1}{g\overline{g}}(g\Gamma \cap \overline{g}\Gamma) = \frac{1}{g\overline{g}}\,\mathrm{lcm}\,(g, \overline{g})\Gamma = \mathbb{Z}[i]$$

since g, $\overline{g}$ are relatively prime. ∎

For $OC(x + \Gamma)$, the situation is more complicated. One can show the following results (see [27]).

Theorem 4. *Let* $\Gamma = \mathbb{Z}[i]$ *and* $x \in \mathbb{C}$.

(i) The set $OC(x + \Gamma)$ *is a subgroup of* $OC(\Gamma)$ *if and only if for any* T_1, $T_2 \in OC(x + \Gamma) \setminus SOC(x + \Gamma)$, $T_1T_2 \in SOC(x + \Gamma)$.

(ii) If $OC(x+\Gamma)$ *contains a reflection* $T \in P(\Gamma)$ *then* $OC(x+\Gamma)$ *is a subgroup of* $OC(\Gamma)$. *Also,* $OC(x + \Gamma) = SOC(x + \Gamma) \rtimes \langle T\rangle$, *where* $\langle T\rangle = \{1, T\} \cong C_2$ *is the group generated by* T.

(iii) Suppose $OC(x + \Gamma)$ *does not contain a reflection* $T \in P(\Gamma)$. *If* $RT_r \in OC(x + \Gamma)$ *where* $R(z, \varepsilon_1) \in SOC(\Gamma)$ *then for any unit* ε_2, $R_2 = R_2(z, \varepsilon_2) \notin SOC(x + \Gamma)$.

When computing for $OC(x + \Gamma)$, we see from Theorem 4 that it is convenient to determine whether there is a reflection $T \in P(\Gamma)$ that is in $OC(x + \Gamma)$. If such a reflection T exists, then $OC(x + \Gamma)$ is a group and it is the semi-direct product of $SOC(x + \Gamma)$ and $\langle T\rangle$. Otherwise, we need to check if $RT_r \in OC(x + \Gamma)$ only for those reflections RT_r for which $R(z, \varepsilon) \in SOC(\Gamma)$ and $R' = R'(z, \varepsilon') \notin SOC(x + \Gamma)$ for all unit ε' holds.

5. Specific examples

For the rest of the discussion, we shall assume that $R(z,\varepsilon) \in SOC(\Gamma)$. The following theorem solves completely the case when x has an irrational component [27].

Theorem 5. *Let $x = a + bi \in \mathbb{C}$. If a or b is irrational then $OC(x+\Gamma)$ is a group of at most two elements. In particular, if*

(i) a is irrational and b is rational then $OC(x+\Gamma) = \begin{cases} \langle T_r \rangle & \text{if } 2b \in \mathbb{Z} \\ \{1\} & \text{otherwise.} \end{cases}$

(ii) a is rational and b is irrational then $OC(x+\Gamma) = \begin{cases} \langle T \rangle & \text{if } 2a \in \mathbb{Z} \\ \{1\} & \text{otherwise,} \end{cases}$

where T is the reflection along the imaginary axis.

(iii) both a and b are irrational, and

(a) a, b are rationally independent then $OC(x+\Gamma) = \{1\}$.

(b) $a = \frac{p_1}{q_1} + \frac{p_2}{q_2}b$ where p_j, $q_j \in \mathbb{Z}$, p_j and q_j are relatively prime (for $j = 1, 2$) with

1. p_2q_2 even, then $OC(x+\Gamma) = \begin{cases} \langle RT_r \rangle & \text{if } q_1 | 2q_2 \\ \{1\} & \text{otherwise,} \end{cases}$

where $R = R(p_2 + q_2 i, 1) \in SOC(\Gamma)$.

2. p_2q_2 odd, then $OC(x+\Gamma) = \begin{cases} \langle RT_r \rangle & \text{if } q_1 | q_2 \\ \{1\} & \text{otherwise,} \end{cases}$

where $R = R(\frac{p_2+q_2}{2} - \frac{p_2-q_2}{2}i, i) \in SOC(\Gamma)$.

Example 1.

(i) Suppose $x = \frac{1}{\sqrt{2}} + \frac{1}{\sqrt{3}}i$. We immediately see that we cannot write $\frac{1}{\sqrt{2}} = c + d\frac{1}{\sqrt{3}}$ where $c, d \in \mathbb{Q}$ since $\sqrt{2} \notin \mathbb{Q}(\sqrt{3})$. Hence, $OC(x+\Gamma) = \{1\}$.

(ii) Let $x = \sqrt{2} - \frac{\sqrt{2}}{2}i$. We have $\sqrt{2} = \frac{0}{1} + \left(\frac{-2}{1}\right)\left(-\frac{\sqrt{2}}{2}\right)$, and since $1|(2\cdot 1)$, $OC(x+\Gamma) = \langle RT_r \rangle$ where $R = R(-2+i, 1) \in SOC(\Gamma)$.

It only remains to consider the case when both a and b are rational. We now consider $x = a + bi \in \mathbb{Q}(i)$ and write $x = \frac{p}{q}$ where p, $q \in \mathbb{Z}[i]$, and p, q are relatively prime (in $\mathbb{Z}[i]$). The following lemma tells us that $SOC(x+\Gamma)$ depends only on the denominator q of x.

Lemma 6. *Let $\Gamma = \mathbb{Z}[i]$, $x = \frac{p}{q} \in \mathbb{Q}(i)$ where p, $q \in \mathbb{Z}[i]$, with p, q relatively prime, and $R(z,\varepsilon) \in SOC(\Gamma)$. Then $R \in SOC(x+\Gamma)$ if and only if q divides $\varepsilon z - \bar{z}$. Furthermore, $SOC(x+\Gamma) = SOC(\frac{1}{q}+\Gamma)$.*

Proof. We know from Lemma 2 that if $R \in SOC(x+\Gamma)$ then $(\varepsilon z - \bar{z})x = \frac{(\varepsilon z - \bar{z})p}{q} \in \mathbb{Z}[i]$. Since p and q are relatively prime, $q|(\varepsilon z - \bar{z})$. The second statement follows from the first. ■

Lemma 7. *Suppose $\Gamma = \mathbb{Z}[i]$ and $x \in \mathbb{C}$. If $x' = Qx$ for some $Q \in P(\Gamma)$ then*

$$OC(x'+\Gamma) = Q[OC(x+\Gamma)]Q^{-1}.$$

Proof. Since $SOC(\Gamma)$ is a normal subgroup of $OC(\Gamma)$, $R \in SOC(\Gamma)$ if and only if $QRQ^{-1} \in SOC(\Gamma)$. Thus, if $R \in SOC(\Gamma)$ then it follows from Theorem 1 that

$$\begin{aligned} R \in OC(x'+\Gamma) &\Leftrightarrow Rx' - x' \in \Gamma + R\Gamma \\ &\Leftrightarrow Q(Q^{-1}RQx - x) \in Q(\Gamma + Q^{-1}RQ\Gamma) \\ &\Leftrightarrow Q^{-1}RQx - x \in \Gamma + Q^{-1}RQ\Gamma \\ &\Leftrightarrow Q^{-1}RQ \in OC(x+\Gamma) \\ &\Leftrightarrow R \in Q[OC(x+\Gamma)]Q^{-1}. \quad ■ \end{aligned}$$

Recall that we only need to consider values of $x = a + bi$ in a fundamental domain of Γ. A fundamental domain of Γ is $\{a + bi \in \mathbb{C} : -\frac{1}{2} \leq a, b < \frac{1}{2}\}$ (see Figure 1). Observe that every point x' in the chosen fundamental domain can be written as $x' = Qx$ where $Q \in P(\Gamma)$ and $x \in \{a + bi \in \mathbb{C} : 0 \leq b \leq a \leq \frac{1}{2}\}$ (a fundamental domain of the symmetry group of Γ which is a crystallographic group of type $p4m$). Hence, it follows from Lemma 7 that we only need to compute $OC(x + \Gamma)$ for values of $x = a + bi$, where $0 \leq b \leq a \leq \frac{1}{2}$ (see Figure 1).

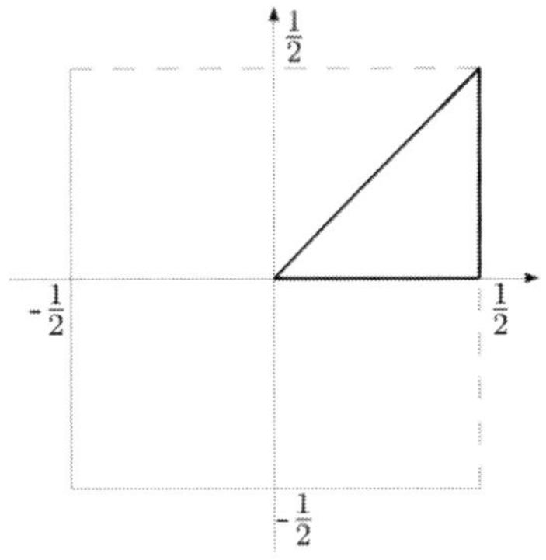

Figure 1. A fundamental domain of Γ, or unit cell, and a fundamental domain of the symmetry group of Γ (black triangle)

Lemma 8. *If $\Gamma = \mathbb{Z}[i]$ and $x = a + bi \in \mathbb{C}$, then $OC(x + \Gamma)$ is a subgroup of $OC(\Gamma)$ if one of the following conditions is satisfied: $a \in \frac{1}{2}\mathbb{Z}$, $b \in \frac{1}{2}\mathbb{Z}$ or $a \pm b \in \mathbb{Z}$. Furthermore, $OC(x + \Gamma) = SOC(x + \Gamma) \rtimes \langle RT_r \rangle$ where $R = R(1, \varepsilon) \in SOC(\Gamma)$, and*

$$\varepsilon = \begin{cases} 1 \text{ if } b \in \frac{1}{2}\mathbb{Z} \\ -1 \text{ if } a \in \frac{1}{2}\mathbb{Z} \\ i \text{ if } a - b \in \mathbb{Z} \\ -i \text{ if } a + b \in \mathbb{Z}. \end{cases}$$

Proof. Consider the reflection $RT_r \in P(\Gamma)$. By Lemma 2, $RT_r \in OC(x + \Gamma)$ if and only if $\varepsilon\overline{x} - x \in \mathbb{Z}[i]$. The result follows by applying Theorem 4. ∎

In particular, for values of a and b for which $0 \leq b \leq a \leq \frac{1}{2}$, $OC(x + \Gamma)$ is a subgroup of $OC(\Gamma)$ when $a = \frac{1}{2}$, $b = 0$, or $a = b$ (the boundaries of the triangle in Figure 1).

Before looking at some examples, we note that given $R(z, \varepsilon) \in SOC(\Gamma)$, we have

$$\varepsilon z - \overline{z} = \begin{cases} 2\,\mathrm{Im}(z) & \text{if } \varepsilon = 1 \\ -2\,\mathrm{Re}(z) & \text{if } \varepsilon = -1 \\ -[\mathrm{Re}(z) + \mathrm{Im}(z)](1 - i) & \text{if } \varepsilon = i \\ -i[\mathrm{Re}(z) - \mathrm{Im}(z)](1 - i) & \text{if } \varepsilon = -i\,. \end{cases} \tag{3}$$

In addition, we see from (1) and (2) that $\mathrm{Re}(z)$ and $\mathrm{Im}(z)$ are relatively prime and of different parity (that is, one is odd and the other is even).

We will also exhibit the number of possible coincidence rotations and CSLs obtained with given index m of the shifted lattice $x + \Gamma$ by means of generating functions. We shall denote by $\hat{f}_x(m)$ the number of coincidence rotations of $x + \Gamma$ of index m, and $f_x(m)$ the number of CSLs of $x + \Gamma$ of index m.

Example 2. $x = \frac{1}{2} + \frac{1}{2}i = \frac{1}{1-i}$

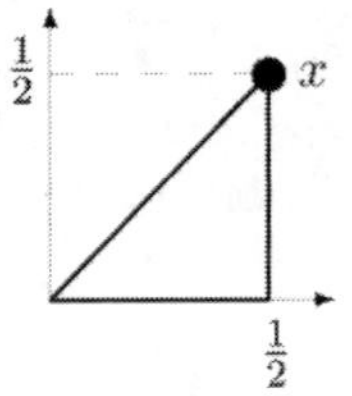

The denominator of x is $q = 1-i$ and we see from (3) that $q|(\varepsilon z - \overline{z})$ for all z, ε. Lemmas 6 and 8 implies that $SOC(x+\Gamma) = SOC(\Gamma) \cong C_4 \times \mathbb{Z}^{(\aleph_0)}$ and $OC(x+\Gamma) = OC(\Gamma) \cong SOC(x+\Gamma) \rtimes C_2$. Clearly, $OC(x+\Gamma)$ is a subgroup of $OC(\Gamma)$ in this case. Also, $\hat{f}_x(m) = \hat{f}(m)$ and $f_x(m) = f(m)$.

These results agree with the results obtained in the Appendix of [9] (just shift the center of the Delaunay cell into the origin).

We also note here that $(S)OC(x+\Gamma) = (S)OC(\Gamma)$ if and only if $x = \frac{m}{2} + \frac{n}{2}i$, where m, n are odd integers.

Example 3. $x = \frac{1}{2}$

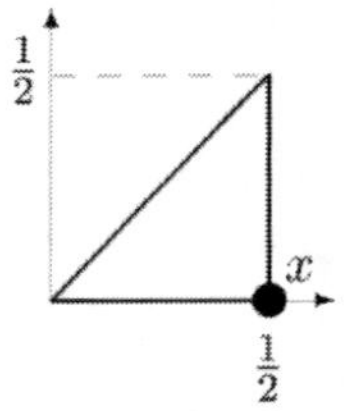

The denominator of x is $q = 2$. Since the sum of $\mathrm{Re}(z)$ and $\mathrm{Im}(z)$ is odd, we obtain from (3) that for all z, $q|(\varepsilon z - \overline{z})$ if and only if $\varepsilon = \pm 1$. Hence, by Lemma 6,

$$SOC(x+\Gamma) = \{R(z,\varepsilon) \in SOC(\Gamma) : \varepsilon = \pm 1\} \cong C_2 \times \mathbb{Z}^{(\aleph_0)}.$$

From Lemma 8, $OC(x+\Gamma) = SOC(x+\Gamma) \rtimes \langle T_r \rangle$ and is a subgroup of $OC(\Gamma)$ of index 2. In this case, we have $f_x(m) = f(m)$ but $\hat{f}_x(m) = 2f_x(m)$.

Example 4. $x_0 = \frac{1}{3}$ and $x_1 = \frac{1}{3} + \frac{1}{3}i$

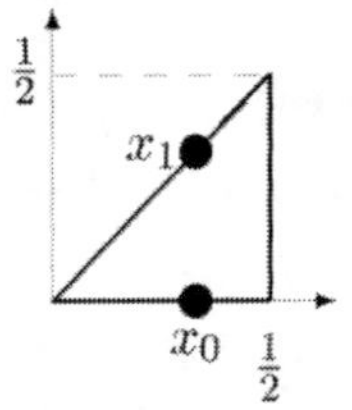

Both x_0 and x_1 have denominator $q = 3$. It is easy to see that if both rational integers m, n are not divisible by q, then either $m+n$ or $m-n$ is divisible by q. Hence, by (3), there is a unique ε so that $q|(\varepsilon z - \overline{z})$ for all z. We conclude from Lemma 6 that $SOC(x_j + \Gamma) \cong \mathbb{Z}^{(\aleph_0)}$ for $j = 0, 1$. In addition, by Lemma 8, $OC(x_j + \Gamma) = SOC(x_j + \Gamma) \rtimes \langle RT_r \rangle$, where $R = R(1, i^j) \in P(\Gamma)$, and $OC(x_j + \Gamma)$ is a subgroup of $OC(\Gamma)$ of index 4. Finally, we have $\hat{f}_{x_j}(m) = f_{x_j}(m) = f(m)$.

Example 5. $x_0 = \frac{1}{5}, \frac{2}{5}$ and $x_1 = \frac{1}{5} + \frac{1}{5}i, \frac{2}{5} + \frac{2}{5}i$

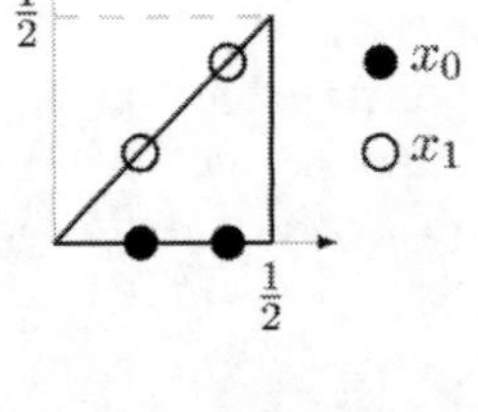

We have the denominator $q = 5$ for x_j, $j = 0, 1$. By considering each possible combination of $\mathrm{Re}(z)$ and $\mathrm{Im}(z)$ modulo q, it can be verified that for all z with $5 \nmid N(z)$, there is a unique ε such that $R(z,\varepsilon) \in SOC(\frac{1}{5}+\Gamma)$. This means that $SOC(x_j+\Gamma) \cong \mathbb{Z}^{(\aleph_0)}$ for $j = 0, 1$ by Lemma 6. Also, it follows from Lemma 8 that $OC(x_j+\Gamma)$ is a subgroup of $OC(\Gamma)$ and $OC(x_j+\Gamma) = SOC(x_j+\Gamma) \rtimes \langle RT_r\rangle$, where $R = R(1, i^j) \in P(\Gamma)$. Furthermore, $\hat{f}_{x_j}(m) = f_{x_j}(m)$ where the Dirichlet series generating function for $f_{x_j}(m)$ is given by

$$\Phi_{x_j}(s) = \sum_{m=1}^{\infty} \frac{f_{x_j}(m)}{m^s} = \prod_{\substack{p\equiv 1(4)\\ p\neq 5}} \frac{1+p^{-s}}{1-p^{-s}}$$
$$= 1 + \tfrac{2}{13^s} + \tfrac{2}{17^s} + \tfrac{2}{29^s} + \tfrac{2}{37^s} + \tfrac{2}{41^s} + \tfrac{2}{53^s} + \tfrac{2}{61^s} + \tfrac{2}{73^s} + \tfrac{2}{89^s} + \tfrac{2}{97^s} + \tfrac{2}{101^s} + \tfrac{2}{109^s} + \tfrac{2}{113^s} + \tfrac{2}{137^s} + \tfrac{2}{149^s} + \tfrac{2}{157^s} + \tfrac{2}{169^s} + \tfrac{2}{173^s} + \tfrac{2}{181^s} + \tfrac{2}{193^s} + \tfrac{2}{197^s} + \tfrac{4}{221^s} + \tfrac{2}{229^s} + \ldots .$$

Example 6. $x = \frac{2}{5} + \frac{1}{5}i = \frac{i}{1+2i}$

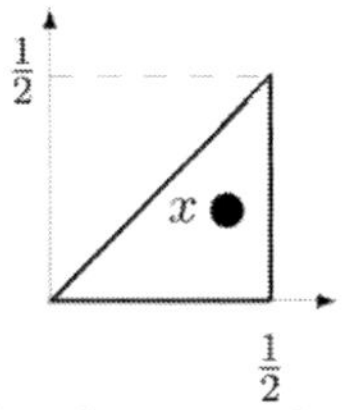

Since the denominator $q = 1+2i$ of x does not divide $1-i$, we see from (3) that $q|(\varepsilon z - \bar{z})$ if and only if $5|(\varepsilon z - \bar{z})$. Hence, $SOC(x+\Gamma) = SOC(\frac{1}{5}+\Gamma) \cong \mathbb{Z}^{(\aleph_0)}$ by Lemma 6. Applying Theorem 4, if $RT_r \in OC(x+\Gamma)$ where $R(z,\varepsilon) \in SOC(\Gamma)$ then $5|N(z)$ because $OC(x+\Gamma)$ does not contain a reflection in $P(\Gamma)$. Observe that given z with $5|N(z)$, we must find either $1+2i$ or $1-2i$ (and not both) in the factorization of z into primes in $\mathbb{Z}[i]$. If $(1-2i)|z$ then $z\bar{x} \in \mathbb{Z}[i]$ and $\varepsilon z\bar{x} - \bar{z}x = \varepsilon z\bar{x} - \overline{z\bar{x}} \in \mathbb{Z}[i]$, $\forall\varepsilon$. Lemma 2 then implies that

$$OC(x+\Gamma) = SOC(x+\Gamma) \cup \{RT_r : R(z,\varepsilon) \in SOC(\Gamma) \text{ with } (1-2i)|z\}.$$

Here, we have an example of a set $OC(x+\Gamma)$ that is not a group. Indeed, let $T_k = R_kT_r \in OC(x+\Gamma)\setminus SOC(x+\Gamma)$ where $R_k = R_k(z,\varepsilon_k) \in SOC(\Gamma)$, for $k = 1, 2$, with $\varepsilon_1 \neq \varepsilon_2$. We obtain that T_1T_2 is not the identity with $T_1T_2 = R_1{R_2}^{-1} \in P(\Gamma)$ which means that $T_1T_2 \notin OC(x+\Gamma)$. By Theorem 4, $OC(x+\Gamma)$ is not a subgroup of $OC(\Gamma)$.

The Dirichlet series generating function for $f_x(m)$ is given by

$$\Phi_x(s) = \sum_{m=1}^{\infty} \frac{f_x(m)}{m^s} = \frac{1}{1-5^{-s}} \cdot \prod_{\substack{p\equiv 1(4)\\ p\neq 5}} \frac{1+p^{-s}}{1-p^{-s}}$$
$$= 1 + \tfrac{1}{5^s} + \tfrac{2}{13^s} + \tfrac{2}{17^s} + \tfrac{1}{25^s} + \tfrac{2}{29^s} + \tfrac{2}{37^s} + \tfrac{2}{41^s} + \tfrac{2}{53^s} + \tfrac{2}{61^s} + \tfrac{2}{65^s} + \tfrac{2}{73^s} + \ldots .$$

Also, if we denote by $\hat{F}_x(m)$ the number of (linear) coincidence isometries of $x+\Gamma$ of index m, we obtain that

$$\hat{F}_x(m) = \begin{cases} f_x(m) \text{ if } 5 \nmid m \\ 4\, f_x(m) \text{ if } 5 \,|\, m. \end{cases}$$

Hence, the Dirichlet series generating function for $\hat{F}_x(m)$ is given by

$$\Psi_x(s) = \sum_{m=1}^{\infty} \frac{\hat{F}_x(m)}{m^s} = \frac{1+3\cdot 5^{-s}}{1-5^{-s}} \cdot \prod_{\substack{p\equiv 1(4)\\ p\neq 5}} \frac{1+p^{-s}}{1-p^{-s}}$$
$$= 1 + \tfrac{4}{5^s} + \tfrac{2}{13^s} + \tfrac{2}{17^s} + \tfrac{4}{25^s} + \tfrac{2}{29^s} + \tfrac{2}{37^s} + \tfrac{2}{41^s} + \tfrac{2}{53^s} + \tfrac{2}{61^s} + \tfrac{8}{65^s} + \tfrac{2}{73^s} + \ldots .$$

6. Conclusion and outlook

We have seen that the coincidence isometries of a shifted lattice are also coincidence isometries of the original lattice. Moreover, the CSLs of the shifted lattice are merely translations of CSLs of the original lattice. Thus, no new values of coincidence indices Σ are obtained by shifting the lattice, and some Σ-values even disappear or their multiplicity is reduced.

The coincidences of a shifted square lattice were examined in this paper by identifying the lattice with the ring of Gaussian integers. The problem was completely solved for the case when the shift consists of an irrational component. For the remaining case, that is, when the shift may be written as a quotient of two Gaussian integers that are relatively prime, one needs to compute the set of coincidence rotations for each possible denominator via some divisibility condition. Partial results are given here and in [27] on how to obtain the coincidence isometries and indices for any given denominator and corresponding numerator. General results in this direction will depend on the arithmetic of the Gaussian integers.

It should be emphasized that the order of rotation and translation of a lattice is in general not interchangeable. In this paper, we compare the shifted lattice with its rotated copy, that is, the translation (say x) comes first before rotation. This corresponds to the situation where the lattice is first rotated by R and is shifted afterwards by the vector $Rx - x$, which is equivalent to a rotation of the lattice about a different point $(-x)$, thus keeping at least one point $(-x)$ fixed. In particular, the CSLs of a shifted lattice are shifted copies of the intersection of a lattice with a rotated, then translated version of the same lattice (see [27]).

The next step is to extend these results to planar modules by identifying the modules with rings of cyclotomic integers ([9]). General results for lattices will of course also hold in three dimensions, but the approach in this case will not be via complex numbers but via quaternions.

Finally, it is expected that the ideas behind the study of a shifted lattice may be applied to crystals where there is more than one atom per primitive unit cell, as described in [25, 26]. A related mathematical problem is the following: Suppose the coincidence problem for a sublattice (of finite index) of a given lattice has already been solved. What can be deduced about the coincidence indices of the the original lattice? A possible approach to answer this question involves looking at the coincidences of the corresponding cosets, which are just shifted copies of the sublattice.

Acknowledgments

The authors are grateful to the referee for his valuable remarks on the manuscript. M. Loquias would like to thank the Deutscher Akademischer Austausch Dienst (DAAD) for financial support during his stay in Germany. This work was supported by the German Research Council (DFG), within the CRC 701.

References

[1] Friedel G 1911 *Leçons de Cristallographie* (Paris: Blanchard)
[2] Kronberg M L and Wilson F H 1949 Secondary recrystallization in copper *Trans. A.I.M.E.* **185** 501-14
[3] Ranganathan S 1966 On the geometry of coincidence-site lattices *Acta Cryst.* **A51** 335-43
[4] Bollmann W 1970 *Crystal Defects and Crystalline Interfaces* (Berlin: Springer)
[5] Pumphrey P H 1976 Special high angle grain boundaries. In *Grain Boundary Structure and Properties* ed G A Chadwick and D A Smith (London: Academic Press Inc.) chapter 5 pp 139-200
[6] Grimmer H, Bollmann W and Warrington D H 1974 Coincidence-site lattices and complete pattern-shift in cubic crystals *Acta Cryst.* **A30** 197-207
[7] Grimmer H 1974 Disorientations and coincidence rotations for cubic lattices *Acta Cryst.* **A30** 685-88
[8] Grimmer H 1974 A reciprocity relation between the coincidence site lattice and the DSC lattice *Scripta Met.* **8** 1221-24
[9] Pleasants P A B, Baake M and Roth J 1996 Planar coincidences for N-fold symmetry *J. Math. Phys.* **37** 1029-58

[10] Baake M 1997 Solution of the coincidence problem in dimensions $d \leq 4$. In *The Mathematics of Long-Range Aperiodic Order* ed R V Moody (Dordrecht: Kluwer) pp 9-44
[11] Warrington D H 1993 Coincidence site lattices in quasicrystal tilings *Mat. Science Forum* **126-128** 57-60
[12] Warrington D H and Lück R 1994 *Proc. Intl. Conf. on Aperiodic Crystals (Les Diablerets)* ed G Chapuis and W Paciorek (Singapore: World Scientific) pp 30-34
[13] Radulescu O, Warrington D H and Lück R 1997 *Proc. Intl. Conf. on Aperiodic Crystals (Alpe d'Huez)* ed M de Boissieu et al (Singapore: World Scientific) pp 783-88
[14] Warrington D H and Lück R 2001 Healing of slip planes and interfaces in quasiperiodic patterns *Ferroelectrics* **250** 357-60
[15] Zeiner P 2005 Symmetries of coincidence site lattices of cubic lattices *Z. Kristallogr.* **220** 915-25
[16] Zeiner P 2006 Coincidences of hypercubic lattices in 4 dimensions *Z. Kristallogr.* **221** 105-14
[17] Baake M and Zeiner P 2008 Coincidences in four dimensions *Phil. Mag.* **88** 2025-32
[18] Baake M, Grimm U, Heuer M and Zeiner P 2008 Coincidence rotations of the root lattice A_4 *European J. Combin.* **29** 1808-19
[19] Zou Y M 2006 Structures of coincidence symmetry groups *Acta Cryst.* **A62** 109-14
[20] Huck C 2009 A note on coincidence isometries of modules in Euclidean space *Z. Kristallogr.* **224** 341-44
[21] Glied S 2008 Similarity versus coincidence rotations of lattices *Z. Kristallogr.* **223** 770-72
[22] Glied S. Similarity and coincidence isometries for modules *Canad. Math. Bull.* in press
[23] Gleiter H and Chalmers B 1972 High-angle grain boundaries. In *Progress in Materials Science* vol 16 ed B Chalmers et al (Oxford: Pergamon Press) chapter 1 pp 1-12
[24] Fischmeister H F 1985 Structure and properties of high angle grain boundaries *J. Phys. Colloques* **46** C4-3-C4-23
[25] Gratias D and Portier R 1982 General geometrical models of grain boundaries *J. Phys. Colloques* **43** C6-15-C6-24
[26] Pond R C and Vlachavas D S 1983 Bicrystallography *Proc. R. Soc. of London* **A386** 95-143
[27] Loquias M J C and Zeiner P. Coincidences of shifted lattices and $\mathbb{Z}$-modules, in preparation

Colour groups in tilings with singularities

Reinhard Lück

Weilstetter Weg 16, D-70567 Stuttgart, Germany

r.v.lueck@web.de

Abstract. The paper is focussed on colour symmetry of regular tilings with singularities. Colour indices of the tilings considered are not only determinded by the symmetry of the tilings but additionally by the rotational symmetry of the singularities. The number of the derived colour indices is found to be limited for a tiling. Coloured examples are presented. The combination of phyllotactic analysis and colour symmetry is discussed. Periodic arrangements of singularities can be introduced to visualize special hyperbolic tilings.

1. Introduction

Colour symmetries have been discussed for Euclidean periodic and quasiperiodic [1, 2] as well as for hyperbolic tilings [3]. The advantage of this research is – beside the aesthetics of the results – a better understanding of atomic superstructures and of coincidence site lattices. Regular tilings with singularities were described by Grünbaum and Shephard [4] in the Section 'Well-behaved tilings' of their textbook. Such tilings are self-similar with a scaling factor like quasiperiodic patterns. There are three types of regular patterns in two dimensions and Euclidean space. These are composed of (i) 4 quadrangles sharing a vertex (4^4), (ii) 6 triangles sharing a vertex (3^6) and (iii) 3 hexagons sharing a vertex (6^3). Tilings of the type (4^4) and the tilings (3^6) and (6^3) will be treated separately. The singularities have a rotational symmetry different from the vertex symmetry; this is similar as for so-called Volterra dislocations or inclinations related to disclinations ([5] and references therein). The colour symmetry of regular patterns without singularities was described by Baake [1] and in other papers [2] and references therein. In the present case, the colour symmetry should be compatible with both sufficient rotational colour symmetry of the undistorted regular pattern and the rotational symmetry of the singularity. In addition, periodic arrangements of singularities will be introduced for the visualization of special hyperbolic tilings. The colour indices of possible colourings in the considered tilings are not only determinded by the symmetry of the tiling – as described in references [1], [2] and [3] – but additionally by the rotational symmetry of the singularities. Examples including phyllotactic analysis will be presented and discusssed.

2. Tilings of the type (4^4)

2.1. Five-fold singular symmetry

There are several tiles of quadrangular shape which form tilings with singularities. The Penrose kite [4] may be regarded as a very common one. As the Penrose kite is related to five-fold symmetry in quasiperiodic patterns, a pattern with five-fold singularity can be designed. The scaling factor of the resulting self-similar tiling is the golden mean $\tau = (5^{1/2} + 1)/2$ as in quasiperiodic patterns. The five-colour (colour index h=5) decoration of the (undistorted) square lattice is well-known from the Σ5 coincidence site lattice (CSL) [2]. The corresponding colouring is not a perfect one, since there is no colour mirror symmetry; it may be considered as semi-perfect. Moreover, an enantiomorphic pair of colourings exists. This colouring can be easily applied on a tiling with five-fold singularity. The colouring of a tiling composed of Penrose kites with a five-fold singularity is demonstrated in Fig. 1. The tiling (4^4) can be decorated with two colours as it is well known from the chess board. This colouring can be transformed to the kite tiling with singularity. A combination of 5 and 2 colours results in 10 colours. A combination of the two different colourings with 5 colours results in a perfect colouring with 25 colours. No other colour indices than 1, 2, 5, 10, 25, and 50 exist for this pattern.

Figure 1. Five-fold pattern composed of Penrose kites. (4^4) tiling with five-fold singular centre. The scaling factor is τ with rotation (e.g. by π) and τ^2 without rotation. Decoration is performed with five colours (h=5). There is no mirror symmetry for colours and therefore no perfect colouring. An enanthiomorphic pair exists for 5 colours. 2, 25 and 50 colours are possible with perfect colouring. An enanthiomorphic pair with 10 colours can be derived.

2.2. Six-fold and three-fold singular symmetries

Kites may be also arranged with other symmetries and/or other scaling factors. In an arrangement with six-fold symmetry eight (Fig. 2) or nine (Fig. 3) colours are possible. The tilings of Figs 2 and 3 are equivalent, but their scaling factors are different. The maximum colour index is 72. The kites may be distorted to asymmetric quadrangles. Figure 4 with nine colours is equivalent to Fig. 3; the pattern in Fig. 4 has the same linear scaling factor $3^{1/2}$. Possible colour indices are 1, 2, 4, 8, 9, 18, 36, and 72. Figure 5 is another example of an asymmetric tiling. The colour index is 5, no other colour indices than 1, 5, 9, and 45 are possible. It should be emphasized that there are many more rotational symmetries of singularities in (4^4) patterns.

Figure 2. 8 colours in a (4^4) tiling with six-fold singular centre.

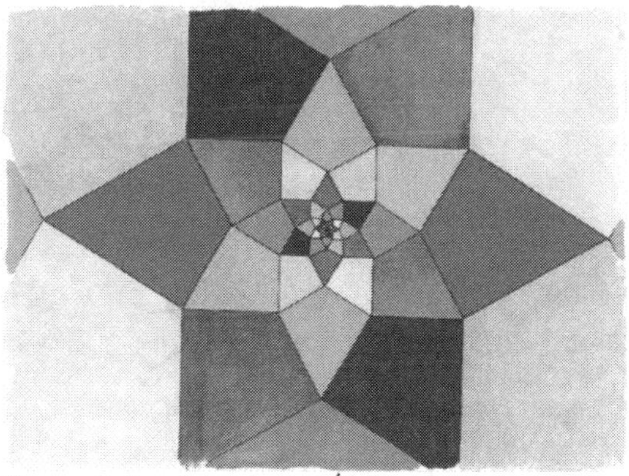

Figure 3. 9 colours in a (4^4) tiling with six-fold singular centre. Symmetric version.

Figure 4. 9 colours in a (4^4) tiling with six-fold singular centre. Asymmetric version.

Figure 5. 5 colours in an asymmetric (4^4) tiling with a three-fold singular centre. Maximum colour index is 45.

3. Tilings with curved lines

Before discussing tilings composed of triangles and hexagons some modifications with curved boundaries should be introduced. From Figs 4 and 5 it may be concluded that spirals are essentials of such tilings. The spirals may be taken directly from the tilings with scaling properties. Another possibility is a conformal transformation of the regular tilings. Conformal transformation or conformal mapping is a standard mathematical procedure described e.g. in textbooks (compare the classical Books by Zeev Negari [6] or by Ludwig Bieberbach [7]) or mathematical handbooks [8].Conformal transformation can be performed by a holomorphic function in the planar representation of complex numbers. Interest in the theory of conformal transformation arose in the second half of the 19th century; applications were developed in the 20th century. For the present purpose it is important that such conformal transformation results in logarithmic spirals. The presented examples are designed using logarithmic spirals. The property of the logarithmic spiral provides that all conformal symmetric (4^4) patterns are constructed with the same spiral. Other spirals (e.g. taken from the self-similar tilings) can be applied to change the scaling factor. We applied both methods, some results will be presented. Figure 6 is analogous to Fig. 1. Five colours are used to decorate. The numbers of left-hand and right-hand spirals are both 5. These numbers determine the maximum colour index. Figure 7 is a conformal

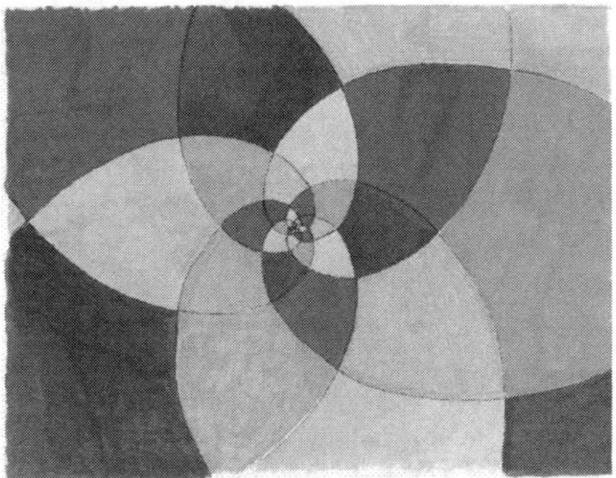

Figure 6. Conformal transformation of (4^4) tiling with a five-fold singular centre. Decoration with 5 colours.

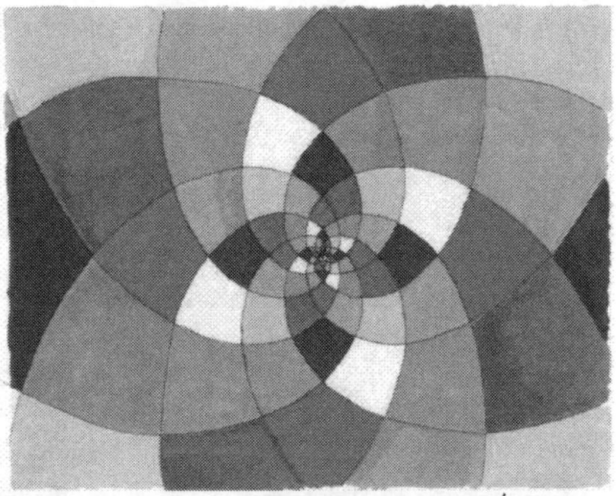

Figure 7. Decoration of an conformal (4^4)tiling with an eight-fold centre with 8 colours.

transformation of a (4^4) pattern with an eight-fold centre. The numbers of spirals are 8 for both orientations. The figure is decorated with 8 colours. Possible colour indices for this tiling are 1, 2, 4, 8, 16, 32, 64, 128. If the numbers of the left-hand and right-hand spirals are respectively m and n, the maximum colour index for (4^4) tilings is

$$h_{\max} = m^2 + n^2 \tag{1}$$

There are more colour indices if the colour group with maximum colour index allows subgroups. The colour indices are exactly the orders of the subgroups of the colour group of maximum colour index; there are no subgroups not corresponding to a colouring.

Figure 8. Four colours in a conformal (3^6) tiling with twelve- singular centre.

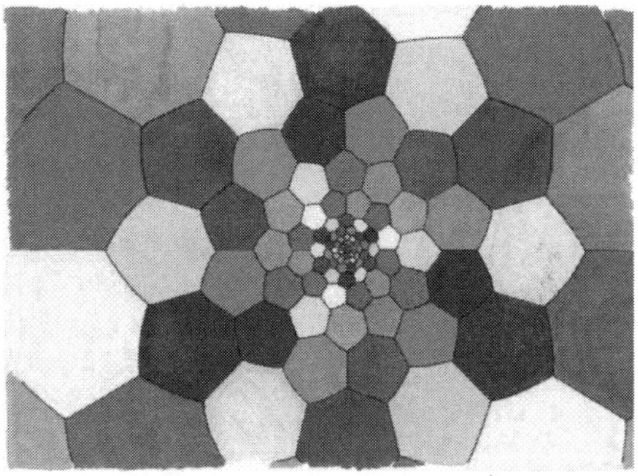

Figure 9. 16 colours in a conformal (6^3) tiling with twelve-fold singular centre.

4. Tilings of the types (3^6) and (6^3)

Tilings composed of triangles or hexagons may be designed by self-similar extension or by conformal transformation. Examples are presented in Figs 8 and 9. A (3^6) tiling with a twelve-fold singular centre is decorated with 4 colours in Fig. 8. Figure 9 represents a (6^3) conformal tiling with a twelve-fold singular centre which is decorated with 16 colours. In case of regular hexagons or vertices of equilateral triangles, the colour indices of undistorted patterns may be generalized by

$$h = M^2 + 3N^2 \tag{2}$$

This equation holds also for the maximum number of colours in six-fold patterns with singularities. The meaning of M and N is determined by the number of spirals crossing perpendicularly edges (M) and vertices (N), respectively. More generally, for colouring hexagons three types of spirals (including straight lines) approaching the singular centre should be considered. If the numbers of these dual lines (i.e. they are crossing centres of hexagons and crossing perpendicularly hexagon edges) are l, m, and n, the maximum colour index was found to be

$$h_{\max} = (l^2 + m^2 + n^2)/2 \tag{3}$$

5. Phyllotactic analysis and colourings

One application of the discussion of tilings with singularities is the phyllotactic analysis of supposed or real botanical materials. Several papers are published on this subject in literature on aperiodics ([9], [10], [11] and references therein) but nothing is said about colours. A short test of several cones (e.g. pine-, scots-pine- or larch-cones) resulted in the combination of consecutive Fibonacci numbers for the spirals involved. For quadrangles, equation (1) should be applied; and the following numbers result for the maximum colour indices $h_{\max}$:

(5, 13, 34,) 89, 233, 610, ..

Numbers within parentheses are examples which were not found in the present analysis of cones. The maximum colour indices are again Fibonacci numbers. In the language of Kepler [12, 13], these are the Fibonacci numbers producing 'males', i.e. the product of consecutive numbers exceeds a square number by one (Kepler wrote 'Einheit'). The skipped Fibonacci numbers produce 'females', where the product of consecutive numbers is less by one than a square number. 5, 13, 89, 233 ... are prime numbers; therefore, these groups do not have subgroups. Important subgroups of the colour indices 34 and 610 are formed by the colour index 2.

For hexagons, however, applying equation (3) the maximum colour indices h_{max} are

(7, 19,) 49, 129, 337, (883, ...)

The tilings can be characterized by three consecutive Fibonacci numbers for *l*, *m*, and *n*:

(1, 2, 3; 2, 3, 5;) 3, 5, 8; 5, 8, 13; 8, 13, 21; (13, 21, 34;....)

7, 19, 337, 883... are prime numbers, the colour groups do not have subgroups. 49 as a square number implies an enantiomorphic pair of subgroups with colour index 7; 129 has subgroups with 3 and 43 as colour index. Pine-cones of the type 5, 8, 13 may be decorated with three colours.

6. Periodic arrangement of singularities and hyperbolic tilings

Coloured hyperbolic tilings were investigated recently [3], [14]. Colour indices were derived for several tilings; perfect and semi-perfect colourings were distinguished [3]. Examples with ten colours were published using the Poincaré model [14]. A disadvantage of the Poincaré model is that the tile size is increasingly shrinking with increasing distance from the centre. Some colour symmetry properties are therefore difficult to be detected. Searching for other possibilities of visualization we introduce a periodic arrangement of singularities. Such figures require the general symmetry of a periodic regular tiling. Another condition is a violation of Euler's equation for Euclidian geometry. The determination of colour index requires the consideration of an additional condition. As a consequence, the number of colour indices is reduced. Figure 10 represents a (4^6) tiling with a periodic arrangement of four-fold singularities (white areas), the pattern is decorated with 6 colours.

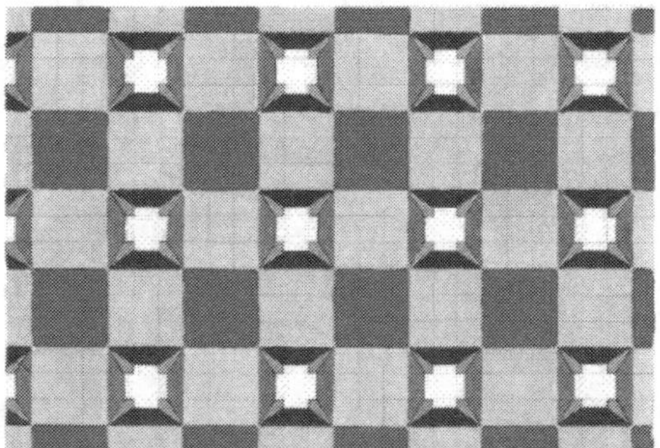

Figure 10. Pattern (4^6) with a periodic arrangement of four-fold singular points, decoration with 6 colours.

Figure 11. A tiling (3^8) with a periodic arrangement of three-fold singular points. It is decorated with 8 colours.

Figure 11 is a representation of the tiling (3^8) with three-fold singular points. This tiling allows a decoration with 2, 4, 8, 16, or 32 colours. Figure 11 is decorated with 8 colours.

It seems to be interesting to transform such figures with periodic arrangements of singularities into tilings with locally regular vertices. A conformal periodic tiling should require that the singularities are formed by closed lines of six-, three-, four-, or two-fold symmetry. A sketched pattern analogous to Fig. 11 is presented in Fig. 12 as an example. An attempt to transform the vertices to a locally regular eight-fold rotational symmetry similar as in the Poincaré model of (3^8) is presented in Fig. 13. Both representations (Figs 12 and 13) are decorated with 16 colours. To visualize a decoration with 4 colours, the 16 colours are organized in four groups (bluish, greenish, yellowish/brownish and reddish) of four different tones. The comparison of the periodic arrangement of singularities and the Poincaré disc is very helpful to understand colour symmetry and colour orbits. Figure 12 demonstrates periodicity of colours, this corresponds to a symmetry operation difficult to detect in the Poincaré disc. However, not all hyperbolical colour symmetries can be transformed to a pattern with a periodic arrangement of singularities. A hyperbolic tiling as it is represented in the Poincaré model can not be completely transformed into a conformal mapping with such a periodic arrangement of singularities; in other words: there are (relatively small) areas in Fig, 13 not represented in Fig. 14.

Figure 12. A tiling (3^8) with periodic arrangement of three-fold singularities and 16 colours.

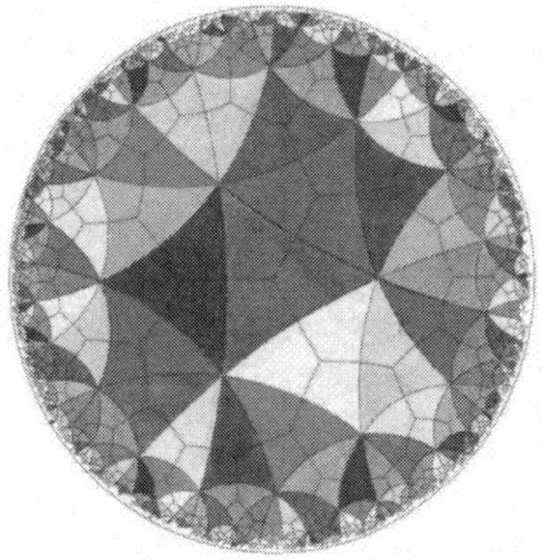

Figure 13. Poincaré disc of (3^8) decorated with 16 colours.

7. Conclusion

Tilings with singularities can be designed by a self-similar procedure or by conformal transformation of a regular tiling. Only a limited number of colourings were found to be possible for regular tilings with a single singularity. The maximum colour index can be easily determined counting the number of differently oriented spirals approaching the singularity. The same procedure can be applied to colour phyllotactic arrangements. Some coloured hyperbolic tilings can be represented by a periodic arrangement of singularities.

8. Acknowledgement

I am grateful to the Bielefeld group, especially to Dirk Frettlöh, for a discussion of the present topic, to Professor Ellen Baake for a discussion on phyllotactic analysis, to David H. Warrington for a critical reading of the manuscript and to an unknown referee for some important proposals.

Journal of Physics: Conference Series **226** (2010) 012027 doi:10.1088/1742-6596/226/1/012027

References

[1] Baake M 1997 Combinatorial aspects of colour symmetry *J. Phys. A: Math. Gen.* **30** 2687-98 mp arc/02-323

[2] Lück R 2008 Colour symmetry of 25 colours in quasiperiodic patterns *Phil. Mag.* **88** 2049-58

[3] Frettlöh D 2008 Counting perfect colourings of plane regular tilings *Z. Krist.* **223** 773-776

[4] Grünbaum B and Shephard C G 1987 Tiling and Patterns (Freeman, New York) 113-164

[5] Kleman M 2009 Interplay between grain boundaries and disclinations in condensed matter physics Int. J. Mat. Res.**100** 1449-1455

[6] Nehari Z 1952 Conformal mapping McGraw-Hill, New York – Toronto – London, reprinted 1975 Dover Publication Inc. New York

[7] Bieberbach L 1956 Einführung in die konforme Abbildung 5th edition, Berlin; 1953 Conformal mapping, translation of the 4th edition which was originally published 1949, Chelsea

[8] Meyers Handbuch über die Mathemtik 1972, H Meschkowski ed. 2nd edition Bibliographisches Institut, Mannheim – Wien - Zürich

[9] Rivier N Pieranski P and Rothen F 1994 Conformal crystals and their defects *Aperiodic'94* (World Scientific, Singapore) Ed Chapuis G and Paciorek W 1211-25

[10] Rivier N and Goldar A 1997 Entropy of aperiodic crystal generated by spirals *Aperiodic'97* (World Scientific, Singapore) Ed de Boissieu M Verger-Gaugry J-L and Currat R 131-135

[11] Jean R V 1992 Nomothetical modelling of spiral symmetry in biology *Fivefold Symmetry* (World Scientific, Singapore) Ed Hargittai I 505-528

[12] Brandmüller J 1992 Fivefold symmetry in mathematics, physics, chemistry, biology and beyond *Fivefold Symmetry* (World Scientific, Singapore) Ed Hargittai I 11- 31

[13] Kepler J 1619 Harmonices Mundi Libri Quinque (Five books on the harmony of the world) Joannis Plancus, Lincii Austria; translated and commented by M Caspar, Weltharmonie (Harmony of the world) R. Oldenbourg-Verlag, Munich 1984 (cited after [12])

[14] Lück R and Frettlöh D 2008 Ten colours in quasiperiodic and regular hyperbolic tilings *Z. Krist.* **223** 782-784

Journal of Physics: Conference Series **226** (2010) 012028 doi:10.1088/1742-6596/226/1/012028

Aperiodic arrays of quantum dots: influence of external magnetic and electric fields

N E Kaputkina [1], **Yu E Lozovik** [2], **R F Muntyanu** [1] **and Yu Kh Vekilov** [1]

[1] State Technological University "Moscow Institute of Steel and Alloys", Moscow, Russia

[2] Institute of Spectroscopy, Academy of Sciences of Russia, Troitsk, Russia

E-mail: kaputkina@mail.ru

Abstract. Electronic and excitonic excitations in aperiodic sequences of quantum dots (Thue-Morse, Cantor, Fibonacci, Double-period) were studied in external electrical and magnetic field. Single-particle and two-particle tunneling probability was taken into consideration. Transmission coefficient was determined using quasi-classical approximation and range of values of confining potential and interdot distances when tunneling is essential was estimated. An external electrical and magnetic field effect on electron localization was taken into consideration, an effective steepness of confining potential in magnetic field was appeared as control parameter of the problem. Energy spectrum of aperiodic quantum dot sequences in external magnetic field was obtained. Possibility to tune the state of the system by magnetic field was studied. The increase of the external electrical field shifts the energy states of the particle in a quantum dot and contributes to particle localization. The localization of the excitations is possible at the finite values of the perturbation in the case of aperiodic sequences of quantum dots (contrary to the case of periodical sequences).

1. Introduction

We investigate the energy spectra and tunneling of electronic and excitonic excitations in aperiodic sequences of quantum dots (QDs). These objects are perspective for nanoelectronics and optoelectronics [1]. The most investigated heterostructures are based on 2 types of semiconductors: GaAs and its solid solutions $Al_xGa_{1-x}As$ or $In_xGa_{1-x}As$ with different composition x = 0.1 – 1. These materials have direct gap structure with energy gap width $E_g \sim 1$ eV. $m_e = 0.07m_0$ for GaAs effective electron mass in the conduction band (m_0 is vacuum electron mass).

One-dimensional periodic structures are of interest as possible photonic crystals [see [2] and Refs therein], waveguides [3-6].

A lot of interesting works were done for quasi-periodic chains of atoms (see [7] and Refs therein). The energy spectrum of horizontal and vertical molecules of QDs were studied in detail [8-11]. Aperiodic chains of QDs have a lot of advantages. With the help of the external parameters it is possible to control such characteristics as the particle localization, tunneling probability, energy spectrum of a quasi-crystalline sequences of QDs. For example, tuning the magnetic field and the value of bias voltage across the double dot it is possible to control the effect of resonance tunneling. This process can occur in the case when the corresponding levels in neighboring QDs align and are situated within the bias window. The transport process is registered by sharp resonance peaks in I-V curves [12]. In the case of slow relaxation the particle is "trapped" within the single QD. We show how such parameters as the steepness of confining parabolic potential, the distance between neighboring QDs, the external transverse magnetic field influence on the localization of excitations and the resonance tunneling. The smoothness of the energy spectra of aperiodic sequences of QDs was studied using the method of level statistics. The energy spectra of Fibonacci (see [13]), Thue-Morse, Cantor, Double-period sequences exhibit singular behavior. The detailed analysis of energy and wave-functions spectra will be published elsewhere.

One can tune the system state by changing controlling parameters, the steepness of confining parabolic potential, the distance between neighboring QDs, external magnetic and electrical fields. The influence of magnetic and electrical field on the localization of excitations, the energy states positions, and the resonance tunneling is analyzed in detail below.

2. Eigenstates and eigenfunctions of excitations in a single QD and the influence of magnetic field.

The particles in QDs are confined by a harmonic potential of steepness α. The Hamiltonian for a single dot is

$$\hat{H} = -\frac{\hbar^2}{2\mu}\frac{d^2}{dx^2} + \alpha x^2 \tag{1}$$

where μ is the effective mass of the particle The energy spectrum and normalized eigenfunctions of a particle in single QD in one-dimensional case are

$$E_n = \hbar\sqrt{\frac{2\alpha}{\mu}}(n+1/2), \tag{2}$$

$$\Psi_n(x) = \left(\frac{\beta}{\pi}\right)^{1/4} \frac{1}{\sqrt{2^n n!}} \exp(-\beta x^2/2) H_n(x\sqrt{\beta}), \tag{3}$$

Where α is the steepness of confining potential, x is the distance between the center of the QD and the particle; $H_n(x\sqrt{\beta})$ are Hermite polynomials, and $\beta = \sqrt{2\mu\alpha}/\hbar$.

It is possible to show that when the external transverse magnetic field is applied the eigenenergies and normalized eigenfuctions have the similar form as in the previous case but α is replaced on the effective steepness of the confinement potential $\lambda = (\mu/2)(eB/\mu)^2 + \alpha$, where B is the external magnetic field.

The wave functions of the Fibonacci generations were obtained by the method of Komoto, Kadanoff, and Tang [14]. Their critical behavior was proved by the analysis of the *2p*-norm[13].

The solution of the secular equation gives the first order corrections to the energy by perturbation theory:

$$\begin{vmatrix} H_{11} - E^{(1)} & H_{12} \\ H_{21} & H_{22} - E^{(1)} \end{vmatrix} = 0 \tag{4}$$

$$E_{1,2}^{(1)} = 0.5(H_{11} + H_{22}) \pm 0.5\left[(H_{11} - H_{22})^2 + 4H_{12}^2\right]^{1/2} \tag{5}$$

where the matrix elements are:

$$H_{11} = \int_{-\infty}^{\infty} \Psi_n^* \hat{V} \Psi_n \, dx, \quad H_{12} = \int_{-\infty}^{\infty} \Psi_n^* \hat{V} \Psi_n \, dx$$

$$H_{21} = \int_{-\infty}^{\infty}\int_{-\infty}^{\infty} \left(\Psi 2_n^{(0)}\right)^{*} H' \Psi 1_n^{(0)} dx_1 dx_2\ ,\quad H_{22} = \int_{-\infty}^{\infty}\int_{-\infty}^{\infty} \left(\Psi 2_n^{(0)}\right)^{*} H' \Psi 2_n^{(0)} dx_1 dx_2\ , \qquad (6)$$

where $\Psi 1_n^{(0)}, \Psi 2_n^{(0)}$ are the normalized eigenfunctions of the particle in the 1st and 2nd QD respectively.

The energy of the particle interaction $H' = 1/(4\pi\varepsilon\varepsilon_0) e^2 / (d + x_2 - x_1)$ is considered as the perturbation. The calculation in the case of excitons is obtained by changing the effective mass from $\mu = 0.07m_0$ to $M = 0.57m_0$ (only heavy holes are considered, the existence of the light holes is neglected).

Now, it is possible to estimate the distance between the centers of QDs when the first order corrections to the energy of the particle become inessential. It is 180Å when $\alpha = 10^{-8} J/m^2$ for the electron. Thus, we can use the pair interaction to obtain the energy spectrum of an aperiodic sequence of QDs. In contrast to aperiodic sequences of atoms, for aperiodic sequences of QDs rather small magnetic fields are essential.

3. The tunneling probability and the influence of magnetic field.

The expression for the single-particle and two-particle tunneling probability was obtained using quasi-classical approximation:

$D=\frac{e^{2L}}{\left(1+\left(2e^{2L}\right)\right)^{2}}$, $L=\frac{1}{\hbar}\int_a^b |p| dx$, $p(x)=\sqrt{2m(E-U(x))}$ is the particle pulse. E is the particle energy. $U(x)$ is the potential energy. a, b are the turning points.

The energy level splitting versus α and d dependence is given by formula:

$$E_2 - E_1 = \frac{\hbar}{m\int_b^a \frac{dx}{p(x)}} \exp\left(-\frac{1}{\hbar}\int_{-a}^{a} |p| dx\right),\quad p(x) = \sqrt{2m(E-U(x))}\ ;\quad \text{a and b are the turning points.}$$

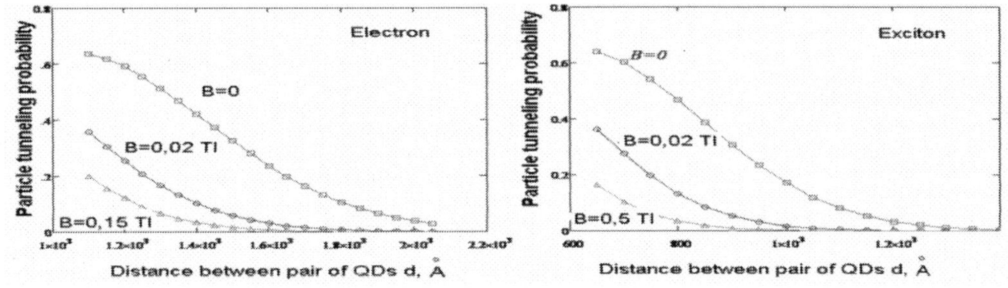

Figure 1. The electron and exciton tunneling probability without magnetic field and in magnetic field *versus* the distance between the centers of the QD pair.

The external magnetic field increasing brings to particle probability peaks close in, thus particle is localized within some QD and B acts like effective steepness potential (see Figure 1).

The expression for the tunneling probability by the perturbation theory is used to show the resonance tunneling of the excitations between the pair of QDs:

$$W_{fi} = \frac{2\pi}{\hbar} \left|V_{fi}\right|^2 \delta\left(E_f - E_i\right) d\nu_f\ , \qquad (7)$$

where $V_{fi} = \int_{-\infty}^{\infty}\int_{-\infty}^{\infty} \left(\Psi 1_f^{(0)}\right) H' \Psi 2_i^{(0)} dx_1 dx_2$ is the matrix element of interaction, and $d\nu_f$ is the density of states.

The particle tunneling is possible only in the case when the quantum states in the neighboring QDs are aligned. The finite perturbation destroys the alignment of quantum states of QDs and localizes the excitations (see [13])

4. The influence of the external electrical field on the energy spectrum of aperiodic sequences of QDs

Let QD is located in the *GaAs* wire and confined by the $Al_xGa_{1-x}As$ layers in two directions (the energy gap of last layer is larger and increases with x). Particles are confined by the parabolic potential in the third direction of the *GaAs* wire (produced, e.g., by lithography.). Only single electron in each QD is considered and spin effects are not taken into account. The parabolic potential is obtained by the gate voltages.

We studied a sequence model based on the pairs of QDs and the distances between their centers as the structural elements of the sequences. The parameters of each QD were identical. Two distances *dA*=400 Å, *dB*=200 Å were used to construct such sequences as Fibonacci, Thue-Morse, Double-period, Cantor, and d = 300 Å was used for the periodic sequence construction. For simplicity we suppose that the electrons in each dot occupy the same level n=3. The number of words in each alphabet equals to 20.

The substitution rules for different types of sequences are following:

Fibonacci sequence: $\sigma_F | A \to AB, B \to A$; Thue-Morse sequence: $\sigma_{TM} | A \to AB, B \to BA$; Period-doubling sequence: $\sigma_{PD} | A \to AB, B \to AA$; Cantor sequence: $\sigma_C | A \to ABA, B \to BBB$.

One-dimensional sequences of QDs are arranged between two leads. At the moment when an electron leaves it's state at the last QD and tunnels into the right lead the resonant tunneling is possible under the influence of the external electrical field when the energy states in the neighboring dots are aligned. The increase of the steepness of the parabolic potential and the external magnetic field contributes to particle localization [13]. The tunneling probability decreases with growth of magnetic field. The energy spectra of the aperiodic sequences are obtained using the approximation of pair-correlation interaction. Coulomb interaction energy and the electrical field acted on charge transfer were considered as the perturbations in the case when their values are much less than the corresponding interlevel distance. The energy spectra were obtained and plotted *versus* the external electrical field.

$E_{n,k} = \hbar\sqrt{\alpha/\mu}(n+1/2) - q^2F_k^2/(2\alpha)$ is the energy of the electron at the n-th state in the QD.

$1 \le n \le N$ is a generation index; N is the number of words in the alphabet; index k denotes the number of the applied electrical field values; α is the steepness of confining potential, and μ is the electron's effective mass in *GaAs*; F is the external electrical field.

It is important to estimate the values of electrical field and the distances when one can consider the the external electrical field as a perturbation. The typical value of electron's energy in the QD is about several meV $(\alpha \sim 10^{-8}\ J/m^2)$, the typical value of the first-order correction to the energy E is $(10^{-4} \div 10^{-5}) \cdot E$. So the values of the electrical field must not exceed $10^4\ V/m$.

According to our calculations the Coulomb energy of interaction can be considered as a perturbation at the distances of order ~550 Å for Fibonacci, Thue-Morse, Double-period and Cantor sequences.

Matrix elements required to determine the first-order corrections to the energy are:

$$H_{i,f} = (4\pi\varepsilon\varepsilon_0)^{-1} \int_{-\infty}^{\infty}\int_{-\infty}^{\infty} \left|\Psi_i(\rho_i)\right|^2 / (d + \rho_i - \rho_f) \overline{\Psi}_f(\rho_f) d\rho_i d\rho_f ; \qquad (8)$$

$$V_{i,f} = \int_{-\infty}^{\infty}\int_{-\infty}^{\infty} \Psi_i(\rho_i) F(d+\rho_i-\rho_f)\bar{\Psi}_f(\rho_f)\, d\rho_i d\rho_f \,, \qquad (9)$$

where d is the distance between the centers of QDs. ρ_i is the distance between the i-th QD and the position of the particle.

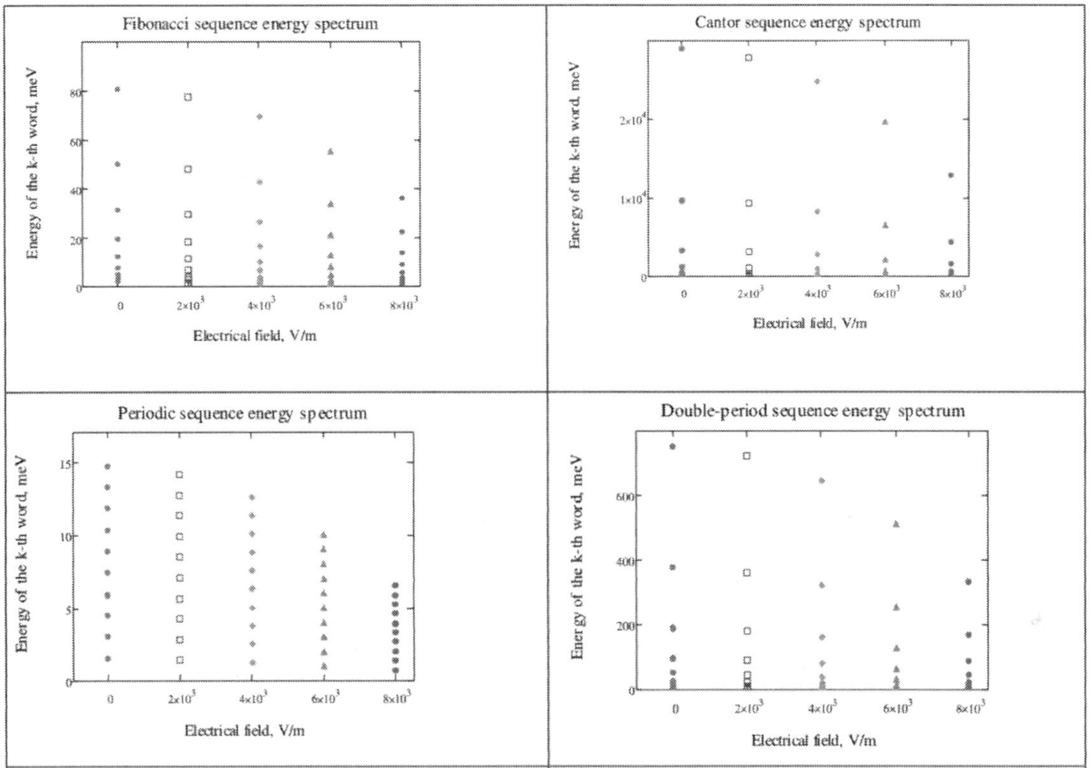

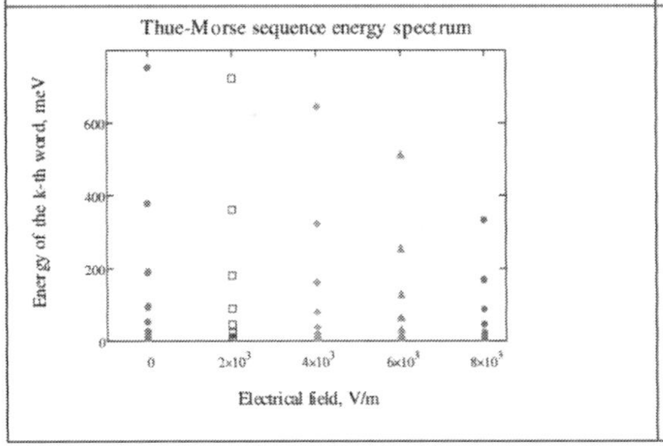

Figure 2 The energy spectrum of Fibonacci, Thue-Morse, Cantor, Double-period, Periodic sequences versus the external electrical field $(0,2,4,6,8)\cdot 10^3$ V/m. The distances dA=400 Å, dB=200 Å were used for Fibonacci, Thue-Morse, Double-period, Cantor sequences, and d = 300 Å was used for the periodic sequence construction. The electrons in each dot occupy the same level n=3. The number of words in each alphabet equals to 20.

The Fibonacci Thue-Morse, Double-period and Cantor sequences are based on the two-letter alphabet (A, B), so the first-order corrections to the energy are:

$$E_{Q_{n,k}}^{(1)} = N_{A_n}\cdot E_{Ak}^{(1)} + N_{B_n}\cdot E_{Bk}^{(1)} + N_{AB_n}\cdot E_{ABk}^{(1)} + N_{AA_n}\cdot E_{AAk}^{(1)} + N_{BB_n}\cdot E_{BBk}^{(1)} \,, \qquad (10)$$

where N_{A_n}, N_{B_n} are the numbers of letters A and B in the n-th word; $N_{AB_n}, N_{AA_n}, N_{BB_n}$ are the numbers of neighboring pairs of letters, $E_{Ak}^{(1)}$, $E_{Bk}^{(1)}$, $E_{ABk}^{(1)}, E_{AAk}^{(1)}, E_{BBk}^{(1)}$ are the first-order corrections to the energy of the word *A*, *B*, and pairs of words AB, AA, BB respectively, calculated under the influence of the *k*-th value of electrical field.

Energy spectrum of aperiodic sequence:

$$E_{Q_{n,k}} = N_{A_n} \cdot E_{An,k} + N_{B_n} \cdot E_{Bn,k} + E^{(1)}_{Q_{n,k}} \qquad (11)$$

Energy spectrum of the periodic sequence:

$$E_{P_{n,k}} = k \cdot E_{n,k} + (k-1) \cdot E^{(1)}_{Pk}, \qquad (12)$$

where $E^{(1)}_{Pk}$ denotes the first-order corrections to the energy of the identical pairs of QDs.

If one compares the dependence of the aperiodic sequences with the dependence of the periodic sequence (see Figure 2) then one can notice that a much less value of the electrical field for the periodic sequence is required to shift the energy of the 10th word to the previous value for example. This supports our idea that the transport blocking occurs at the infinitively small values of perturbation in the case when periodic sequences are considered, and finite values of perturbations are needed to block the transport in the case of aperiodic sequences.

5. Conclusions

The energy spectra of aperiodic sequences Thue-Morse, Cantor, Fibonacci, Double-period were obtained in the external electrical and magnetic field. The increase of the external electrical field shifts the energy states of the particle in a QD and contributes to particle localization. The tunneling of the excitations is possible only in the case when the quantum states of neighboring QDs are aligned. The transport blockage of the excitations is possible at the finite values of the perturbation in the case of aperiodic sequences of QDs while relatively small values of perturbations (external fields etc.) are required to block the transport in the case of periodical sequences. External magnetic field influences also on the wave functions overlapping and tunneling probability. It is possible to tune the state of the system by magnetic field.

Acknowledgments This work was supported by the Russian Foundation for Basic Research (grants 08-02-01461-Ф, 09-02-001447-a, 08-02-00685-a) and the Russian Government Program "Development of High School Potential" (grant 2.1.1.1552).

References

[1] Kulbachinskii V A 2003 *Semiconductor physics and technics*, 2003, vol 37, issue 1
[2] Busch K, von Freymann,G, Linden S, Mingaleev S F, Tkeshelashvili L, Wegener M 2007 *Physics reports* vol 444 pp. 101-202
[3] Jones A L 1965 *J.Opt.Soc.Am.* vol 55 p 261
[4] Somekh S, Garmire E, Yariv A, Garvin H L, Hunsberger R G 1973 *Appl.Phys.Lett.* vol 22 p 46
[5] Stegeman G I, Segev M 1999 *Science*, vol 286 p 1518
[6] Christodoulidis D M, Lederer F, Silverberg I 2003 *Nature* vol 424 p 817
[7] Albuquerque E L, Cottam M G 2003 *Physics reports* vol 376 pp. 225-337
[8] Kaputkina N E, Lozovik Yu E 1998 *Physics of the Solid State* vol 40 pp 1929-1934
[9] Kaputkina N E, Lozovik Yu E 1998 *Physics of the Solid State* vol. 40 pp 1594-1599
[10] Lozovik Yu E, Kaputkina N E 1998 *Physica Scripta* vol 57 p 542
[11] Kaputkina N E, Lozovik Yu E 2006 *J. Phys.: Condens. Matter*, vol 18 pp S2169–S2174
[12] Van der Wiel V G, De Franceschi S, Elzerman J M, Fujisawa T, Tarucha S, Kouwenhoven L P 2003 *Reviews of modern physics* vol 75 pp 1-21
[13] Kaputkina N E, Lozovik Yu E, Muntyanu R F, Vekilov Yu Kh 2008 *Philosophical Magazine*, vol 88 issue 13 pp 2253-2259
[14] Komoto M, Kadanoff L P, and Tang C 1983 *Phys. Rev. Lett.* vol 50 p 1870

Similarity of eigenstates in generalized labyrinth tilings

Stefanie Thiem and Michael Schreiber

Institut für Physik, Technische Universität Chemnitz, D-09107 Chemnitz, Germany

E-mail: {stefanie.thiem, michael.schreiber}@physik.tu-chemnitz.de

Abstract. The eigenstates of d-dimensional quasicrystalline models with a separable Hamiltonian are studied within the tight-binding model. The approach is based on mathematical sequences, constructed by an inflation rule $\mathcal{P} = \{w \rightarrow s, s \rightarrow sws^{b-1}\}$ describing the weak/strong couplings of atoms in a quasiperiodic chain. Higher-dimensional quasiperiodic tilings are constructed as a direct product of these chains and their eigenstates can be directly calculated by multiplying the energies E or wave functions Ψ of the chain, respectively.

Applying this construction rule, the grid in d dimensions splits into 2^{d-1} different tilings, for which we investigated the characteristics of the wave functions. For the standard two-dimensional labyrinth tiling constructed from the octonacci sequence ($b = 2$) the lattice breaks up into two identical lattices, which consequently yield the same eigenstates. While this is not the case for $b \neq 2$, our numerical results show that the wave functions of the different grids become increasingly similar for large system sizes. This can be explained by the fact that the structure of the 2^{d-1} grids mainly differs at the boundaries and thus for large systems the eigenstates approach each other. This property allows us to analytically derive properties of the higher-dimensional generalized labyrinth tilings from the one-dimensional results. In particular participation numbers and corresponding scaling exponents have been determined.

1. Introduction

Since the discovery of quasicrystals research aims at understanding the physical properties of these materials. Today many exact results are known for one-dimensional quasicrystals [1, 2] but the characteristics of wave functions in two or three dimensions have been clarified to much lesser degree. In particular, they have mainly been studied by numerical calculations of finite systems or periodic approximants due to the missing translational symmetry of these materials. For instance numerical results for the Penrose tiling, the Ammann-Kramer-Neri tiling or octagonal tiling models have been determined but for relatively small systems making it difficult to estimate the behavior for macroscopic system sizes [3, 4]. Further, it is possible to calculate some exact eigenstates of the tight-binding Hamiltonian on the Penrose tiling [5, 6], which is a first step to understand the nature of eigenstates and properties of quasicrystals in higher dimensions.

In order to get a better insight into the properties of wave functions in quasicrystals, another approach employs models of d-dimensional quasicrystals, which can be built from d separate one-dimensional systems yielding straightforward solutions. This allows us to study very large systems in higher dimensions based on the solutions in one dimension [7]. Unfortunately, these structures do not occur in real quasicrystals, however, they can be artificially constructed. Examples are conducting nanowires, coupled nanomechanical resonators, or photonic quasicrystals [7, 8]. The best studied model within this approach is based on the Fibonacci sequence corresponding to the golden mean [1, 7], whereas in this paper we concentrate on the other metallic mean quasiperiodic sequences [9, 10, 11], which have been less investigated so far.

At first we introduce the quasicrystal model based on metallic mean sequences and discuss the symmetry properties of the eigenstates in Sec. 2. This approach is expanded to higher dimensions for the labyrinth tiling in Sec. 3. In general, the grid splits into 2^{d-1} different tilings in d dimensions but the results indicate that the wave functions of these tilings approach each other for large systems. We give a reasonable argument for this behavior as well. This property can be used to calculate or even analytically derive characteristics for generalized labyrinth tilings. As an example, Sec. 4 comprises results for the participation ratio of the eigenstates and their scaling behavior in one, two, and three dimensions. The paper is concluded in Sec. 5.

2. Eigenstates and wave functions of metallic mean chains

The metallic mean sequences for a parameter b are defined by the inflation rule $\mathcal{P} = \{w \rightarrow s, s \rightarrow sws^{b-1}\}$ based on an alphabet containing the symbols $\{s, w\}$ with the starting symbol w. After a iterations we obtain the ath order approximant $\mathcal{C}_a$ of the quasiperiodic chain. Regarding the length f_a of this approximant $\mathcal{C}_a$ we can also find a recursive rule $f_a = bf_{a-1} + f_{a-2}$ with $f_0 = f_1 = 1$. The ratio of the lengths of two successive iterants in the limit $\lim_{a\to\infty} f_a/f_{a-1} = \lambda$ is given by different metallic means depending on b with a continued fraction representation $\lambda = [\bar{b}] = [b, b, b, ...]$. This leads to the well known Fibonacci sequence for $b = 1$ with the golden mean $\lambda_{\mathrm{Au}} = [\bar{1}] = (1 + \sqrt{5})/2$, while $b = 2$ results in the octonacci sequence with silver mean $\lambda_{\mathrm{Ag}} = [\bar{2}] = 1 + \sqrt{2}$ and $b = 3$ corresponds to the bronze mean $\lambda_{\mathrm{Bz}} = [\bar{3}] = (3 + \sqrt{13})/2$ [11].

For the orthogonal basis states $|l\rangle$ the tight-binding Hamiltonian can be written as

$$\mathcal{H} = \sum_{l=1}^{f_a} |l\rangle\, t_{l,l+1} \langle l+1| + \sum_{l=1}^{f_a+1} |l\rangle\, \varepsilon_l \langle l| \; . \tag{1}$$

This can be interpreted as an electron on a discrete grid which can hop from one vertex of this graph to any of its neighboring vertices and the aperiodicity of the grid is given by the quasiperiodic sequence with the symbols w and s corresponding to weak and strong bonds. Hence, the hopping parameter t is determined according to the sequence $\mathcal{C}_a$ with $t_s = 1$ for a strong bond and $t_w = v$ for a weak bond $(0 \leq v \leq 1)$. Further, usually one assumes zero on-site potentials $(\varepsilon_l = 0)$ for these models so that no vertex is energetically preferred [3, 10].

The discrete energy values E^i and corresponding wave functions $|\Psi^i\rangle = \sum_{l=1}^{f_a+1} \Psi_l^i\, |l\rangle$ are solutions of the time-independent Schrödinger equation for the quasiperiodic systems

$$\mathcal{H}\,|\Psi^i\rangle = E^i\,|\Psi^i\rangle \Longrightarrow E^i\Psi_l^i = t_{l-1,l}\Psi_{l-1}^i + t_{l,l+1}\Psi_{l+1}^i \; . \tag{2}$$

The results show that the eigenvalues are symmetric with respect to 0. However, we have to distinguish between even and odd system sizes $N_a = f_a + 1$: for even N_a all energy values E have a symmetric value $-E$ but for odd N_a there is one state $E^M = 0$, which has no corresponding state. Additionally, also the eigenfunctions show a symmetry. The eigenstates Ψ with an eigenvalue E and $\widetilde{\Psi}$ with the opposite eigenvalue $-E$ only differ by an alternating sign depending on the position l with

$$\widetilde{\Psi}_l = (-1)^l \Psi_l \; . \tag{3}$$

Further, for odd $N_a = 2M - 1$ the eigenvector Ψ_l^M associated to the eigenvalue $E^M = 0$ has a special structure, namely Ψ_l^M vanishes either on all odd or on all even sites l (cp. [10]), i.e.,

$$\Psi^{M-}_{l \bmod 2 \equiv 0} = 0 \qquad \text{or} \qquad \Psi^{M+}_{l \bmod 2 \equiv 1} = 0 \; . \tag{4}$$

3. Eigenstates and wave functions of the generalized labyrinth tiling

The generalized labyrinth tiling is based on two quasiperiodic chains perpendicular to each other. Hence, the labyrinth tiling can be expressed as the Euclidean product $\mathcal{L}_a = C_a \times C_a$ of the two chains. The bonds are given by the diagonals of the constructed grid and depending on the starting point the grid decomposes into two separate ones, cp. Fig. 1(a). Connecting all vertices with even index sum, we obtain the finite ath order approximant $\mathcal{L}_a$ of the generalized labyrinth tiling $\mathcal{L}$. Analogously, all vertices with odd index sum connected by diagonal bonds also form a labyrinth tiling $\mathcal{L}_a^\star$ [9, 10]. Both tilings are shown for different inflation rules in Fig. 1. For the octonacci sequence they are identical if $\mathcal{L}_a$ is mirrored at the centre of the x- or y-axis or rotated by 90°, i.e. $\mathcal{L}_a^\star$ is dual to $\mathcal{L}_a$. For other inflation rules $\mathcal{L}_a$ and $\mathcal{L}_a^\star$ are slightly shifted against each other and mainly differ at the boundaries as shown in Fig. 1(b) and 1(c). These similarities originate from hidden mirror symmetries in the quasiperiodic chains. While the octonacci chain is perfectly mirror symmetric, for the other chains mirror symmetry can be achieved by neglecting a few symbols at one end of the sequence (e.g. for the Fibonacci chain the last two symbols on the right) and by interchanging two consecutive bonds for the case $b > 2$ (i.e. a phason flip).

The eigenstates of the generalized labyrinth tiling are constructed as the product of the eigenstates of two one-dimensional chains, i.e. $E^{ij} = E^{1i}E^{2j}$ and $\Phi_{lm}^{ij} \sim \Psi_l^{1i}\Psi_m^{2j}$ [10]. The indices i and j enumerate the eigenvalues E in ascending order and the corresponding wave functions Ψ, and l and m represent the coordinates of the vertices. However, due to the symmetries of the eigenfunctions of Eq. (3), some of the wave functions Φ_{lm}^{ij} and related eigenvalues E^{ij} are identical. Then only one of them is allowed to be considered. In two dimensions all combinations

$$\left\{E^{ij} \,|\, 1 \le i \le N_a/2 \wedge 1 \le j \le N_a\right\} \tag{5}$$

are valid for even chain lengths N_a. Further, the wave functions Φ_{lm}^{ij} have to be normalized. For odd N_a the selection of the eigenstates is more complicated due to the special structure of the eigenfunction for $E^M = 0$. This results in the two-dimensional energy values

$$\left\{E^{ij} \,|\, (1 \le i < M \wedge 1 \le j \le N_a) \vee (i = M \wedge 1 \le j \le M)\right\} \,. \tag{6}$$

In three dimensions we follow the same considerations as in two dimensions. The grid now decomposes into four different grids depending on the choice of the starting point because the

(a) silver mean model $\mathcal{L}_3^{\mathrm{Ag}\star}$ (b) golden mean model $\mathcal{L}_7^{\mathrm{Au}}$ (c) golden mean model $\mathcal{L}_7^{\mathrm{Au}\star}$

Figure 1. Labyrinth tilings $\mathcal{L}_a$ and $\mathcal{L}_a^\star$ for different inflation rules: (a) visualizes the construction of $\mathcal{L}_a$ and $\mathcal{L}_a^\star$ for the octonacci chain; (b) shows $\mathcal{L}_7^{\mathrm{Au}}$ and (c) $\mathcal{L}_7^{\mathrm{Au}\star}$ for the golden mean model, where $\mathcal{L}_a$ is mirrored at the centre of the y-axis to show the similarities of both grids.

bonds correspond to one of the four main diagonals of a cuboid. Without loss of generality we consider only the labyrinth tiling $\mathcal{L}_a^{3\mathrm{d}} = \mathcal{C}_a \times \mathcal{C}_a \times \mathcal{C}_a$, for which all vertices have either all even or all odd indices l, m, and n.

We analogously compute the eigenvalues and vectors in three dimensions. Here they are based on three independent quasiperiodic chains, where each one is perpendicular to the other two chains. The wave functions are now given as a product of three one-dimensional eigenfunctions with $\Phi_{lmn}^{ijk} \sim \Psi_l^{1i}\Psi_m^{2j}\Psi_n^{3k}$, where i, j, and k label the energy values and l, m, and n the vertices of the tiling. For even N_a the eigenvalues $E^{ijk} = E^i E^j E^k$ of the generalized labyrinth tiling $\mathcal{L}_a^{3\mathrm{d}}$ in three dimensions are given by the set

$$\left\{E^{ijk} \,|\, 1 \le i \le N_a/2 \wedge 1 \le j \le N_a/2 \wedge 1 \le k \le N_a\right\} . \tag{7}$$

This equation results from the symmetry of the wave functions from Eq. (3). Due to the structure of the eigenfunctions to the eigenvalue $E^M = 0$, for odd N_a the allowed combinations are given by

$$\begin{aligned}\{E^{ijk} \,|\, &(1 \le i < M \wedge 1 \le j < M \wedge 1 \le k \le N_a) \vee \\ &(i = M \wedge 1 \le j \le M \wedge 1 \le k \le M) \vee (1 \le i < M \wedge j = M \wedge 1 \le k \le M)\} .\end{aligned} \tag{8}$$

The characteristics of the wave functions Φ for the different tilings are interesting. Figure 2 shows several wave functions of the generalized labyrinth tilings $\mathcal{L}_a$ and $\mathcal{L}_a^\star$ in two dimensions. For the silver mean model the wave functions are identical for $\mathcal{L}_a$ and $\mathcal{L}_a^\star$ if one of them is mirrored at the centre of the x- or y-axis, cp. Fig. 2(a) and 2(d). Both wave functions are also identical if rotated by 90° taking into account that the indices i and j of one of the eigenstates

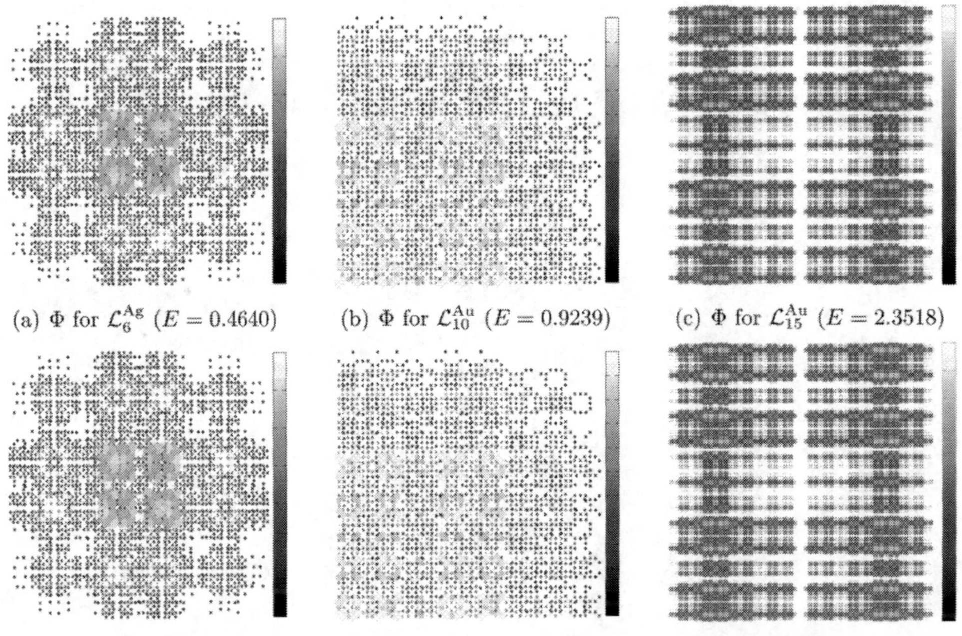

(a) Φ for $\mathcal{L}_6^{\mathrm{Ag}}$ ($E = 0.4640$) (b) Φ for $\mathcal{L}_{10}^{\mathrm{Au}}$ ($E = 0.9239$) (c) Φ for $\mathcal{L}_{15}^{\mathrm{Au}}$ ($E = 2.3518$)

(d) Φ for $\mathcal{L}_6^{\mathrm{Ag}\star}$ ($E = 0.4640$) (e) Φ for $\mathcal{L}_{10}^{\mathrm{Au}\star}$ ($E = 0.9239$) (f) Φ for $\mathcal{L}_{15}^{\mathrm{Au}\star}$ ($E = 2.3518$)

Figure 2. Similarities of wave functions for the two different generalized labyrinth tilings $\mathcal{L}_a$ and $\mathcal{L}_a^\star$ shown for the silver mean model and for two different approximants of the golden mean model. The point size and color encode the probability $|\Phi|^2$ with a logarithmic scale, cp. [6].

have to be interchanged. For other inflation rules the wave functions are not necessarily identical for small approximants due to the different boundaries of the lattice. Nevertheless, they are quite similarly distributed over the lattice as e.g. displayed in Fig. 2(b) and 2(e) and several of the wave functions show even a symmetry with respect to the x- or y-axis. For larger approximants the wave functions approach each other because the influence of the boundaries and of the phason flip for $b > 2$ vanish, see Fig. 2(c) and 2(f). This characteristic can be helpful for analytical considerations for properties of the general labyrinth tiling.

4. Scaling behavior of participation ratios

The participation ratio p reflects the fraction of the total number of sites for which the probability measure of the wave function $|\Phi_{\mathbf{r}}|^2$ is significantly different from zero and, hence, provides information about the degree of localization of the wave function $\Phi_{\mathbf{r}}$. For the discrete positions $\mathbf{r}$ (all valid combinations of the indices l, m, and n) it is defined via the inverse participation number P and the number of states V_a $(= N_a^d/2^{d-1}$ for even $N_a)$ in d dimensions as

$$p(\Phi, V_a) = \frac{P(\Phi)}{V_a} = \frac{1}{V_a}\left[\sum_{\mathbf{r}} |\Phi_{\mathbf{r}}|^4\right]^{-1} . \tag{9}$$

The participation ratio $p(\Phi, V_a)$ of a state scales with a power law in the number of sites as $p(\Phi, V_a) \sim V_a^{-\gamma(\Phi)}$, where the exponent γ is connected to the nature of the eigenstates [3]. The wave function of a localized state is characterized by $\gamma = 1$, and $\gamma = 0$ corresponds to an extended state. For intermediate values of γ $(0 < \gamma < 1)$ we obtain fractal eigenstates, which are neither extended over the whole system nor completely localized at a certain position and show self-similar patterns [6, 12, 13].

Usually, the participation ratios in a small energy interval are distributed over a certain range resulting in a snow-flurry-like appearance when plotted. However, the average participation ratio is almost identical for states at the centre and the edge of the energy spectrum for the present

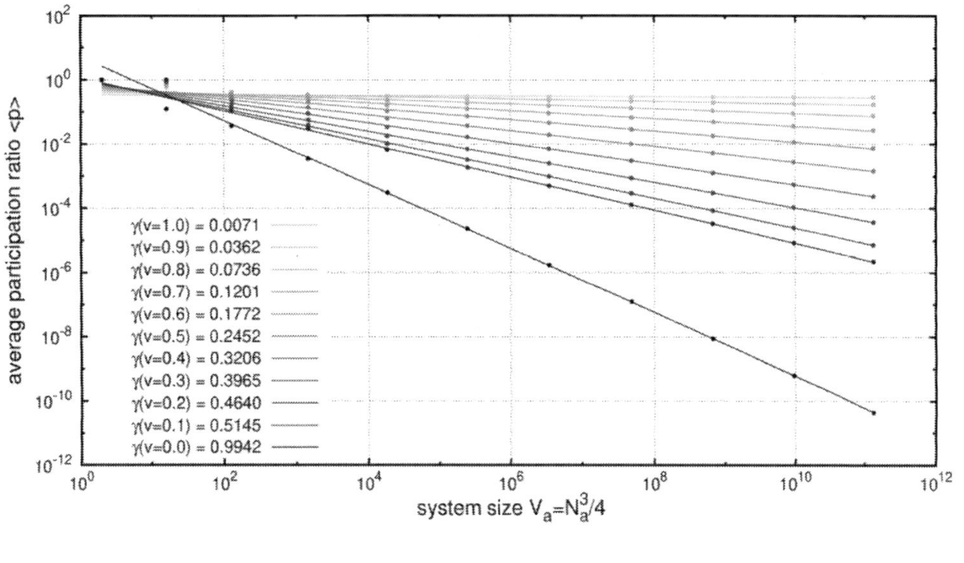

Figure 3. Scaling behavior of $\langle p \rangle$ for the silver mean model in $d = 3$. The dots indicate $\langle p \rangle$ for different system sizes and the lines correspond to a mean least squares fit, where the slope equals the scaling exponent γ.

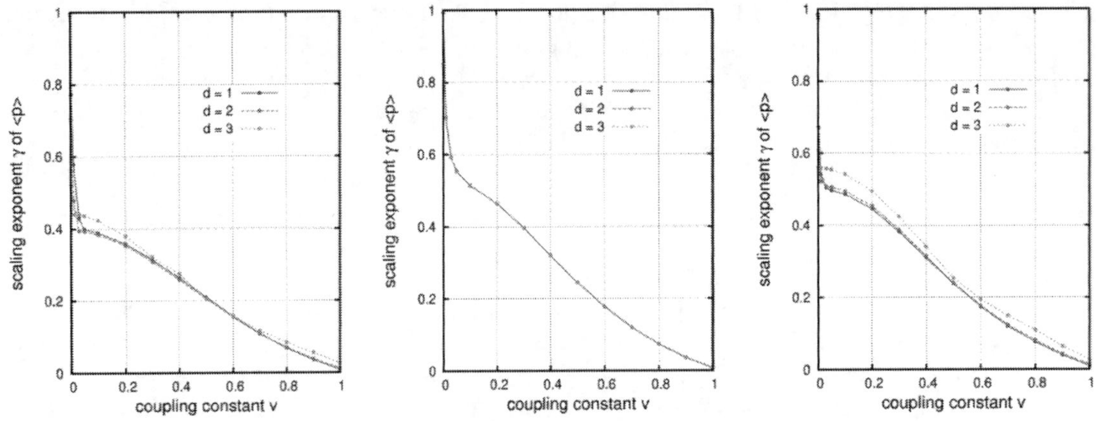

(a) golden mean model (b) silver mean model (c) bronze mean model

Figure 4. Scaling exponents γ of the average participation ratio $\langle p \rangle$ over the complete energy spectrum in one, two, and three dimensions for different metallic mean models.

model [12]. For other quasiperiodic tilings differences can be observed, e.g. the participation ratios of eigenstates of the Penrose tiling, the generalized Rauzy tiling and octogonal tiling models are smaller at the centre than at the edges of the spectrum, whereas e.g. the Ammann-Kramer-Neri tiling shows an inverse behavior [3, 4, 14]. In order to compare the characteristics of the eigenstates for different dimensions we compute the scaling behavior for the average participation ratio $\langle p(\Phi, V_a)\rangle = \frac{1}{V_a}\sum_{\mathbf{s}} p(\Phi^{\mathbf{s}}, V_a)$. Results for the silver mean model for three dimensions are shown in Fig. 3. The average participation ratio scales very well with the system size. In Fig. 4 we show the dependency of the scaling exponent γ on the coupling constant v for different metallic means in one, two, and three dimensions. The plot indicates that the scaling exponent γ is independent of the dimension, which can be also analytically proven for the limit of infinite systems. The deviations for the different dimensions, especially for $d = 3$, for the golden and bronze mean model are caused by the higher computational efforts allowing only the consideration of comparatively small systems. For the octonacci chain participation ratios of higher-dimensional tilings are given as a product of the participation ratios of the chains [15].

The similarities of the wave functions of the different general labyrinth tilings allow us to find good approximations for the calculation of certain properties such as participation ratios. In Fig. 5 we compare the approximate results obtained by the product approach with the exact results and find that the deviations become very small already for system sizes with $N_a \approx 20$.

5. Conclusion

In this paper we investigated the eigenstates for the generalized labyrinth tilings. We presented the construction rules for the eigenstates in two and three dimensions and found that for inflation rules with $b \neq 2$ the wave functions of the 2^{d-1} different tilings in d dimensions approach each other with increasing system sizes in analogy to the occurrence of identical wave functions for $b = 2$ for the different grids. This property is useful for approximate calculations of certain properties as shown for the participation ratios but also for analytical considerations.

Acknowledgment

The authors like to thank the conference organizers for financial support to one of them (S. T.) by the International Union of Crystallography to attend the conference Aperiodic'09 and the Stiftung der Deutschen Wirtschaft for funding the research.

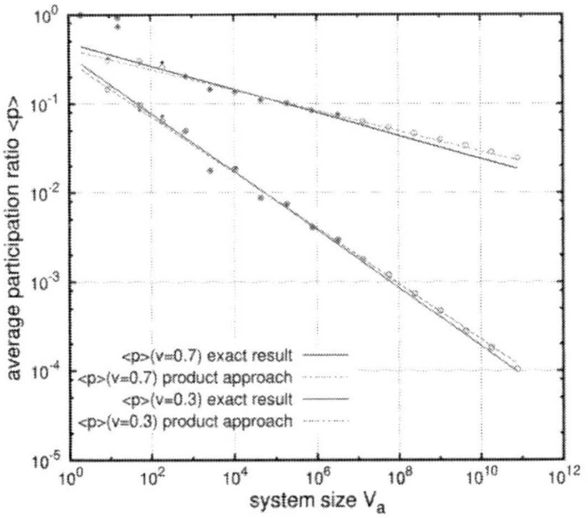

(a) golden mean model ($b = 1$)

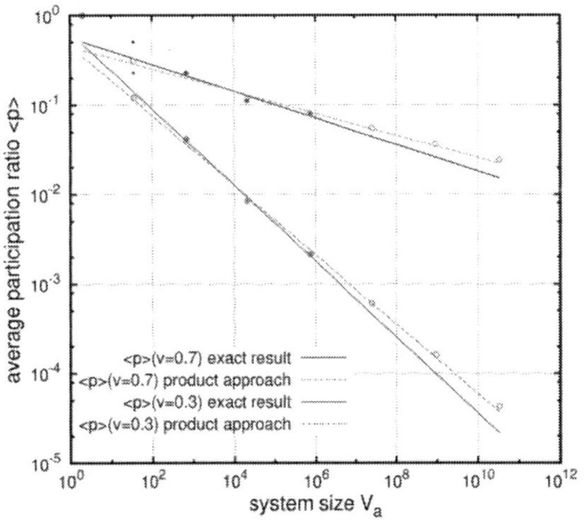

(b) bronze mean model ($b = 3$)

Figure 5. Difference of the average participation ratio for the exact numerical result with the results obtained using a product approach for different generalized labyrinth tilings in three dimensions.

References

[1] Sire C and Mosseri R 1989 *J. Phys. France* **50** 3447
[2] Damanik D 2000 *Directions in Mathematical Quasicrystals* ed Baake M and Moody R V (American Mathematical Society, Providence) pp 277–304
[3] Grimm U and Schreiber M 2003 *Quasicrystals: Structure and Physical Properties* ed Trebin H R (Wiley-VCH, Berlin) pp 210–35
[4] Passaro B, Sire C and Benza V G 1992 *Phys. Rev. B* **46** 13751
[5] Rieth T and Schreiber M 1995 *Phys. Rev. B* **51** 15827
[6] Repetowicz P, Grimm U and Schreiber M 1998 *Phys. Rev. B* **58** 13482
[7] Ilan R, Liberty E, Mandel S E D and Lifshitz R 2004 *Ferroelectrics* **305** 15
[8] McGrath R, Grimm U and Diehl R D 2004 *Phys. World* **17** (12) 23
[9] Sire C, Mosseri R and Sadoc J F 1989 *J. Phys. France* **50** 3463
[10] Yuan H Q, Grimm U, Repetowicz P and Schreiber M 2000 *Phys. Rev. B* **62** 15569
[11] Gumbs G and Ali M K 1989 *J. Phys. A* **22** 951
[12] Thiem S 2008 *Electronic States in Quasicrystal Models* Master Thesis Chemnitz University of Technology
[13] Kohmoto M, Sutherland B and Tang C 1987 *Phys. Rev. B* **35** 1020
[14] Triozon F, Vidal J, Mosseri R and Mayou D 2002 *Phys. Rev. B* **65** 220202
[15] Cerovski V Z, Schreiber M and Grimm U 2005 *Phys. Rev. B* **72** 054203

6th International Conference on Aperiodic Crystals (APERIODIC'09) IOP Publishing
Journal of Physics: Conference Series **226** (2010) 012030 doi:10.1088/1742-6596/226/1/012030

Electronic Structures and Stability of Ag-In-Ca Surfaces

K.Nozawa[1] **and Y. Ishii**[2]

[1]Graduate School of Materials Science, University of Hyogo, Hyogo 678-1297, Japan
[2]Department of Physics, Chuo University, Tokyo 112-8551, Japan

E-mail: ishii@phys.chuo-u.ac.jp

Abstract. Surface properties and stability of quasicrystal-related complex metallic compounds Ag-In-Ca has been studied theoretically. Optimal structures of surface is obtained by relaxing the atoms in a slab to their force-free positions by the first-principles method. It is found that, after relaxation, In is at topmost position near the vacuum, Ag intrudes into the bulk and Ca is located at intermediate layers between In and Ag. To simulate STM images for the slab model, we calculate the charge distribution constructed from the electronic states in an appropriate energy range. We find that bright contrast is observed on different subsets of the In sites on surface for the charge distributions calculated from occupied and unoccupied states. Roles of the low-lying unoccupied Ca-3d states and the sp-d hybridization on surface electronic structures are discussed.

1. Introduction

Quasicrystals (QCs) are long-range ordered materials with rotational symmetry incompatible with translational one. The Fermi energy of QC-related compounds is usually located at a pseudogap in the electronic density of states and hence minimizing the band energy [1]. Such characteristic electronic structures should affect chemical properties of QC surfaces. Extensive studies on clean surfaces of Al-based QCs have shown that Al-rich surfaces are more stable than those with higher concentration of transition elements [2]. This seems reasonable if we assume that surface dangling bonds associated with transition d-states cost larger energy although the dangling bond picture is not so obvious for intermetallic compounds. One may interpret stability of the Al-rich surface as a consequence of the sp-d hybridization, which is considered important for the pseudogap formation. To understand a role of the sp-d hybridization on surface properties, it is particularly of interest to study how the selectivity and surface properties are modified for a family of Cd-based QCs [3], for which the sp-d hybridization is again considered to play an essential role [4]. In this paper, we shall investigate electronic structures of a slab model for an approximant of Cd-based QCs. Since experimental data are not available for surface of Cd-based compounds, we study electronic structures of a slab of Ag-In-Ca cubic approximant.

In the next section, we explain a slab model studied here and calculation method. In section 3, we calculate charge density distributions simulating STM images for relaxed surface and discuss possibility obtaining different STM images by changing sample bias voltage. Roles of the low-lying unoccupied Ca-3d states and the sp-d hybridization on surface electronic structures will be discussed.

2. Models and Calculation

A cubic approximant of Cd-based QCs is found in many systems such as Cd-Yb [5], Cd-Ca [6], Zn-Sc [7], Ag-In-Yb [8], Cu-Al-Sc [9] and so on. Structure of the cubic approximant is composed of icosahedral clusters on a vertex and a body-center of a cubic unit cell [3]. The core of the cluster is an atomic shell of non-icosahedral symmetry, presumably of tetrahedral shape. The second and third atomic shells are a dodecahedron of 20 atoms and an icosahedron of 12 atoms, respectively. The fourth shell is an icosidodecahedron obtained by placing 30 atoms on the edge of the third shell. A few more atoms are placed between the clusters and can be interpreted as forming a triacontahedral fifth shell interpenetrating with the neighbouring cluster [10].

Gómez has studied chemical ordering in Ag-In-Yb cubic approximant [8]. Alloy composition of the sample is $Ag_{39}In_{47}Yb_{14}$. The innermost core and the fourth shell consist of In atoms whereas the second one consists of mainly Ag. The third shell is occupied by Yb atoms as in a case of Cd-Yb and others. Here we note that the In-rich composition leads to the valence electron density (the number of valence electrons per atom) $e/a \approx 2.1$. For ternary systems with trivalent Sc such as Cu-Al-Sc, the monovalent Cu-rich composition is reported and again $e/a \approx 2.1$ is realized [9]. It may be reasonable to assume that the innermost tetrahedral core consists of Cu in this case.

We assume a chemically ordered structural model for the cubic approximant according to Gómez's observation. Starting from atomic positions for cubic Cd_6Ca, we relaxed the atoms to their optimal positions under symmetry constraint that the space group is $I23$. A bulk model obtained here is used for constructing a slab model. A lattice constant a is chosen to be 15.4118 Å as reported for Ag-In-Yb by Gómez [8].

Five-fold surface of icosahedral Ag-In-Yb has been studied by Sharma and co-workers [11]. Comparing the scanning tunneling microscope (STM) images with the structural model by Takakura *et al.* [10], they concluded that observed wide terraces are Yb-rich flat surfaces with high atomic density. For the two-fold (001) surface of the cubic approximant, a flat plane intersecting the cluster center is one with high atomic density. A slab model is constructed by terminating the bulk structure at a flat (001) plane intersecting the cluster center. Thickness of the slab is the same as the lattice constant of the cubic approximant and the slab is separated by a vacuum layer with the same thickness as the slab. In lateral direction, we assume the same periodicity as the cubic approximant. So size of the unit cell for calculation is $a \times a \times 2a$ with a being a lattice constant of cubic approximant. The slab is in a range of $|z| \leq 0.25$ and the total number of atoms in the cell is about 190.

The cluster centers in the present slab model are at $(x, y, z) = (0, 0, \pm 0.25)$ and $(0.5, 0.5, 0.0)$. Since the atoms in the tetrahedral core are not on the flat surface intersecting the cluster center, we have some ambiguity about how to construct flat surface models. In this paper, we shall investigate three models with and without the tetrahedral core at surface: Model A has no tetrahedral core at the cluster center and hence a large hole there. Model B1 has In_4 tetrahedron at the cluster center at surface and the In_4 tetrahedron at surface is replaced by Ag_4 in model B2.

Electronic structure calculations and structural relaxation have been done with the ab-initio total-energy and molecular-dynamics program VASP (Viena Ab-initio Simulation Package) [12, 13] with the projected-augmented wave (PAW) method [14]. Calculations are based on the local density approximation (LDA) in the density-functional theory. The wave functions are expanded with a plane-wave basis set up to a kinetic energy cut-off of 250 eV. The Brillouin zone is sampled with a single k-point (the Γ point).

Electronic structure calculations within the density-functional theory give a narrow $4f$ band of Yb at 0.1-0.2 eV below the Fermi energy E_F [4]. The position of the 4f band is, however, not accurate and, in fact, the photoemission experiments show that it is deeper than the calculated position [15]. This is due to well-known self-interaction error in the LDA [16], which is considerably large for localized 4f states. To avoid such subtle aspects of the 4f states, we shall study here ternary Ag-In-Ca slab models because Ca, which has no f-states near the Fermi energy, is chemically similar to Yb.

3. Results and Discussions

The slab models are constructed by terminating the bulk structure at a flat (001) plane intersecting the cluster center. To obtain optimal structures of surface, we relax the atoms to their force-free positions. Atoms in a thin layer near the surface (up to 1.5 Å in depth) are considered for relaxation and those atoms outside of this layer are fixed to avoid unphysical deformation of bulk layers. We have checked, however, that general trends discussed below are not essentially changed if we consider more atoms for relaxation. Atomic positions after relaxation are shown in Fig.1 where Ag, In and Ca are represented in red, green and blue, respectively. Although the unrelaxed surface is flat except for the tetrahedral cluster, small corrugation is observed for the relaxed surface. If we look more carefully, we can see general trends that In is at topmost position near the vacuum, Ag intrudes into the bulk and Ca is located at intermediate layers between In and Ag.

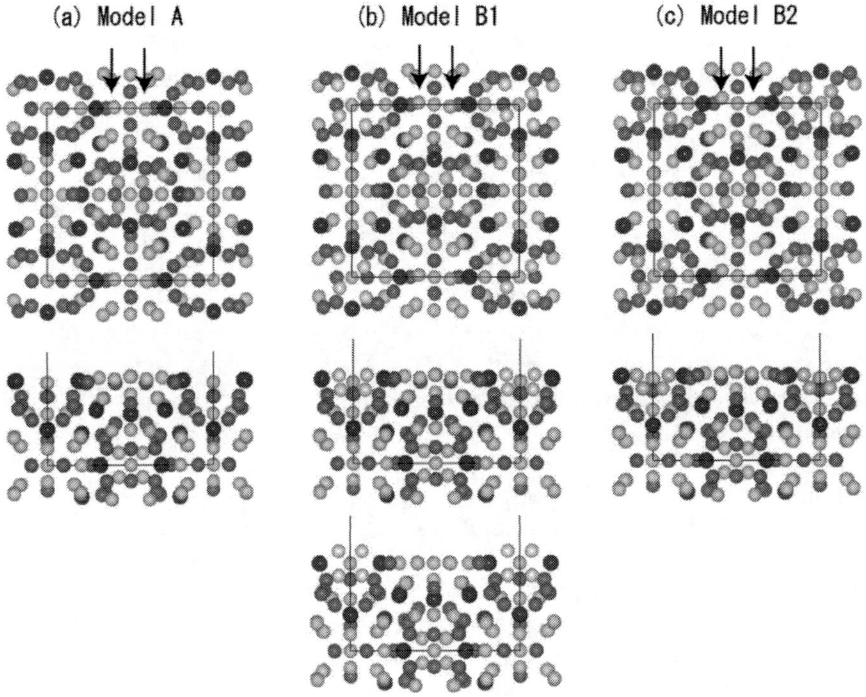

Figure 1. Atomic positions after structural relaxation seen from (001) (top) and (010) (middle) directions. Atomic positions for the unrelaxed slab is shown in a bottom panel of (b). Ag, In and Ca are represented in red, green and blue, respectively. Atoms shown in light color (green or red) in (b) and (c) are those in the tetrahedral clusters at the surface.

The tetrahedral clusters at the surface in models B1 and B2, which are shown in light green and light red in Fig.1, are deformed significantly. The tetrahedral cluster protrudes slightly to the vacuum in the unrelaxed slab model but, after relaxation, the tetrahedron is entirely below the topmost In site. The Ag_4 tetrahedron is deformed and intrudes into the bulk more than the In_4. This is consistent with the general trends mentioned above. Those In atoms on surface indicated by arrows in Fig. 1 are also significantly displaced in the models B1 and B2. Because of slightly larger atomic radius for Ag, the intruding Ag_4 cluster causes larger displacements of atoms in the surface layer.

We compare total energies of the slab models in Table 1. Here the total energy is defined as difference from that for isolated atoms. Total energies for the unrelaxed B1 and B2 are lower than that for the model A by 30 eV. In the models B1 and B2, two tetrahedral clusters (one cluster on each side of the slab) are added to the model A. Total energies of regular tetrahedral In_4 and Ag_4 clusters with the same size as that in the bulk model are calculated as -5.9 and -4.7 eV, respectively. So bonding of the tetrahedral clusters onto the surface yields considerable stabilization. After the relaxation, total energies of the models B1 and B2 are reduced more than the model A. This implies that deformation around the tetrahedral core contributes to stabilize the surface. The Ag_4 cluster in the model B2 is deformed more than the In_4 in the model B1 and hence larger reduction of the total energy in the model B2. Since the Brillouin zone sampling in the present calculation is done at a single k-point, the total energies may not be very accurate. So it would not be safe to discuss relative stability of the models B1 and B2. But we may conclude that the flat surface missing the tetrahedral core at the cluster center is less stable than that with the significantly deformed tetrahedral core.

Table 1. Total energies of the slab models in eV. ΔE is reduction of the total energy after relaxation.

	model A	model B1	model B2
unrelaxed	-618.4	-650.3	-648.0
relaxed	-622.1	-655.4	-654.4
ΔE	-3.7	-5.1	-6.4

To simulate STM images for the slab model, we calculate the surface local density of states, which is the charge distribution constructed from the electronic states in an appropriate energy range, at constant height [17]. In STM experiments, electrons tunnel from occupied states of a sample to unoccupied states of a tip with negative sample bias voltage $V_b < 0$. To simulate STM images for negative bias $V_b < 0$, the charge distribution is calculated from the occupied states in the energy window $E_F - |V_b| < \epsilon < E_F$. For positive bias $V_b > 0$, the charge distribution is calculated from the unoccupied states in the energy window $E_F < \epsilon < E_F + |V_b|$. The charge distribution calculated from the electronic states in a particular energy range is also of interest because the states near the Fermi energy are affected by surface formation and relaxation.

In Fig. 2, we show the charge distribution on a plane parallel to the surface at $z = 0.3$, which is about 1.4 Å above the topmost atoms. The charge distribution is shown in a 2x2 cell with atoms on the surface layer in a single unit cell. The charge distribution in Fig. 2 is calculated from the occupied states within 1 eV below E_F, representing the STM image for negative bias. One can see that bright contrast, which corresponds to the high electron concentration, is observed on some of In atom sites. It should be noted that the bright contrast is not obtained near all of the In sites on the surface. Also the bright contrast is not observed at the Ag and Ca sites. This is because Ag and Ca atoms on the surface layer intrude slightly into the bulk in comparison with In. For the model B1, we have additional bright contrast on In atoms in the tetrahedral cluster on the surface. We expect that the STM study with atomic resolution may clarify atomic arrangements at the cluster center more precisely.

Figure 3 shows the charge distribution calculated from the unoccupied states within 1 eV

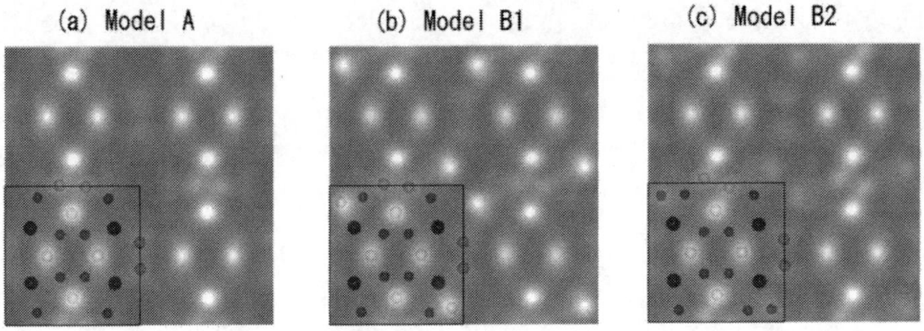

Figure 2. Charge distribution calculated from the occupied states within 1 eV below E_F, representing the STM image for negative bias.

above E_F, representing the STM image for positive bias. Bright contrast is observed on a different subset of the In sites from those giving the bright contrast for negative bias. In addition to point-like contrast corresponding to a single In atom, we have textured weak contrast. This contrast is made from the charge distribution around Ca and Ag sites in the surface layer. If one chooses the bias voltage larger, more strong contrast is obtained near the Ca sites because of large contribution from the low-lying unoccupied Ca-3d states.

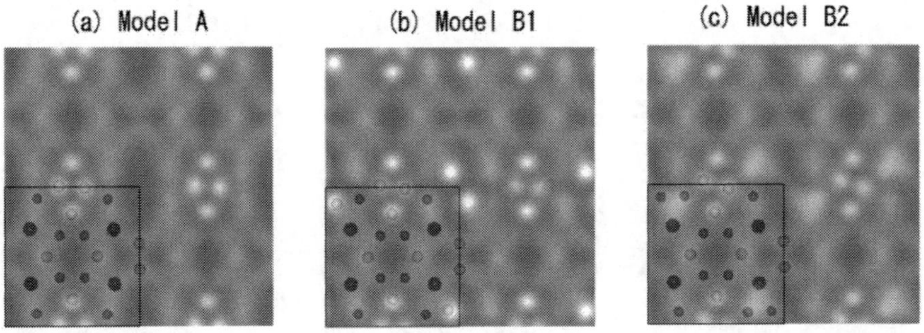

Figure 3. Charge distribution calculated from the unoccupied states within 1 eV above E_F, representing the STM image for positive bias.

The charge distribution of the slab model on a vertical section is shown in Fig. 4. The charge density is calculated from the occupied states within 1 eV below E_F on a plane at $x = 0.5$ for the model B1. Electron cloud extends towards the vacuum around the topmost In on the surface and this gives bright contrast in Fig. 2. The electron distribution is also enhanced between atoms inside the slab and such rearrangement of the charge density may contribute to stabilization of the surface. For cubic approximants Cd_6M (M=Yb, Ca), it is known that the low-lying unoccupied d-states of the second element M are indispensable for compound formation and the sp-d hybridization between the Cd-5p states and the low-lying d-states of M plays an essential role in cohesion [4]. We compare the charge distribution with that for a hypothetical slab model where Ca is replaced with Mg, which has no low-lying unoccupied atomic d-states (Fig. 4). One can see that the characteristic charge distribution between atoms near the surface is significantly reduced in the hypothetical slab with Mg. This suggests that the low-lying unoccupied Ca-3d states contribute to the occupied states stabilizing the surface.

Journal of Physics: Conference Series **226** (2010) 012030 doi:10.1088/1742-6596/226/1/012030

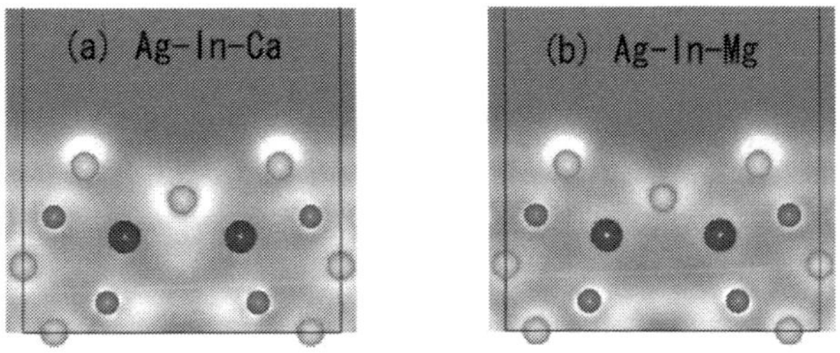

Figure 4. Charge distribution calculated from the occupied states on a plane at $x = 0.5$ for (a) Ag-In-Ca and (b) hypothetical Ag-In-Mg. Atoms on the plane are shown in red, green and blue, respectively, for Ag, In and Ca in (a) and substituted Mg is shown in purple in (b).

4. Conclusion

Surface properties and stability of QC-related complex metallic compounds has been studied theoretically. Electronic structures are calculated by the first-principles method for a slab of Ag-In-Ca cubic approximant. A slab model is constructed by terminating the bulk structure of a cubic approximant at a flat (001) plane intersecting the cluster center. Optimal structures of surface is obtained by relaxing the atoms to their force-free positions. It is found that, after relaxation, In is at topmost position near the vacuum, Ag intrudes into the bulk and Ca is located at intermediate layers between In and Ag. To simulate STM images for the slab model, we calculate the charge distribution from the electronic states in an appropriate energy range. We find that bright contrast is observed on different subsets of the In sites for the charge distributions calculated from occupied and unoccupied states. This suggests that STM images change depending on sample bias voltage but they must be carefully interpreted because the difference is not necessarily due to differences in atomic species or height of the atomic positions. By comparing the charge distribution with a hypothetical Ag-In-Mg slab model, we conclude that the low-lying unoccupied Ca-3d states contribute to stabilizing the surface.

Acknowledgments

The authors appreciate discussions with M.Shimoda, H.R.Sharma, C.P.Gómez and A.P.Tsai. Figures of crystal structure and charge density in this paper are drawn by using VESTA developed by Momma and Izumi [18].

References

[1] Ishii Y and Fujiwara T 2008 *Quasicrystals* ed Fujiwara T and Ishii Y (Amsterdam: Elsevier) chap 6 pp 171–208
[2] Sharma H R, Shimoda M and Tsai A P 2007 *Adv. Phys.* **56** 403
[3] Tsai A P, Guo J Q, Abe E, Takakura H and Sato T J 2000 *Nature* **408** 537
[4] Ishii Y and Fujiwara T 2001 *Phys. Rev. Lett.* **87** 206408
[5] Palenzona A 1971 *J. Less-Common Metals* **25** 367
[6] Bruzzone G 1972 *Gazzeta Chimica Italiana* **102** 234
[7] Andrusyak R I, Kotur B Y and Zavodnik V E 1989 *Soviet Phys. Crystallogr.* **34** 600
[8] Gómez C P *private communication*
[9] Ishimasa T and Honma T 2009 *preprint*
[10] Takakura H, Goméz C P, Yamamoto A, de Boissieu M, Tsai A P 2007 *Nature Materials* **6** 58
[11] Sharma H R, Shimoda M, Sagisaka K, Takakura H, Smerdon J A, Nugent P J, McGrath R, Fujita D, Ohhashi S, Tsai A P 2009 *Phys. Rev.* B **80** 12140(R)
[12] Kresse G and Hafner J 1993 *Phys. Rev.* B **47** RC558
[13] Kresse G and Furthmüller J 1996 *Phys. Rev.* B **54** 11169
[14] Kresse G and Jourbert D 1999 *Phys. Rev.* B **59** 1758

[15] Tamura R, Murao Y, Takeuchi S, Kiss T, Yokoya T and Shin S 2002 *Phys. Rev.* B **65** 224207
[16] Perdew J P and Zunger A 1981 *Phys. Rev.* B **23** 5048
[17] Tersoff J and Hamann D R 1985 *Phys. Rev.* B **31** 805
[18] Momma K and Izumi F 2008 *J. Appl. Crystallogr.* **41** 653

First-principles study for effect of lattice defects on the crystal structure of the Zn-Sc cubic crystalline approximant

Kazuki Nozawa[1], **Yasushi Ishii**[2] **and Kenji Makoshi**[1]

[1] Graduate School of Material Science, University of Hyogo, Kouto, Hyogo 678-1297, Japan
[2] Department of Physics, Chuo University, Kasuga, Tokyo 112-8551, Japan

E-mail: nozawa@sci.u-hyogo.ac.jp

Abstract. Effects of impurity and vacancy introductions are investigated by the first-principles structural relaxation. Studied concentration of the lattice defects is 0.6 at%. As the impurity, a Cu atom is substituted for a Zn atom in Zn_6Sc. For this concentration, the plausible candidate site of Cu substitution is the glue site distributing in the interstitial region between icosahedral clusters. The most stable vacancy site is the core of the icosahedral cluster. Some of the introduced vacancies migrate several angstrom and reach the core. A special site being outside the core plays an important role for the vacancy migration. The vacancy introduced near this site migrates to the core. Significant Zn-Zn bonds are found in the charge density map. The stability of the defect-introduced structure is discussed in terms of a breaking of this bond.

1. Introduction

The binary quasicrystals $Cd_{5.7}Yb/Ca$ are valuable samples for discussing the origin of the quasiperiodicity, because those are free from chemical disorder [1, 2]. The approximant crystals of those intermetallic compounds, Cd_6Yb/Ca and Zn_6Sc, have very close composition and structure to the quasicrystalline phase [1, 3]. Due to these similarities, the approximant phases are intensively investigated as a target of comparative studies for the quasicrystals. The approximant phases of this system have a notable feature that an order-disorder transition found around 100 K [4, 5, 6]. The order-disorder transition is considered to be attributed to orientational changes in the tetrahedral clusters located at the center of an icosahedral cluster. The icosahedral cluster, which is called Tsai-cluster, consists of four layered atomic shells: tetrahedron, dodecahedron, icosahedron and icosidodecahedron from inner to outer. The Tsai-cluster is a building block of not only the approximants but also quasicrystals [3]. But, interestingly, the transition is not found in the quasicrystalline phase. Although the reason why the quasicrystalline phase does not exhibit the transition has not been clarified yet, the transition would be incompatible with the quasiperiodicity. In this sense, the order-disorder transition may provide a clue to understand the quasiperiodic structure.

Recently, Yamada and Tamura investigated effects of lattice defects on the order-disorder transition in Zn-Sc system [7]. They found that Cu substitution of 0.1 at% for the Zn site lowers the transition temperature from 155 K to 100 K. Moreover the transition disappears by 0.5 at% substitution. They speculated that the ordering of the tetrahedral shell is suppressed by the distortion of outer shells resulting from the Cu substitution. The atomic sites having a significant influence on the ordering of the tetrahedral shell are considered to be the tetrahedral shell itself or the dodecahedral second shell being just outside of the tetrahedral core. Lin and Corbett studied the crystal structure of $Zn_{18-y}Cu_ySc_3$ (y=0.65, 1.48 and 2.19), and concluded that Cu atoms occupy the dodecahedral shell [8], though the Cu concentration is rather higher than the experiment by Yamada and Tamura [7](y=0.65 corresponds 3.1 at%). In this paper, we study the substitution site by Cu in Zn-Sc systems using the first principles electronic structure calculation. If the distortion of the outer cluster suppresses the ordering of the tetrahedral core, atomic vacancies may do so as well. The atomic site capable to be a vacant site is also studied in this paper.

2. Calculation methods and systems

The density functional method within the local density approximation is used to calculate relaxed structures and those total energies. Calculations are done with a program package VASP [9, 10, 11, 12, 13, 14]. The ionic potentials are approximated by the ultrasoft pseudopotentials [15]. In this study, we treat the Zn-Sc, Zn-Cu-Sc systems. The number of electrons treated as the valence electrons are 12, 9 and 11 for Zn, Sc and Cu. The electronic wavefunctions are expanded by plane-wave basis set up to a kinetic energy of 209.6 and 233.8 eV for Zn-Sc and Zn-Cu-Sc systems, respectively. A cubic cell including 168 atoms is adopted as the unit cell. The unit cell contains two Tsai-clusters (2×66 atoms) and 36 glue Zn atoms. The Brillouin zone is sampled by $3\times3\times3$ Monkhorst-Pack grids (14 irreducible k-points). The atomic positions are fully relaxed according to the evaluated force obtained in each self-consistent calculation step.

One of the 168 sites in Zn_6Sc is substituted by Cu, i.e., the concentration of Cu is 0.6 at% comparable to the experiment by Yamada and Tamura (0.5 at%) [7]. The substitution site can be classified according to the symmetry of the unit cell. The structure of the tetrahedral cluster determines the set of atomic sites for substitution, because it has the lowest symmetry. As the tetrahedral cluster, we adopted the one obtained as the most stable structure for Cd_6Ca by first-principles calculations [16]. It is also a candidate of the stable structure in Zn-Sc system [17]. The adopted tetrahedron is an appropriate subset of the 24g sites reported by a X-ray diffraction experiment for Zn_6Sc [8]. We assume that two tetrahedra in the unit cell, which are on the vertex and body-center of the cubic cell, face to the opposite direction. This orientational configuration is denoted as LCI structure in the references [16, 17]. Using the symmetry of the tetrahedral core, the substitution sites are classified into 28 inequivalent sites. Note that the icosahedral third shell, which is occupied by second elements (Sc) in pure system, is not assumed to be a substitution site, because the atomic size of Cu(1.28 Å) is much smaller than Sc(1.68 Å).

After the structural relaxation of Cu-introduced system (Zn-Cu-Sc), Cu atom is removed from the system in order to investigate the vacancy introduced system. In other words, the relaxed Cu-introduced systems are used as starting structures of structural relaxations for vacancy-introduced Zn-Sc systems. Thus the case where the vacancy is at the third shell is not studied. However it seems to be reasonable to consider that the majority element (Zn) is removed more easily than the minority one (Sc).

3. Results and discussions

3.1. Copper substitution

One of the Zn atoms in the cubic Zn_6Sc is replaced by Cu and the positions of the 168 atoms are fully relaxed according to the force obtained as a derivative of the calculated energy. The obtained total energies are listed in Table 1. The second column of the table denotes the total energy, and in the third column the relative energy from the most stable one is given. The most stable site for Cu substitution is the third glue-24g site, which is denoted as glue-24g-3 hereafter.

Note that the Wyckoff sites referred in this paper are the ones with Im$\bar{3}$ symmetry and only for comparison with experiments. In the calculation, no constraint for symmetry is adopted and the system does not have Im$\bar{3}$ symmetry even for the starting structure because the fractional sites in the first shell of Im$\bar{3}$ model cannot be treated in the first-principles calculation. The glue-sites are the sites distributed in the interstitial region between Tsai-clusters. The stabilization mechanism of each site has not been understood well yet. Generally an impurity at the glue-24g sites gives lower energies. Some exceptions are found for glue-12e-3, glue-24g-5 and 2nd-24g-1.

From the obtained list of the total energies, we can say that the most favorable site for Cu substitution is the glue site in this concentration. Lin and Corbett performed the X-ray diffraction analysis for Zn-Cu-Sc system and concluded that the sites occupied by Cu are the 24g sites in the second shell [8]. But the concentration they studied is rather higher than in our case. The lowest concentration they studied includes about 2.6 atoms per Tsai-cluster (3.1 at%), but 0.5 (0.6 at%) in the present study. The favorable substitution site probably depends on the concentration. It would be interesting to compare these calculations with the ones done for higher concentrations.

3.2. Vacancy introduction

The Cu atom considered in the previous section is removed to describe the vacancy- introduced system. Accordingly, the concentration of the vacancy is also 0.6 at%. The atomic position of the rest of the 167 atoms in the cubic unit cell is re-relaxed. The evaluated total energy for each system is listed in Table 2 in the same manner as for the Cu substitution. One can find that the distribution of the relative energy is obviously different from that of Cu substitution. The glue sites which are concluded as the plausible candidates for the Cu substitution belong to the highest energy group. The lowest energy site is the 2nd-16f-1. But some sites give similar energies: 1st-2, 2nd-16f-2, 2nd-24g-2, -3, 5, 4th-48h-2, -3, -4 and -6. The relaxed structures of these ten cases are in fact the same. The energy difference between them is therefore considered as numerical errors. The most stable structure for the vacancy-introduced Zn-Sc system has only three Zn atoms in the cluster center forming a regular-triangle perpendicular to the threefold axes of the unit cell (the directions from the cluster center to the 2nd-16f sites). It is surprising that the vacancy introduced into the 4th shell migrates to the cluster center. If the vacancy is introduced into the second shell, one of the atoms of the first shell migrates to fill up the *hole*, and the rest three atoms of the first shell forms the regular-triangle. The hole in the 4th-shell is also filled up using this process. The 2nd-16f atom near the vacancy fills up the hole in the 4th-shell, then the first-shell's atom goes outside to behave as the missing 2nd-16f. The obtained results can be classified into three types:

- Type A; the above-mentioned type, the introduced vacancy of which is filled up by neighboring atoms and the vacancy migrates to the first shell at the final. The stable structure of the first shell is the regular triangle facing to a 2nd-16f site.
- Type B; the vacancy migrates to the first shell, but the first shell does not reach the stable structure (the regular triangle).
- Type C; the vacancy remains at the initial position.

The classification is noted in the table. The vacancy introduced in the 2nd-24g-1 is not occupied by the first-shell's atom completely. The structural relaxation may be stuck in the metastable state during migration of the first-shell's atom to the 2nd-24g-1 site. Therefore a proper description is that an intermediate state between B and C.

Except for a few cases, the vacancy introduced in the first, second or fourth(48h) shells migrates to the first shell finally. The exceptions might be originated from the distortion of the Tsai-cluster. As shown in the Fig. 1, the second shell is distorted by the first shell. There is a trend that the vacancy introduced in the right-hand region in the figure seems to be easily

Journal of Physics: Conference Series **226** (2010) 012031 doi:10.1088/1742-6596/226/1/012031

Table 1.
The total and relative total energies for Cu substitution. The unit of energy is eV.

Sites	Total energy	Relative energy
1st		
1	-.48220041E+03	0.167
2	-.48218426E+03	0.183
2nd-16f		
1	-.48216444E+03	0.203
2	-.48218063E+03	0.187
2nd-24g		
1	-.48230281E+03	0.065
2	-.48222623E+03	0.141
3	-.48224785E+03	0.120
4	-.48209150E+03	0.276
5	-.48222710E+03	0.141
4th-12d		
1	-.48215608E+03	0.212
2	-.48219498E+03	0.173
3	-.48222511E+03	0.143
4	-.48224693E+03	0.121
4th-48h		
1	-.48213196E+03	0.236
2	-.48216059E+03	0.207
3	-.48213638E+03	0.231
4	-.48216652E+03	0.201
5	-.48212565E+03	0.242
6	-.48216336E+03	0.204
glue-12e		
1	-.48222113E+03	0.147
2	-.48210742E+03	0.260
3	-.48232904E+03	0.039
4	-.48222139E+03	0.146
glue-24g		
1	-.48230305E+03	0.065
2	-.48231948E+03	0.048
3	-.48236760E+03	0.000
4	-.48233763E+03	0.030
5	-.48225418E+03	0.113

Table 2. The total and relative total energies for the vacancy introduction. The final structures classified into three types are also presented.

Sites	Total energy	Relative energy	Type
1st			
1	-.47648571E+03	0.142	B
2	-.47662634E+03	0.001	A
2nd-16f			
1	-.47662732E+03	0.000	A
2	-.47662258E+03	0.005	A
2nd-24g			
1	-.47626236E+03	0.365	B
2	-.47662720E+03	0.000	A
3	-.47662555E+03	0.002	A
4	-.47563540E+03	0.992	C
5	-.47662571E+03	0.002	A
4th-12d			
1	-.47617290E+03	0.454	C
2	-.47619388E+03	0.433	C
3	-.47628535E+03	0.342	C
4	-.47617324E+03	0.454	C
4th-48h			
1	-.47554261E+03	1.085	C
2	-.47662653E+03	0.001	A
3	-.47662391E+03	0.003	A
4	-.47662682E+03	0.001	A
5	-.47560934E+03	1.018	C
6	-.47662720E+03	0.000	A
glue-12e			
1	-.47613425E+03	0.493	C
2	-.47607770E+03	0.550	C
3	-.47629842E+03	0.329	C
4	-.47615365E+03	0.474	C
glue-24g			
1	-.47578792E+03	0.839	C
2	-.47570832E+03	0.919	C
3	-.47572245E+03	0.905	C
4	-.47569326E+03	0.934	C
5	-.47572637E+03	0.901	C

filled up compared with that introduced in the left-hand region. We should pay attention to the energies of the 2nd-24g-4, 4th-48h-1 and 4th-48h-5. Those are the ones classified as type C, the vacancy of which does not migrate to the first shell. When the vacancy remains at these sites, the system is considerably less stable than others (4th-12d and glue-12e sites). Fig. 2 is a charge density map of the ideal Zn_6Sc without defects. The displayed charge density is the difference charge density: $\rho = \rho(\mathrm{SCF}) - \rho(\mathrm{atom})$, where $\rho(\mathrm{SCF})$ is the charge density obtained by the SCF calculation and $\rho(\mathrm{atom})$ is the one given as a superposition of the charge of isolated atoms. Thus in a rough description the positive value of the difference charge density means a bonding trend around the region. The positive value is shown only in the figure. The red, blue, green and black balls in the figure denote the atoms of the first, second, third and fourth shells. One can find a significant increase of the difference charge density on lines connecting the blue balls (second-shell's atom). The charge clouds forming triangles around the 2nd-16f sites are originated from the bond between the the second- and fourth-shell's atoms. By the replacement of these atoms with a vacancy, the significant bond is broken. Present results imply that the energy loss of the bond-breaking is rather large, because the total energy of the 2nd-24g-4, 4th-48h-1 and 4th-48h-5 is twice as large as the other cases of the 4th-12d and glue-12e.

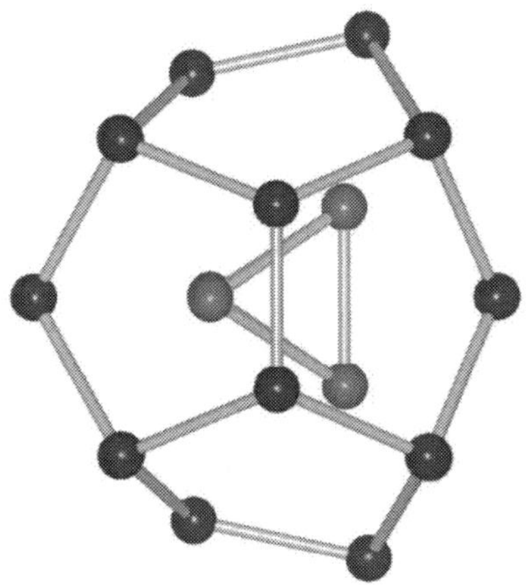

Figure 1. The Tsai-cluster is distorted by the tetrahedral core. The red and blue balls represent the atoms of first and second shells.

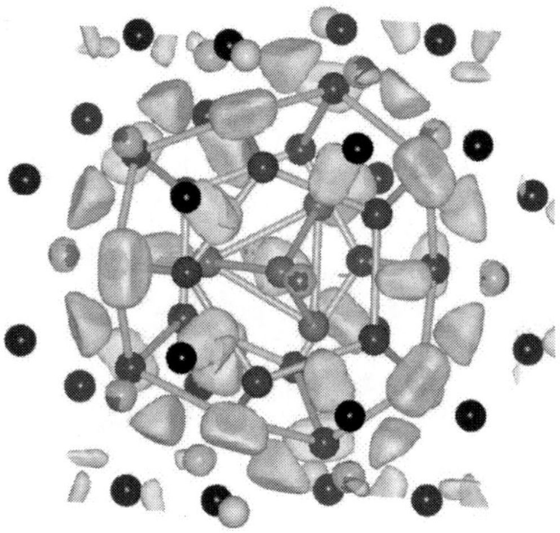

Figure 2. The difference charge density map in complete Zn_6Sc. The significant increase in the charge density is found on the line connecting the second shell.

The vacancies introduced at the glue or 4th-12d sites do not migrate to other sites. This is because the 2nd-16f sites are not around these sites. In the case that the 4th-48h sites are filled up successfully, these sites have a 2nd-16f site as the nearest neighbor (NN) or next nearest neighbor (NNN), and the atoms at the 2nd-16f sites fill the vacancies without exception. The NN and NNN of 4th-12d sites are the 2nd-24g or other 4th-12d sites. As for the glue sites, those are other glue sites. Although the reason of that has not been understood, we can say that the 2nd-16f site would be easily migrate outer shells in comparison with the 2nd-24g sites.

4. Summary

We performed first-principles structural relaxations to investigate the crystal structure of lattice-defects-introduced Zn-Sc cubic approximants. The lattice defects studied are the atom (Cu) substitution and vacancy introduction. The concentration of the impurities is 0.6 at%, which means a single impurity in the cubic unit cell. The most plausible site for Cu substitution is concluded as the glue sites for this concentration. As for the vacancy introduction, some vacancies introduced in the second and fourth shells migrate to the first shell. It may imply the existence of a strong bond between the second and fourth shells. The total energies of the case that the vacancies remain at the introduced sites are explained by breaking of the significant bonds. Two trends for the mobility of the vacancies are found. One is concerned with the distortion of the Tsai-cluster. The vacancy introduced in a distorted part is easily filled up by a neighboring atom. Another one may indicate a special character of the 2nd-16f site. The vacancies introduced in the fourth shell are filled up by an atom at 2nd-16f sites without exception. The vacant sites far from the 2nd-16f site are not filled up. The 2nd-16f site seems to migrate easily from the initial position in comparison with the 2nd-24g site. In the quasicrystalline phase, the dodecahedral shell is not distorted [3]. The 16f site is no longer a particular site in that phase, because the 16f and 24g sites discussed here is equivalent. Therefore the special character of the 16f site, or the threefold linkage between Tsai-clusters, in the approximant phase may be deeply related with the origin of the order-disorder transition, which has not been found in the quasicrystalline phase.

Acknowledgments

The authors would like to thank T. Ishimasa for his important suggestions for vacant sites in the Zn-Sc system. The authors also express gratitude to fruitful discussions with R. Tamura and T. Yamada. Figures of the crystal structures and charge densities are drawn by VESTA developed by K. Momma and F. Izumi [18]. Calculations are partly carried out on SR11000 at the Supercomputer Center, Institute of Solid State Physics, University of Tokyo, and SX8 at YITP, Kyoto University.

References

[1] Tsai A P, Guo J Q, Abe E, Takakura H and Sato T J 2000 *Nature*(London) **408** 538
[2] Guo J Q, Abe E, and Tsai A P 2000 *Phys. Rev.* B **62** R14605
[3] Takakura H, Gómez C P, Yamamoto A, de Boissieu M and Tsai A P 2007 *Nature Mater.* **6** 58
[4] Tamura R, Murao Y, Takeuchi S, Ichihara M, Isobe M and Ueda Y 2002 *Jpn. J. Appl. Phys.* **41** L524
[5] Tamura R, Edagawa K, Murao Y, Takeuchi S, Suzuki K, Ichihara M, Isobe M, and Ueda Y 2004 J. Non-Crystal. Solids. bf 334, 335 173
[6] Tamura R, Nishimoto K, Takeuchi S, Edagawa K, Isobe M, and Ueda Y 2005 *Phys. Rev. B* **71** 092203
[7] Yamada T and Tamura R 2009 *Z. Krist.* **224** 101
[8] Lin Q and Corbett J D 2004 *Inorg. Chem.* **43** 1912
[9] Hohenberg P and Kohn W 1964 *Phys. Rev.* **136** B864
[10] Kohn W and Sham L J 1965 *Phys. Rev.* **140** A1133
[11] Kresse G and Hafner J 1993 *Phys. Rev.* B **47** 558; Kresse G and Hafner J 1994 *Phys. Rev.* B **49** 14251
[12] Kresse G and Furthmüller J 1996 *Comput. Mat. Sci.* **6** 15
[13] Kresse G and Furthmüller J 1996 *Phys. Rev.* B **54** 11 169
[14] Kresse G and Hafner J 1994 *J. Phys.: Condens. Matt.* **6** 8245
[15] Vanderbilt D 1990 *Phys. Rev.* B **41** 7892
[16] Nozawa K and Ishii Y 2008 *J. Phys.: Condens. Matter.* **20** 315206
[17] Hatakeyama T, Nozawa K and Ishii Y 2008 Z. Krist. **223** 830
[18] Momma K and Izumi F 2006 *Commission on Crystallogr. Comput., IUCr Newslett.* No. 7 106

Hall effect in the Y-Al-Ni-Co and *o*-$Al_{13}Co_4$ decagonal approximants

J Ivkov[1], M Komelj[2], A Smontara[1], J Dolinšek[2] and P Gille[3]
[1]Institute of Physics, Laboratory for the Study of Transport Problems, Bijenička 46, POB 304, HR-10001 Zagreb, Croatia.
[2]Jožef Stefan Institute, University of Ljubljana, Jamova 39, SI-1000 Ljubljana, Slovenia
[3]Ludwig-Maximilians-Universität München, Department of Earth and Environmental Sciences, Crystallography Section, Theresienstrasse 41, D-80333 München, Germany.

E-mail: ivkov@ifs.hr

Abstract. The Hall effect in Y-Al-Ni-Co and *o*-$Al_{13}Co_4$ single crystals has been investigated for all combinations of the electrical current and magnetic field directions and in the temperature interval from 90 to 380 K. In these intermetallics the Hall coefficient R_H exhibits well defined anisotropy and only weak temperature dependence. R_H is negative electron-like or zero for the magnetic field perpendicular to the plane that corresponds to the quasiperiodic plane in decagonal quasicrystals, and is positive hole-like or zero for the magnetic field parallel to this plane. The results for the anisotropy of R_H are well correlated to the anisotropy of R_H in *d*-Al-Ni-Co and *d*-Al-Cu-Co quasicrystals. Therefore, the anisotropy in both crystalline and quasicrystaline samples originates from the specific stacked-layer structure and the chemical decoration of the lattice. The investigated anisotropic R_H in Y-Al-Ni-Co was reproduced theoretically by calculation using Boltzmann transport theory, *ab-initio* calculated anisotropic Fermi surface, and isotropic relaxation time approximation.

The monoclinic Y-Al-Ni-Co phase and orthorhombic $Al_{13}Co_4$ phase belong to the $Al_{13}TM_4$ (TM = transition metal) group of complex metallic compounds and are approximants to the decagonal quasicrystals (*d*-QCs). While decagonal quasicrystals can be structurally viewed as a periodic stack of quasiperiodic atomic planes, their approximant phases are characterized by large unit cells which periodically repeat in space with the atomic decoration closely resembling that of *d*-QCs. Moreover, atomic layers of the approximant phases correspond to the quasiperiodic atomic planes of *d*-QC in the sense that they show locally similar patterns, which means that their structures on the scale of near-neighbor atoms closely resemble each other. Further, the periodicity lengths along the stacking direction of these layers in the approximant phases are almost identical to those along the periodic direction of *d*-QCs. Decagonal approximants thus offer valid comparison to the *d*-QCs.

A consequence of the anisotropic structure of both *d*-QCs and their approximants are anisotropic electrical and thermal transport properties (electrical resistivity, thermoelectric power, Hall coefficient, thermal conductivity) when measured along different crystalline directions [1]. The anisotropy of the Hall coefficient R_H of *d*-QCs [2,3] is especially intriguing, being positive hole-like ($R_H > 0$) for the magnetic field lying in the quasiperiodic plane, whereas it changes sign to negative ($R_H < 0$) for the field along the periodic direction, thus becoming electron-like. This R_H anisotropy was reported for the *d*-Al-Ni-Co, *d*-Al-Cu-Co and *d*-Al-Si-Cu-Co and is considered to be a universal feature of *d*-QCs.

However, the lack of translational periodicity within the quasiperiodic layers prevents any quantitative theoretical analysis of these anisotropic properties in *d*-QCs. At the same time, for the approximant phases to the decagonal phase theoretical simulations are straightforward to perform and they can give us information for the microscopic origin of this anisotropy in general. Therefore, in this article we present a brief overview of our experimental results and *ab-initio* calculation obtained up to now for the anisotropic Hall coefficient for the Y-Al-Ni-Co [1,4] and o-$Al_{13}Co_4$ [5] complex metallic alloys.

The single crystals used in our study were grown by the Czochralski technique [6]. The structure of the Y-Al-Co-Ni ($Al_{76}Ni_{22}Co_2$) matched well to the monoclinic unit cell of the Zhang *et al.* model [7] and the structure of the *o*-$Al_{13}Co_4$ matched well to the orthorhombic unit cell of the Grin *et al.* model [8]. The monoclinic Y-Al-Ni-Co phase has two atomic layers within one periodic unit of ≈ 0.4 nm along the stacking [010] direction and a relatively small unit cell, comprising 32 atoms, and *o*-$Al_{13}Co_4$ complex metallic compound has four atomic layers within one periodic unit of ≈ 0.8 nm along the [100] stacking direction and a unit cell comprising 102 atoms.

In order to perform crystalline-direction-dependent studies, we have cut from the ingots of each alloy three bar-shaped samples of dimensions 1×1×7 mm^3, with their long axes along three orthogonal directions. For each sample the long sides of the bar are perpendicular to other two orthogonal directions. The three orthogonal directions are equal to *b* and *c* crystallographic directions for both alloys, and to *a* direction for orthorhombic $Al_{13}Co_4$ alloy. For monoclinic Y-Al-Co-Ni the third orthogonal direction denoted as *a** lies in the monoclinic plane at an angle 26° with respect to *a*, and is perpendicular to *c* and *b*.

The Hall effect measurements were performed by the five-point method using standard ac technique in magnetic fields up to 1 T and in the temperature interval from 90 to 380 K. The temperature-dependent Hall coefficient $R_H = E_y/j_xBz$ of Y-Al-Ni-Co and *o*-$Al_{13}Co_4$, for all combinations of the current and magnetic field directions, is displayed in figure 1.

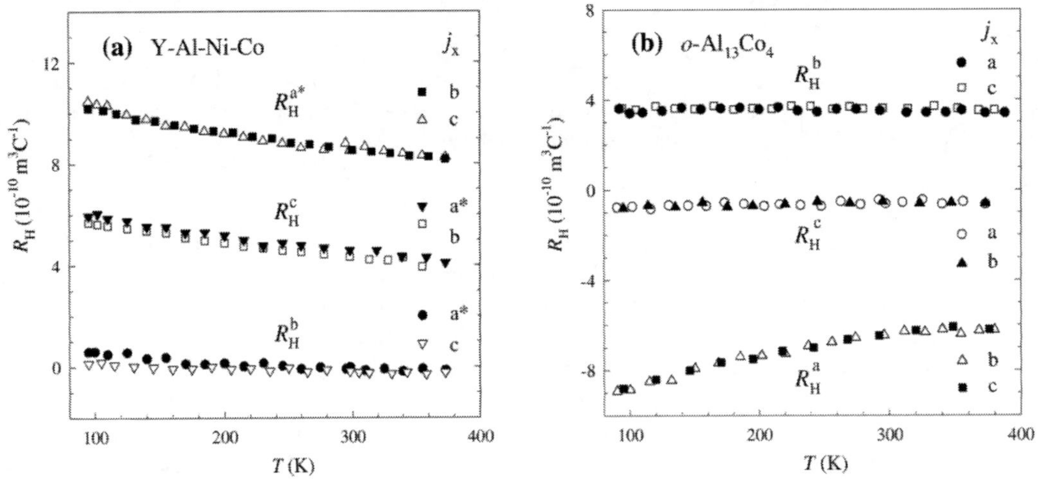

Figure 1. Anisotropic temperature-dependent Hall coefficient $R_H = E_y/j_xB_z$ of monoclinic Y-Al-Ni-Co (a) and o-$Al_{13}Co_4$ (b) for different combinations of the current j_x and magnetic field B_z. The superscript on R_H denotes the direction of magnetic field. Note that the stacking direction for Y-Al-Ni-Co and o-Al13Co4 are *b* and *a* directions, respectively.

The R_H values are typical metallic in the range 10^{-10} m^3C^{-1} (with the experimental uncertainty of ±0.1×10^{-10} m^3C^{-1}). For each alloy the six R_H sets of data form three groups of two practically identical R_H curves, where the magnetic field in a given crystallographic direction yields, in accordance with the Onsager relations [9], the same R_H for the current along the other two crystallographic directions in the perpendicular plane. For each alloy R_H exhibits a pronounced anisotropy and a weak temperature dependence that tends to disappear at higher temperatures, as is typical for the metallic systems. The anisotropic values of R_H at 300 K are collected in table 1. The comparison of these results to the results for *d*-QCs will be discused later after the presentation of a theoretical analysis of the Hall effect.

Table1. Experimental and calculated values of the anisotropic Hall coefficient $R_H=E_y/j_xB_z$ of Y-Al-Ni-Co and o-$Al_{13}Co_4$ for various directions of the current j_x and magnetic field B_z at 300 K. Average values for a particular field and two current directions are given.

Y-Al-Ni-Co				o-$Al_{13}Co_4$			
B_z	j_x	R_H (10^{-10} m^3C^{-1}) Experiment	Theory	B_z	j_x	R_H (10^{-10} m^3C^{-1}) Experiment	Theory
*a**	*b*, *c*	8.5	8.8	*a*	*b*, *c*	-6.5	-18.2
b	*a**, *c*	0	-0.3	*b*	*a*, *c*	3.5	16.2
c	*a**, *b*	4.5	5.5	*c*	*a*, *b*	-0.6	14.2

A quantitative theoretical analysis of the anisotropic Hall coefficient requires knowledge of the anisotropic electronic band structure i.e. the Fermi surface. The calculations of the band structure were performed for two similar structural models (original [7] and relaxed [4]) for Y-Al-Co-Ni and for two similar structural models (original [8] and new [5]) for o-$Al_{13}Co_4$. It is out of scope of the present article to go into minor differences connected with the different models, and here we present the results for relaxed Y-Al-Ni-Co and original o-$Al_{13}Co_{14}$. The *ab-initio* calculation of the band structure was performed within the framework of the density functional theory by applying Wien97 code [10] and ABINIT code [11]. The calculated Fermi surfaces visualized by using the *xcrysden* program [12], are presented in figure 2. It is transparent that the Fermi surfaces are highly anisotropic for both Y-Al-Co-Ni and o-$Al_{13}Co_4$, which is at the origin of the experimentally observed anisotropy in the electronic transport coefficients along different crystallographic directions.

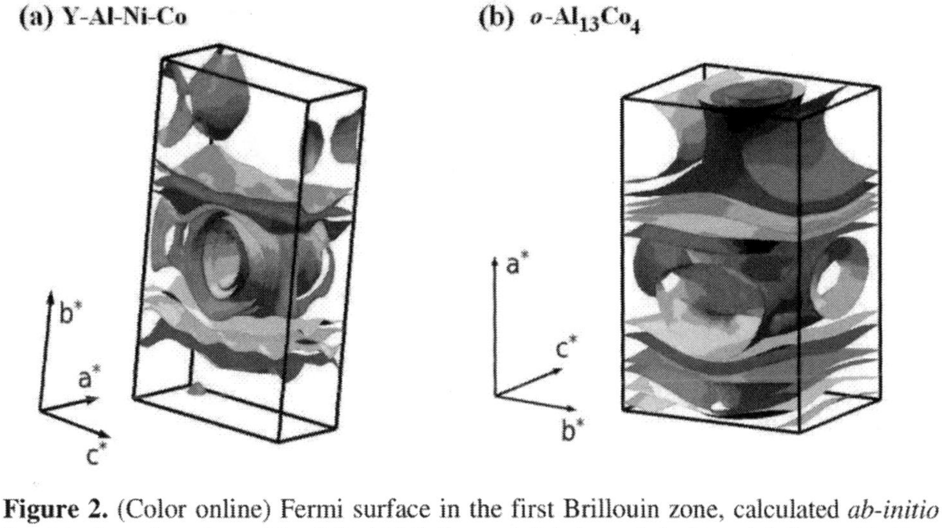

Figure 2. (Color online) Fermi surface in the first Brillouin zone, calculated *ab-initio* for (a) the relaxed model of Y-Al-Ni-Co and (b) for original model for o-$Al_{13}Co_4$. Orientation of the reciprocal-space axes *a**, *b** and *c** is also shown.

The theoretical Hall coefficient was calculated by means of the Boltzmann semiclassical theory as implemented in the BoltzTraP code [13] by applying the band structure calculated *ab-initio* as described above, and for the same set of combinations of the current and field directions as the experimental ones in figure 1. The results are shown in figure 3 and table 1. For both complex intermetallic compounds, the six theoretical R_H sets of data form three groups of two similar R_H curves, where the magnetic field in a given crystalline direction yields similar values of the coefficient for the current along the other two crystalline directions in the perpendicular plane. For Y-Al-Ni-Co there exist some temperature-dependent differences within each group, where the temperature dependence originates from the Fermi-Dirac function. The main objection to the theoretical analysis is that it assumes a single (band independent) and isotropic scattering time which can be canceled out in the actual calculations.

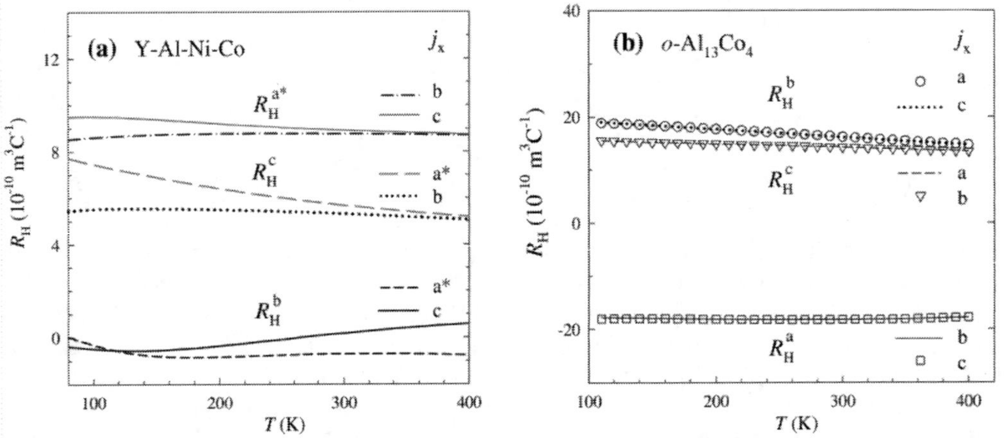

Figure 3. (Color online) Theoretical anisotropic Hall coefficient, calculated *ab-initio* for (*a*) the relaxed Y-Al-Ni-Co model [4] and (*b*) the old o-$Al_{13}Co_4$ model [7]. R_H was calculated for the same set of current and field directions as employed in the experimental measurements of the Hall coefficient shown in figure 1. The superscript on R_H denotes the direction of the magnetic field. Note that the stacking direction for Y-Al-Ni-Co and o-$Al_{13}Co_4$ are b and a directions respectively

From figures 1 and 3 and table 1 it can be concluded that the theoretical Hall coefficient, calculated for the relaxed Y-Al-Ni-Co phase reproduces well the experimental features of the anisotropic Hall coefficient in this alloy. For o-$Al_{13}Co_4$ compound the theory gives too large values for all combinations of directions. However, the theory reproduces well the order of appearance of the anisotropic Hall coefficient (the theoretical order $R_H^b > R_H^c > R_H^a$ matches the experimental one) and the crystallographic-direction-dependent change of sign $R_H^b > 0$ and $R_H^a < 0$.

In conclusion, the origin of the anisotropic Hall coefficient in Y-Al-Ni-Co and o-$Al_{13}Co_4$ is the anisotropic Fermi surface, the anisotropy of which in turn originates from the specific stacked-layer structure and the chemical decoration of the lattice. The theory reproduces the experimental results quantitatively for Y-Al-Ni-Co and only qualitatively for o-$Al_{13}Co_4$. There is almost complete analogy between the R_H anisotropy of Y-Al-Ni-Co and o-$Al_{13}Co_4$ on the one side and d-QCs on the other. In d-QCs $R_H > 0$ for the field lying in the quasiperiodic plane (corresponding to the in-plane coefficients $R_H^{a*}, R_H^c > 0$ of Y-Al-Ni-Co and $R_H^{a*}, R_H^c \geq 0$ of o-$Al_{13}Co_4$), whereas R_H changes sign to negative for the field along the periodic direction (corresponding to $R_H^b \approx 0$ of Y-Al-Ni-Co and $R_H^b < 0$ of o-$Al_{13}Co_4$). This suggests that long-range quasiperiodicity of the structure is of marginal importance (if it is of any at all) for the anisotropy in d-QCs, which originates from the stacked-layered structure and the complex local atomic order on the scale of nearest-neighbor atoms.

Acknowledgements
This work was done within the activities of the 6th Framework EU Network of Excellence "Complex Metallic Alloys" (Contract No. NMP3-CT-2005-500140), and has been supported in part by the Ministry of Science, Education and Sports of the Republic Croatia through the Research Project No. 035-0352826-2848.

References

[1] Smontara A *et al* 2008 *Phys. Rev.* B **78** 104204, and references therein
[2] Zhang D L, Lu L, Wang X M, Lin S Y, He L X and Kuo K H 1990 *Phys. Rev.* B **41** 8557
[3] Wang Y P, Zhang D L and Chen L F 1993 *Phys. Rev.* B **48** 10 542
[4] Komelj M, Ivkov J, Smontara A, Gille P, Jeglič P and Dolinšek J 2009 *Solid State Commun.* **149** 515
[5] Dolinšek J *et al* 2009 *Phys. Rev.* B **79** 184201
[6] Gille P and Bauer B 2008 *Cryst. Res. Technol.* **43** 1161
[7] Zhang B, Gramlich V and Steurer W 1995 Z. *Kristallogr.* **210** 498
[8] Grin J, Burkhardt U, Ellner M and Peters K 1994 *J. Alloys Compd.* **206**, 243
[9] Smith A C, Janak J F and Adler R B 1967 *Electronic Conduction in Solids* (New York: McGraw-Hill) p 226
[10] Blaha P, Schwarz K, Sorantin P and Trickey S B 1990 *Comput. Phys. Commun.* **59** 399
[11] Gonze X *et al.* 2002 *Comput. Mater. Sci.* **25** 478, the ABINIT computer program is a common project of the Université Catholique de Louvain, Corning Incorporated, and othe contributors (URL: http://www.abinit.org).
[12] Kokalj A Comp. 2003 *Mater. Sci.* **28** 155; code available from http://www.xcrysden.org.
[13] Madsen G K H and Singh J 2006 *Comput. Phys. Commun.* **175** 67

Stability of mechanically alloyed vacancy ordered phase in $Al_{70}Cu_{15}Ni_{15}$ alloy during annealing

Thakur Prasad Yadav[1], Nilay Krishna. Mukhopadhyay[2], Radhey Shyam Tiwari[1], and Onkar Nath Srivastava[1*]
[1]Department of Physics, Banaras Hindu University, Varanasi-221 005 (INDIA)
[2]Department of Metallurgical Engineering, Institute of Technology, Banaras Hindu University, Varanasi-221 005, (INDIA)

E-mail: hepons@yahoo.com (ONS), yadavtp@gmail.com (TPY)

Abstract. A nano τ_3 vacancy-ordered phase in the Al-Cu-Ni alloy system has been synthesized with a composition close to $Al_{70}Cu_{15}Ni_{15}$ by mechanical alloying a mixture of elemental powder in a high-energy ball mill by varying milling time from 10 to 100 hours. The stability of nano-crystalline τ_3 vacancy-ordered phase has been studied under thermal annealing in vacuum as well as in air. The x-ray diffraction and transmission electron microscopy techniques were employed for characterization of the milled and annealed samples. The powder after 100 h of milling was found to contain mostly nano τ_3 phase with the partial ordering, and with crystallite sizes in the range of 10-20 nm along with a lattice strain of ~0.67 %. The milled powder, after annealing in vacuum at 700 °C for 60 h, revealed the formation of a strain-free and ordered τ_3 phase with a crystallite size of 80 nm, indicating grain coarsening. It is interesting to note that the milled powder annealed in air at 700 °C for 60 h showed the formation of $(Cu,Ni)Al_2O_4$ type spinel phase with the lattice parameter of 8.1 Å and the lattice strain as 0.52 %. The average grain size of spinel phase was found to be ~ 40 nm.

1. Introduction

In quasicrystal forming alloy systems, there always exist crystalline phases with the same or similar local structures as those of the quasicrystals. These crystalline phases are referred to as approximants and usually they are formed with large unit cells and complicated structures. The complex crystalline phases in metallic alloys including vacancy ordered phase often co-exist with quasicrystals or incommensurate crystal in many Al-based alloy systems [1]. The vacancy ordered phases (VOPs) being an interesting class of intermetallic phases comes under the broad category of structures with ordered defects [2]. It has been shown that the vacancy ordered phase can be considered as an approximant phase to one-dimensional as well as two-dimensional decagonal quasicrystals [3-5]. VOPs represent a class of structure where compositions and diffraction effects display a close relationship with the golden mean, τ (where, $\tau=1/2(\sqrt{5}+1)$). These structures can be described in terms of basic B2 (CsCl) type of cell with the ordering among transition metals and vacancy along the [111] direction. Dong et al. [6] have carried out extensive investigations on the occurrence of B2 phase, which resulted in the realization that the valence electron concentration of the B2 phase is similar to that of a quasicrystal. The B2 phase and related VOPs were found to co-exist extensively with the quasicrystalline phases, particularly in Al-Cu-TM (Ni, Co, TM= transition metal) systems and they follow a well-defined orientation relationship [7-10]. VOPs and disordered B2 phases are of considerable interest for the understanding of the stability and transformation behavior of quasicrystalline phases during nonequilibrium processing [10-11]. The τ_3 VOP (Al_3Ni_2 type) exhibits rhombohedrally distorted B2 lattice (hexagonal). Thus it can be considered as a superstructure of B2 phase due to ordering among transition metal atoms and vacancy along [111] direction in the (111) layers. It has been suggested that Al-Cu-Ni system has an arrangement of vacant and filled sites in the truncated Fibonacci sequence along [111] quasiperiodic Fibonacci chain [11]. During complete disordering, the structure becomes bcc (i.e disordered B2). The τ_3 or any other higher order VOPs can be formed because of ordering among the vacancies and the transition metal atoms along [111] directions. This structural description using distorted bcc lattice is convenient for understanding the order-disorder transformation qualitatively.

In the last two decades, a class of materials with a nanometer-sized microstructure have been synthesized and studied for their useful properties and possibility for technological applications [12]. Nanocrystalline phases were found to exhibit size dependent properties, such as lower melting points, higher energy gaps, and stability of thermodynamically non-equilibrium structures [13-15]. In comparison to macro-scale phases, increased ductility and toughness of the brittle materials was observed [16]. Among various processing techniques, high energy ball milling processes have been widely used for producing the bulk nanocrystalline and amorphous materials in complex metallic alloy systems [17-19]. Mechanical alloying (MA) is now recognized as a versatile process for synthesizing nanocrystalline powders [20].

More recently, there have been studies on the formation of spinel phases from quasicrystalline materials in Al-Co-Ni and Al-Ni-Fe systems [21-22]. The synthesis of aluminum–transition metal nanocrystalline spinel has been investigated intensively in the recent years due to their unique potential applications [23-24]. The spinel structure is characterized by a simultaneous occupation of tetrahedral and octahedral positions by metallic cations in an FCC oxygen lattice. Aluminum based spinels constitute an important class of advanced ceramic materials with a wide variety of technological applications. The $CoAl_2O_4$, $NiAl_2O_4$ and, $CuAl_2O_4$, possess interesting electronic, magnetic and catalytic properties and are used in industry as ceramic pigments, coatings and catalysts [25-26]. The spinel materials can be prepared by many methods such as solid-state reaction, hydrothermal, alkoxide hydrolysis, sol-gel method and microwave-induced methods [27-28]. The most general method is the solid-state reaction, which involves the mixture of metal oxides followed by sintering at high temperature. Recently, the nanospinel materials have been synthesized by milling quasicrystalline precursor, followed by annealing [29]. It is not clear in any of the prior work whether true thermodynamic equilibrium of the spinel phases could be achieved.

Here we reported the stability of nano τ_3 vacancy-ordered phase formed by MA in the $Al_{70}Ni_{15}Cu_{15}$ alloy system in order to understand the microstructural evolution during annealing in vacuum and air. No other VOPs or QC were detected during the investigation. The $(Ni,Cu)Al_2O_4$ spinel phase with nano size microstructure was found to form during air annealing.

2. Experimental procedure

Powder mixture of 10 g containing 70 at% Al, 15 at% Cu and 15 at% Ni respectively were used in Szegvari Attritor Ball mill with a ball to powder ratio of 80: 1 and with a speed of 400 rpm. The attritor consists of a cylindrical stainless steel tank of inner diameter of 13 cm. Hardened steel balls of diameter of 6 mm were used. The milling operation was conducted for 10 to 100 hours in a hexane medium. Isothermal heat treatment of milled sample was carried out at 700 0C for 60 h in vacuum as well as in air. The mechanically milled and annealed samples were characterized by powder X ray diffraction (XRD), Philips PW 1710 diffractometer using CuKα radiation (λ=1.5418 Å). Transmission electron microscopy (TEM) was carried out in a Tecnai G20 at 200 kV electron microscope employing imaging and diffraction modes. A chemical analysis using EDX attached with TEM exhibited that the contamination of Fe (0.6 wt %) in the alloy are within the acceptable limit and thus it can be negligible. The grain size and the lattice strain of the sample have been calculated from the integral width of the physical broadening profile. Cauchy and Gaussian components are obtained from the ratio of full width at half maximum intensity (2ω) and integral breadth (β). In a single line analysis the apparent crystallite size 'D' and strain 'e' can be related to Cauchy (βc) and Gaussian (βG) widths of the diffraction peak at the Bragg angle θ [30].

3. Results and Discussion

Figure 1 (a-c) shows the XRD patterns of elemental Al, Cu & Ni peaks as function of milling time, indicating gradual alloying and the evolution of τ_3 phases after 100 h of milling. There is no significant change during milling up to 50 h except peak shift and peak broadening (Fig.1 (b)). After 100 h of milling (Fig.1(c)), the peak corresponding to any pure elements cannot be identified, suggesting that a single phase has formed. It is difficult from the XRD patterns alone to ascertain whether disordered B2 (bcc) or disordered VOPs have formed, because the superlattice peaks corresponding to VOPs cannot be identified easily except in case of 100 h milling where the superlattice peak of (100) can be discerned. It is important to point out that (100) peak shown on the patterns (Fig.1c) confirms the existence of τ_3 phase as this peak cannot be indexed due to B2 phase though other peaks can be indexed based on the B2 phase.

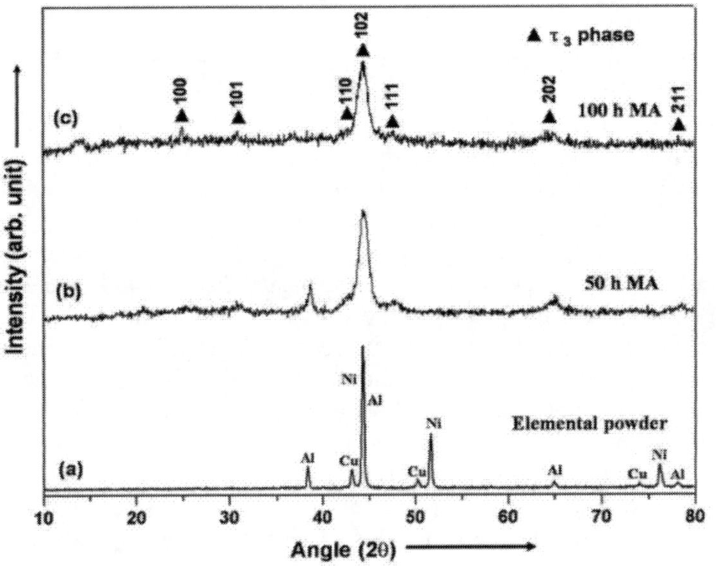

Figure1. X-ray diffraction patterns of elemental powderwith composition $Al_{70}Ni_{15}Cu_{15}$ (a), as well as milled powder for different milling durations (b-c). The gradual evolution of τ3 phase can be noticed.

The (100) peak can certainly confirm the formation of τ_3 phase. However, the presence of B2 phase as a minor one cannot be ruled out. The intensity of the other superlattice peaks is expected to be too weak to be detected due to defects, strain and nanocrystallinity effects. However with the help of TEM investigation it has been confirmed that the nano τ_3 phase is a major phase in the milled sample. The peak positions and the corresponding lattice parameters (hexagonal, a=4.1050 Å, c= 4.9541 Å) of the milled samples are found to agree closely with JCPDS data of the Al_3Ni_2 type τ_3 phase (a=4.065 Å, c=4.906 Å) [31]. The increase in lattice parameters can be interpreted due to Cu atoms and nanocrystallinity effect. In the initial stages of milling, the formation of metastable disordered Al-Ni (Cu) solid solution (B2 type) and/or disordered VOP appears to form. In the final stage these phases accommodate residual pure aluminum and give rise to partially ordered τ_3 phase. It can be argued that the intensity ratio of (100) or (001) peaks with (102) peak (most intense peak) can be approximately correlated with the degree of superlattice ordering of the τ_3 phase. In the disordered condition this will be zero and the structure could be related to partially ordered B2 phase. It should be emphasized that τ_3 phase can be considered as superstructure of B2 phase with an introduction of structural vacancy along [111]. The effect of the reduced size of the crystallites and lattice strain can be observed on the overall peak intensities and broadening, which is obvious after 100h of milling (Fig.1b). By analyzing the peak broadening, the grain size and lattice strain were found to vary continuously with milling time. In case of 100 h milling, the crystallite size and the lattice strain were found to be ~ 10-20 nm and ~ 0.675 % respectively.

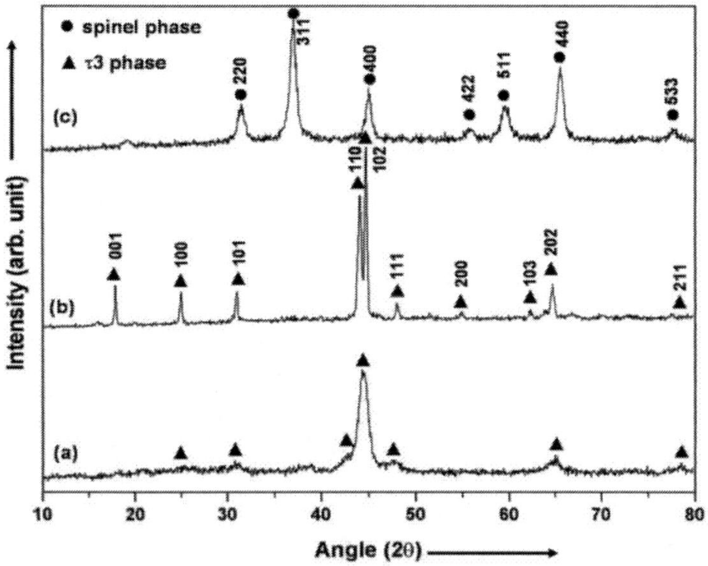

Figure 2. XRD patterns obtained from 100h MM (a), after annealing in vacuum (b) and in air (c) at 700 °C for 60h respectively. In case of vacuum annealing the peak broadening is found to be reduced compared to that in the as-milled condition, where as in case of air annealing the $(Ni,Cu)Al_2O_4$ spinel phase has been observed.

Figure 2 (a-c), show XRD patterns obtained from 100 h milling followed by annealing at 700 °C for 60 h in vacuum as well as air respectively. In case of vacuum annealing, the peak broadening was found to reduce after annealing (Fig.2 (b)). The sharpness and intensity of the peaks were improved significantly. This can be attributed to the ordering, annihilation of defects, strain relaxation, and grain coarsening. The degree of superlattice ordering which is found to enhance can be estimated from the ratio of the (100) or (001) peak with (102) peak. All the peaks in the diffraction patters were indexed according to the Al_3Ni_2 structure type. The splitting of (110) peak of B2 phase into (110) and (102) peaks of τ_3 phase is an indicative that the phase transformation towards ordered structures has taken place. It can be concluded that the partially ordered τ_3 phase obtained after milling gave rise to a perfect and long range ordered τ_3 phase after 60 h of annealing. It is known that the high-energy ball milling technique introduces defects, disordering and reduction of grain size, and thereby destabilizes the ordered phase. However in the present case, it is interesting to note that the vacancy ordered phase has been formed even after MA of 100 h. The mechanical energy imparted during milling could not completely destroy the ordering between the vacancy and Cu/Ni atoms. It can be stated that τ_3 is more stable than B2 phase in this composition during milling though the disordered B2 phase can be considered as the parent phase from which τ_3 phase can be obtained as a superstructure due to the ordering of vacancy. The XRD pattern of 100 h mechanically milled and air annealed sample for 60h at 700 °C shows the formation of $(Ni,Cu)Al_2O_4$ spinel phase. The XRD pattern (Fig. 2(c)) shows the peaks corresponding to the spinel structure with the lattice parameter (FCC, a=8.15±0.04 Å). From these results, it is obvious that high rate of oxygen diffusion through the nano-VOP phases has lead to the formation of spinel structures since the kinetics depends on the particle size and activated surface.

TEM study revealed many interesting features of microstructural evolution in the milled and annealed samples. Fig 3(a) displays the bright field image obtained from the powder particle after 100 h of milling. The equiaxed nanocrystalline grains can be observed clearly in the microstructure. The grains are found to have sizes ranging from 10 to 20 nm. The corresponding SAD pattern (Fig.3 (b)) was indexed based on τ_3 phase.

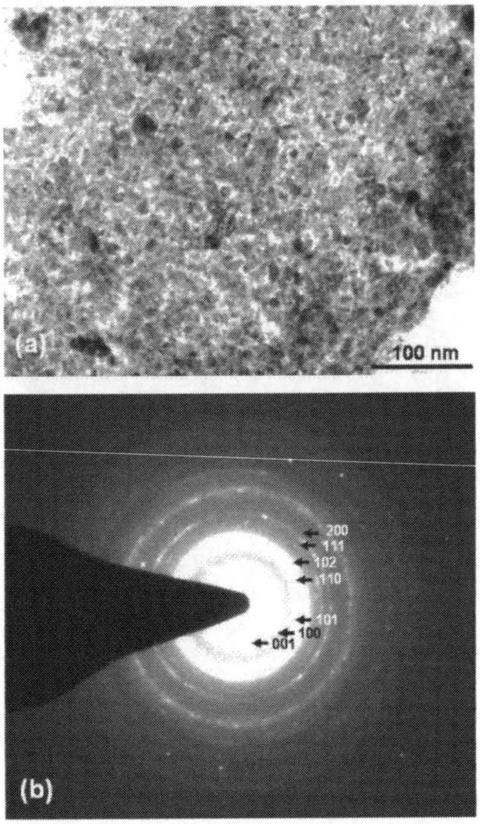

Figure 3. (a) Bright field images from 100h milled powder and (b) corresponding diffraction pattern showing a spotty ring pattern corresponding to the τ3 phase.

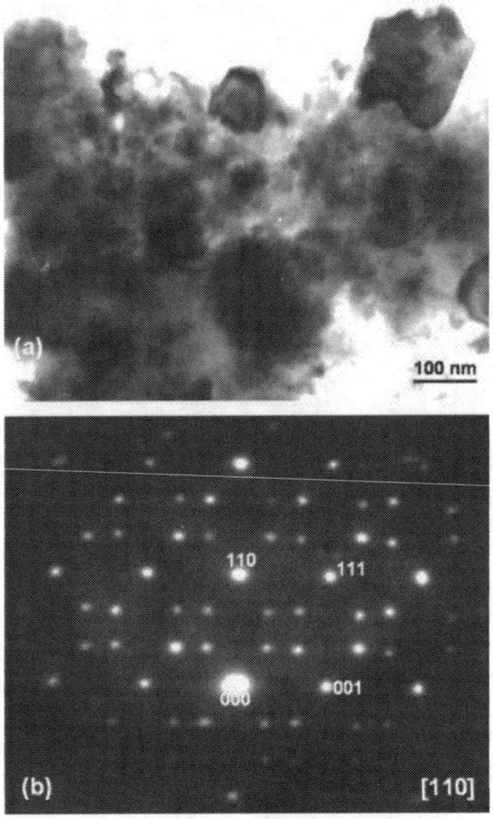

Figure 4. (a) The bright field image showing the microstructure of powder subjected to 100h milling followed by vacuum annealing at 700 °C for 60h, (b) corresponding diffraction patterns. The indexing of the diffraction spots reveals the presence of τ3 phase of [110] zone axis.

The lattice parameters obtained are consistent with that obtained from XRD patterns. The TEM evidence suggests that the τ3 phase is a major phase during milling. The existence of B2 phase could not be established in the milled samples. Figs. 4 (a) shows the microstructures of coarser grains of τ3 phase obtained from the powder sample subjected to 100 h milling followed by annealing at 700 °C for 60 h. The corresponding diffraction patterns along various zone axis can be indexed based on τ3 phase Fig. 4(b) showing the fundamental and superlattice spots (weak intensity). The indexing of the fundamental spots have been shown based on the B2 structure (a=2.94 Å) for clear distinction between superlattice and fundamental spots. In Fig.4 (b) the diffraction spot (111) is clearly divided into three equal spacing, confirming the structure to be τ3 phase. The spot 1/3 (111) is a superlattice peak and equivalent to (001) of τ3 phase. It is clear that perfectly ordered τ3 phase has formed after annealing as the diffraction pattern does not show any kind of diffuse scattering. In addition to the ordering the grain coarsening was found to be significant grains have grown from sizes of 10-20 nm to ~80-100 nm. The bright field microstructures of 100 h MM and air annealed sample at 700 °C for 60 h are shown in Fig. 5 (a). The average particle size appears to be consistent with data obtained from XRD. It is clearly visible that grains are having size ranging in between 40-80 nm. The corresponding selected area diffraction pattern (Fig. 5(b)) confirms the presence of spotty rings due to the (Ni,Cu)Al2O4 spinel phase with lattice parameter 8.1 Å of FCC lattice.

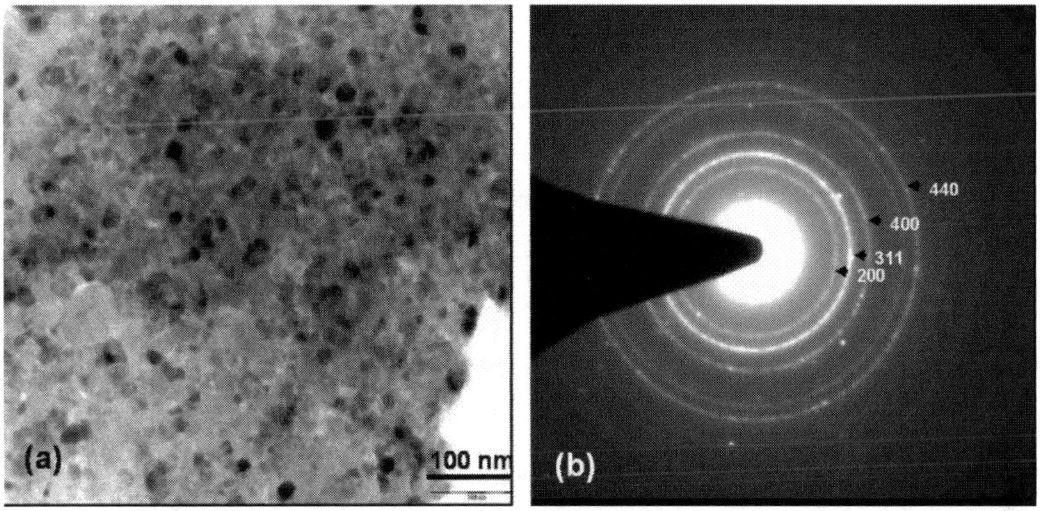

Figure 5. (a) Bright field images of powder sample after 100 h of milling and 60 h of annealing in air at 700 °C, showing the aggregates of nano spinel of the order of 60-80 nm in size, (b) the corresponding diffraction ring pattern shows the formation of spinel phase.

In the present milling experiments τ_3 phases was found to be evolving after 100 h of milling. Earlier the formation of micron size grains of τ_3 phase by rapidly solidification in Al-Cu-Ni system have reported and it was found difficult to obtain the nanometric grain size [32]. The present study demonstrates that by high energy ball milling and controlled annealing nano τ_3 phase can be synthesized. It appears that the ordering tendency among the vacancy and the metallic atoms are so strong that the complete disordering cannot be achieved even after 100 h of milling. Furthermore, it will be worth studying the evolution of superlattice ordering as a function of further milling as well as annealing. The formation of spinel phase was possible due to profuse oxidation during air annealing. However the nanocrystalline phase has also played an important role for the formation of spinel as it has enhanced the oxidation kinetics. The nano VOP phase obtained after milling of the elemental powder has reduced the activation energy for the diffusion of oxygen atoms due to the strained and nanoscale microstructure. The oxidation reaction being exothermic in nature it generates heat, which further increases the rate of oxidation. This is very much similar to the process called as self propagating high temperature synthesis (SHS), reported for synthesizing spinels by solid state reactions [33-34]. Thus the mechanically activated powder surface along with the nano scale microstructures, and exothermic reaction, has caused the easy formation of nanospinel even at moderate annealing temperature. It is interesting to point out that the significant grain growth of the spinel has not taken place. This could be understood due to the fact that the phase transformation needs to be advanced by the restructuring of the resultant spinel phase. The micromechanisms of spinel phase formation as well as the evolution of nano VOPs phase requires further investigation.

4. Conclusions

In the present investigation formation of nano vacancy ordered phase in Al-Cu-Ni system by mechanical alloying of elemental powders has been achieved. The stability of nano VOPs has been studied under vacuum and air annealing. The ordering, strain relaxation and grain coarsening have been observed after annealing of the nano VOPs in vacuum at 700 °C for 60 h. The crystallite size of the vacuum annealed nano $Al_{70}Cu_{15}Ni_{15}$ powder is found to be ~ 100 nm. It has been found, that for same annealing condition in air, a nano (Ni,Cu)Al_2O_4 spinel phase has formed. The minimum annealing temperature and time were found to be 700 °C and 60 h respectively.

Acknowledgement
The authors would like to thank Dr. M.A.Shaz for many several discussions. The authors also gratefully acknowledge the Department of Science and Technology: UNANST (DST), New Delhi and Ministry of New and Renewable Energy (MNRE), New Delhi, India for financial support.

References

[1] Sun W, Chen Y H andZhang Z 2007 *Philosophical Magazine* ***87*** 2815
[2] Van Sande M, Ridder R De, Landuyt J Van. et al. 1978 *Physica Stat. Sol (a)* ***50*** 587
[3] Chattopadhyay K, Lele S, Ranganathan S and Thangaraj N 1987 *Acta Metall.* ***35*** 727
[4] Tendeloo G Van, Heurck C V and Amelinckx S, 1999 *Solid State Communications* ***71*** 705
[5] Dong, C J *1995 Phys. I. France* ***5*** 1625
[6] Dong, C, Zhang Li-M, Zhau Q G et al. 1999 *Bull. Mater Sci.* ***22*** 465
[7] Grushko B, Urban K and Freiburg C 1991 *Scripta Met. Mater.* **25** 2533
[8] Dong C, Perrot A, Dubois J M and Belin E 1994 *Material Sci Forum* ***403*** 150
[9] Zhang Z and Urban K 1989 *Scripta Met.* ***23*** 767
[10] Dong C 1995 *Scripta Metall Mater* ***33*** 239
[11] Subramaniam A and Ranganathan S 2004 *Journal of Non-Crystalline Solids* ***334&335*** 114
[12] Tjong S C, Chen Haydn 2004 *Materials Science and Engineering* R ***45*** 1
[13] Sanders P G, Barker J G and Weertman J R 1996 *J. Mater. Res.* ***11*** 311
[14] Castro T, Reifenberger R, Choi E and Andres R P 1990 *Phys. Rev. B* ***42*** 8548
[15] Huang Y K, Menovsky A A and Boer de F R 1993 *Nanostruct. Mater.* ***2*** 505
[16] Nieman G W, Weertman J R and Siegel R W 1991 *J. Mater. Res.* ***6*** 1012
[17] Murty B S and Ranganathan S 1998 *International Materials Reviews* ***43*** 101
[18] Calka A and Wexler D 2002 *Nature* ***419*** 147
[19] Koch C C 1989 *Ann. Rev. Mater Sci.* ***18*** 121
[20] Suryanarayana C 2001 *Progress in Materials Science* ***46*** 1
[21] Yadav T P, Mukhopadhyay N K, Tiwari R S and Srivastava O N 2007 *J. of Nano Sci.& Tech.* ***7*** 575
[22] Yadav T P, Mukhopadhyay N K, Tiwari R.S and Srivastava O N 2008 *Phil. Mag.* ***88*** 2227
[23] Meyer F, Hempelmann Rolf, Mathur S and Veith, Michael 1999 *J. Mater. Chem.* ***9*** 1755
[24] Woelk H J, Hoffmann B, Mestl G and Schloegl R 2002 *J. Am. Ceram. Soc.* ***85*** 1876
[25] Shimizu K, Maeshima H, Yoshida H et al. 2000 *Phys. Chem. Chem. Phys.* ***2*** 2435
[26] Cho W S and Kakihana M 1999 *Journal of Alloys and Compounds* ***287*** 87
[27] Jeevanandam P, Koltypin Yu and Gedanken A. 2002 *Mater Sci Eng. B* ***90*** 125
[28] Peelamedu R, Agrawal D and Roy R 2002 *Materials Lett.* ***55*** 234
[29] Yadav T P, Mukhopadhyay N K, Tiwari R S and Srivastava O N 2008 *J. of Applied Ceramic Tech.* ***5*** 449
[30] Yadav T P, Mukhopadhyay N K, Tiwari R S and Srivastava O N 2005 *Mater. Sc. Eng. A* ***393*** 366
[31] *Powder Diffraction File* Swarthmore, PA, **1990,** No. 60696
[32] Yadav T P, Shaz M A,Tiwari R S and Srivastava O N 2003 *Z. für Krystallographic* ***218*** 12
[33] Li Y, Zhao J and Han J 2002 *Bull. Mater. Sci.* ***25*** 263
[34] Radishevskaya N I, Kasatskii N G, Chapskaya, et al. 2006 *Glass and Ceramics* ***63*** 52

Journal of Physics: Conference Series **226** (2010) 012034 doi:10.1088/1742-6596/226/1/012034

Thermal conductivity of Taylor phase T-$Al_{73}Mn_{27}$ complex metallic alloy

D Stanić[1,2], P Popčević[1], I Smiljanić[1], A Bilušić[1,3], I Batistić[4], J Ivkov[1] and A Smontara[1]

[1]Institute of Physics, Laboratory for the Study of Transport Problems, Bijenička 46, POB 304, HR-10001 Zagreb, Croatia.
[2]Department of Physics, University of Osijek, Gajev trg 6, HR-31000 Osijek, Croatia
[3]Department of Physics, Faculty of Natural Sciences of the University of Split, Nikole Tesle 12, HR-21000 Split, Croatia
[4]Faculty Science, University of Zagreb, Bijenička c. 32, HR-10000 Zagreb, Croatia

E-mail: bilusic@pmfst.hr

Abstract. Thermal conductivity coefficient, κ, of Taylor phase T-$Al_{73}Mn_{27}$ complex metallic alloy with giant unit cell has been studied in the temperature interval from 2 K to 300 K. κ is relatively small in magnitude (comparable to that of thermal insulators), have a slope-change at about 50 K and increases up to 300 K. Such a low κ originates in complex structure: aperiodic in short length scale, that leads to frequent electron scattering and consequent low electron contribution to the thermal conductivity, while large lattice constant defines a small Brillouin zone that enhances *umklapp* scattering of extended phonons, and suppressing lattice thermal conductivity. Above ~100 K hopping of localized lattice vibrations gives an additional heat-carrying channel.

Complex metallic alloys (CMA) denote intermetallic phases with a giant unit cell containing from tens up to thousands atoms. The atoms inside the unit cell are arranged in clusters with the local icosahedral symmetry. In this paper we study thermal conductivity of the orthorhombic $Al_{73}Mn_{27}$ compound first discovered by Hofmann [1] and later studied by Taylor [2,3] , so it is often referred as the Taylor (T) phase. The structure is built of atomic layers stacked along *b* crystallographic axis, along which pentagonal columnar clusters are formed [4]. Therefore, they are considered to be approximants of the decagonal (*d*) Al-Mn phases. The unit cell of the T-$Al_{73}Mn_{27}$ phase contains 156 atoms with many of the sites having either fractional occupation or mixed Al/Mn occupation, so that a great inherent chemical disorder exists on the lattice [5].

This study is a continuation of our previous investigation of the thermal conductivity of different families of CMAs [6-10], from which two conclusions can be drawn: (*i*) the thermal conductivity of CMAs is unusually low as compared to that of typical metallic alloys and (*ii*) the contribution of electrons to the thermal transport is significantly lower than that due to phonons over a wide temperature range. At the same time this study is a part of systematic investigation of thermal and electrical properties of *T*-$Al_{73}Mn_{23-x}(Pd,Fe)_x$ ($0 < x < 6$) alloys which are convenient for the research due to the strong variations of these properties with the composition.

In what follows we first present some experimental details and the results for the thermal conductivity in T-$Al_{73}Mn_{27}$ phase. Next, we separate the thermal conductivity into electron and lattice parts. There, the much smaller electron contribution was determined by the spectral conductivity model. Further, the lattice contribution is analyzed by considering propagation of long-wavelength phonons within the Debye model and by a hopping of localized vibrations of a short wavelength. Finaly, the results of a fitting procedure to experimental data are show.

Polygrain ingot sample of desired composition was produced from constituent elements by levitation induction melting in a water-cooled copper crucible under an argon atmosphere. To obtain T-phase, parts of the sample was annealed in argon at 900 ^{0}C for up to 312 h, additionally at 930 ^{0}C for 3 h and subsequently quenched into water [5]. The annealed sample was single phase.

Thermal conductivity κ has been measured in the temperature interval from 2 to 300 K in liquid helium cryostat using an absolute steady-state heat-flow method. The thermal flux through the sample was generated by a 1 kΩ RuO_2 chip-resistor, glued to one end of the sample, while the other end was attached to a copper heat sink. The temperature gradient across the sample was monitored by a chromel – (gold+0.07 at.% Fe) differential thermocouple with the wires diameter of 25 μm. Electrical resistivity ρ (conductivity $\sigma = 1/\rho$) and the thermoelectric power S were measured in the same temperature range: ρ using the standard four-terminal technique, while S by a differential method with two identical thermocouples (chromel-gold + 0.07 at.% iron, 50 μm of wire thickness), attached to the sample with silver paint.

The temperature dependence of the thermal conductivity $\kappa(T)$ for the T-$Al_{73}Mn_{27}$ is shown in figure 1. $\kappa(T)$ shows characteristic behaviour for complex metallic alloys (e.g. $Mg_{32}(Al,Zn)_{49}$ [11], $Al_{74}Pd_{22}Mn_4$, β-Al_3Mg_2 [12] and ε-phases (Al-Pd-transition metal) [8-10]: relatively small in magnitude, change of slope at about 50 K and increases up to 300 K.

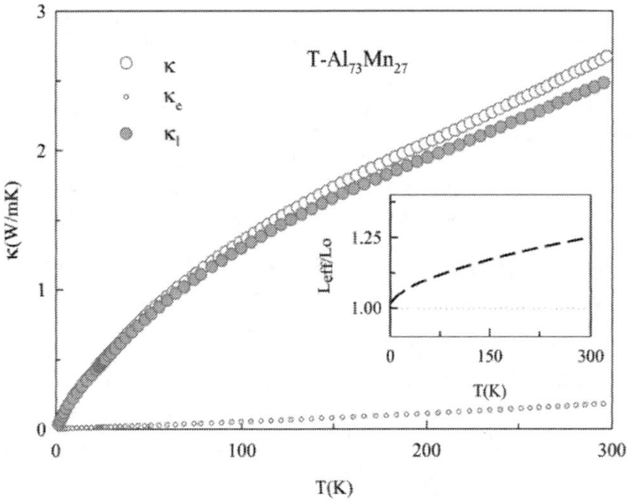

Figure 1. Temperature dependence of the thermal conductivity $\kappa(T)$ between 2 and 300 K in T-$Al_{73}Mn_{27}$, electronic contribution, κ_e, calculated by the model of spectral conductivity, and estimated lattice contribution $\kappa_l = \kappa - \kappa_e$. The inset shows temperature dependence of the ratio of effective Lorentz number L_{eff} to the Lorentz number L_0.

At room temperature the magnitude of κ is 2.7 W/mK. Such small magnitude of thermal conductivity is the characteristic of thermal insulators like SiO_2 [13] and Zr/YO_2 [14]. The small value of κ has been found also in icosahedral quasicrystals *i*-Al-Pd-Mn [15,16] which is explained by (a) the low electronic density of states at Fermi level E_F (small contribution of electrons to thermal conductivity), and (b) lattice aperiodicity (that suppresses the lattice contribution to κ). Similarly, the electrical conductivity of T-$Al_{73}Mn_{27}$ is rather low compared to simple metals. In that sense, we apply the thermal conductivity model appropriate to systems with the long-scale lattice periodicity and the short-scale atomic clustering structure that has been described in details in ref. [6].

To estimate the electronic part to the thermal conductivity (κ_e) we apply analysis based on the Kubo-Greenwood response theory [17,18]. The central quantity of this formalism is the spectral conductivity function $\sigma_S(E)$ that incorporates both the band structure and the transport properties of the system. The analysis of the temperature dependence of the electrical conductivity $\sigma(T)$ and the thermoelectric power $S(T)$ gives the shape of $\sigma_S(E)$ in the vicinity of E_F. The details of the analysis, which it is a modified version of the procedure originally developed by Landauro and Macia [19,20] and adjusted to suite the experimental data in this class of compounds, can be found in the reference [21]. Once calculated, $\sigma_S(E)$ can be used to determine κ_e (for room-temperature values see table 1).

Table 1. Values at room temperature of: the electrical σ, and thermal conductivity, κ, the electronic contribution κ_e calculated by the model of spectral conductivity and the fit parameters for the lattice thermal conductivity $\kappa_l=\kappa_D+\kappa_H$ (see text).

σ ($10^4\ \Omega^{-1}\mathrm{m}^{-1}$)	κ (W/mK)	κ_e (W/mK)	A ($10^7\ \mathrm{s}^{-1}\mathrm{K}^{-2}$)	κ_H^0 (W/mK)	E_a (meV)
1,97	2,69	0,18	5,6	4,8	17,7

It is interesting to consider the validity of the Wiedemann–Franz law (WFL) $\kappa_{e0}= L_0T\sigma$, where $L_0=$ 2,44×10^{-8} $\Omega\mathrm{K}^{-2}$ is the Lorenz number. Additionally, from the spectral conductivity model we can calculate the effective Lorentz number $L_{eff}=\kappa_e/T\sigma$. The ratio L_{eff}/L_0 is given in the inset of Fig. 1: one can see that WFL is valid at the limit T→0 only. Similar deviation from WFL in icosahedral quasicrystals is observed in ref. [22].

The lattice contribution, $\kappa_l(T) = \kappa(T) - \kappa_e(T)$ is analyzed by considering propagation of long-wavelength phonons within the Debye model (κ_D) and hopping of localized vibrations of a short wavelength (κ_H). This picture assumes that large atomic clusters of icosahedral symmetry strongly suppress propagation of phonons in the lattice of complex metallic alloys, except acoustic phonons with long wavelength for which this material is an elastic continuum. Additionally, localized vibrations within the cluster substructure can participate in the heat transfer via thermally activated hopping. In the simplest model, hopping of localized vibrations is described by the single activation energy E_a, yielding a contribution to the thermal conductivity $\kappa_H = \kappa_H^0 \exp(-E_a/k_BT)$ [7], where κ_H^0 is a constant.

The thermal conductivity within Debye model is given by [6]:

$$\kappa_D = C_D T^3 \int_0^{\theta_D/T} \tau(x) \frac{x^4 e^x}{(e^x-1)^2} dx \tag{1}$$

where $C_D = k_B^4/2\pi^2\bar{v}\hbar^3$, $\bar{v}$ is the average sound velocity, θ_D the Debye temperature, τ the phonon relaxation time, $x = \hbar\omega/k_BT$, and $\hbar\omega$ is the phonon energy. Since the Debye temperature of the investigated T-$Al_{73}Mn_{27}$ is not known, we have used $\theta_D = 500\ K$ reported for the related icosahedral *i*-Al–Pd–Mn quasicrystals [23]. Since our $\kappa(T)$ data are available only up to 300 K, it turns out that the fit is insensitive to a slight change of θ_D. The Debye constant C_D was also not taken as a free parameter, but was instead calculated by using $\bar{v}$ = 4000 ms^{-1}, a typical value for such complex structure [6]. The different phonon-scattering processes are incorporated into the relaxation time $\tau(x)$ and we assume that Matthiessen's rule is valid, $\tau^{-1} = \sum \tau_j^{-1}$, where τ_j^{-1} is a scattering rate related to the *j*-th scattering channel. In analogy to the ε-phases in Al–Pd–Mn [6], we consider two dominant scattering processes in the investigated temperature interval. First, scattering of phonons on structural defects of stacking-fault type with the scattering rate

$$\tau_{sf}^{-1} = \frac{7}{10}\frac{a^2}{\bar{v}}\gamma^2 N_s \omega^2 \equiv A\omega^2 \tag{2}$$

where a is a lattice parameter, γ is the Grüneisen parameter and N_s is the linear density of stacking faults. The second scattering mechanisms are the *umklapp* processes with the phenomenological form of the scattering rate pertinent to complex metallic alloys [15], $\tau_{um}^{-1} \propto \omega^{\alpha} T^{4-\alpha}$. Thus for the total scattering rate we get $\tau^{-1} = \tau_{sf}^{-1} + \tau_{um}^{-1}$.

Results of a fitting procedure of a relation $\kappa_l(T) = \kappa_D(T) + \kappa_H(T)$ to experimental data are shown in figure 2. Parameters of the fitting procedure are given in table 1. From the figure 2, it is easy to see

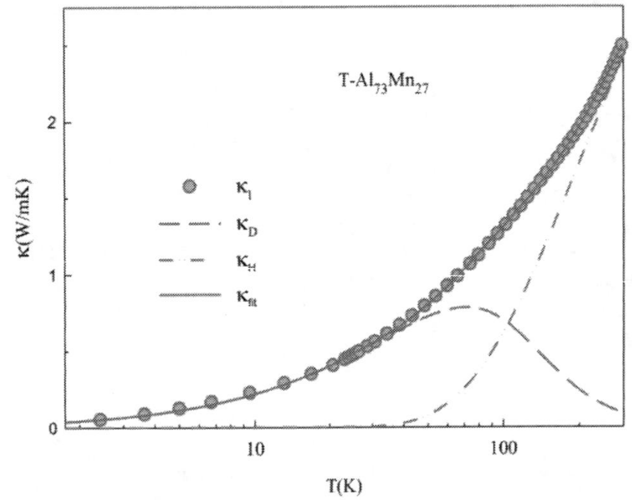

Figure 2. Temperature dependentce of the phonon thermal conductivity κ_l together with the Debye contribution κ_D and hopping contribution κ_H of T-$Al_{73}Mn_{27}$ shown in log scale.

that Debye contribution $\kappa_D(T)$ has a maximum at about 50 K while it becomes smaller at higher temperatures, a characteristic for the periodic structures that originates in phonon-phonon *umklapp* scattering processes. From the fitting parameter A we can obtain N_s: for typical values for the lattice parameter $a \approx 1.5$ nm and the Grüneisen parameter $\gamma \approx 2$, we get $N_s = 2\ \mu m^{-1}$. This micrometer-scale N_s value is comparable to those reported for ψ-Al-Pd-Mn [6], *i*-Al-Pd-Mn [24], and decagonal *d*-Al-Mn-Pd [25]. The parameter α defines phonon scattering by *umklapp* processes in a phenomenological way. Fitting procedure gives $\alpha = 1$, so the frequency and temperature dependence of the *umklapp* term is $\tau_{um}^{-1} \propto \omega T^3$. Hopping contribution κ_H becomes significant above 100 K; obtained activation energy E_a is 17.7 meV and is smaller by factor 2 than E_a of ξ-Al-Pd-Mn [6]. This smaller E_a value reflects the considerably less steep $\kappa(T)$ increase at temperatures above 100 K for our samples compared to ξ-Al-Pd-Mn. On the other side, the above E_a values correlate with the inelastic neutron and X-ray [6] scattering experiments on *i*-Al-Pd-Mn quasicrystals, where dispersionless vibrational states were identified for energies higher than 12 meV. Such states indicate localized vibrations and are considered to be a consequence of a dense distribution of energy gaps in the phonon excitation spectrum. Therefore, localized vibrations also appear to be present in the T-$Al_{73}Mn_{27}$, where their origin may be attributed to the cluster substructure of the giant unit cell.

To conclude, we have investigated thermal conductivity of *T*-$Al_{73}Mn_{27}$ sample which show behavior typical for complex metallic alloys: relatively small value and change of slope at about 50 K. We have separated $\kappa(T)$ into electron $\kappa_e(T)$ and lattice (phonon) $\kappa_l(T)$ parts. Electron contribution was determined by the spectral conductivity model, and it is much smaller than the lattice one, due to rather low electrical conductivity. The low level of the lattice thermal conductivity is determined by enhanced *umklapp* processes of phonon scattering (caused by large lattice constant and, consequently small Brillouin zone) and by inherent short- and middle scale size aperiodicity of the structure.

Acknowledgements

This work was done within the activities of the 6th Framework EU Network of Excellence "Complex Metallic Alloys" (Contract No. NMP3-CT-2005-500140), and has been supported in part by the Ministry of Science, Education and Sports of Republic of Croatia through the Research Projects: 035-0352826-2848; 035-0352826-2847 and 119-1191458-0512. We are especially grateful to M Feuerbacher (Jülich) for the provision of the sample.

References

[1] Hofmann W 1938 *Aluminium* (Berlin) **20** 865
[2] Taylor M A 1960 *Acta Metall.* **8** 256
[3] Taylor M A 1961 *Acta Cryst.* **14** 84
[4] Klein H, Boudard M, Audier M, deBoissieu M, Vincent H, Beraha L and Duneau M 1997 *Philos. Mag. Lett.* **75** 197
[5] Balanetskyy S, Meisterernst G, Heggen M and Feuerbacher M 2008 *Intermetallics* **16** 71
[6] Dolinšek J, Jeglič P, McGuiness P, Jagličić Z, Bilušić A, Bihar Ž, Smontara A, Landauro C, Feuerbacher M, Grushko B and Urban 2005 *Phys. Rev.* B **72** 064208
[7] Bihar Ž, Bilušić A. Lukatela J, Smontara A, Jeglič P, McGuiness P, Dolinšek J, Jagličić Z, Janovec J, Demange V, Dubois J.M 2006 *J. Alloys Compd.* **407** 65
[8] Smontara A, Smiljanić I, Bilušić A, Grushko B, Balanetskyy S, Jagličić Z, Vrtnik S and Dolinšek J 2008 *Journal of Alloys and Compounds* **450** 92
[9] Smiljanić I, Smontara A, Bilušić A, Lukatela J, Stanić D, Barišić N, Dolinšek J, Feuerbacher M and Grushko B 2008 *Philosophical Magazine* **88** 2155
[10] Smontara A, Bilušić A, Bihar Ž and Smiljanić I 2009 *Properties and Application of complex Metallic Alloys* ed. E Belin-Ferre (London: World Scientific) p. 113
[11] Smontara A, Smiljanić I, Bilušić A, Jagličić Z, Klanjšek M, Dolinšek J, Roitsch S and Feuerbacher M 2007 *Journal of Alloys and Compounds.* **430** 29
[12] Dolinšek J, Apih T, Jeglič, P, Smiljanić I, Bihar Ž, Bilušić A, Smontara A, Jagličić Z and Feuerbacher M 2007 *Intermetallics.* **15** (2007) 1367-1376 .
[13] Zhu D M 1994 *Phys. Rev.* B **50** 6053
[14] Mevrel R, Laizet J C, Azzopardi A, Leclercq B, Poulain M, Lavigne O and Demange D 2004 *J Eur. Cer. Soc.* **24** 3081
[15] Bilušić A, Budrović Ž, Smontara A, Dolinšek J, Canfield P C and Fisher I. R 2002 *J. Alloys Compd.* **342** 413
[16] Bilušić A, Smontara A, Dolinšek J, McGuiness P and Ott H R 2007 *J. Alloys Compd.* **432** 1
[17] Kubo R 1957 *Journal of the Physical Society of Japan* **12** 570
[18] Greenwood D A 1958 *Proceedings of the Physical Society* **71** 585
[19] Landauro C V, Maciá E and Solbrig H 2003 *Phys. Rev.* B **67** 184206
[20] Maciá E, Takeuchi T and Otagiri T 2005 *Phys. Rev.* B **72** 174208
[21] Batistić I, Stanić D, Tutiš E and Smontara A 2010 *Croatica Chemica Acta* **83** in press
[22] Maciá E and Rodríguez-Oliveros R 2007 *Phys. Rev.* B **75** 104210
[23] Wälti C, Felder E, Chernikov M A, Ott H R, de Boissieu M and Janot C 1998 *Phys. Rev.* B **57** 10504
[24] Legault S, Ellman B, Strom-Olsen J, Taillefer L, Lograsso T and Delaney D 1998 *Quasicrystals: Proc. 6th Int. Conf. (Tokyo, 1997)* (Singapore: World Scientific) p. 475
[25] Matsukawa M, Yoshizawa M, Noto K, Yokoyama Y and Inoue A 1999 *Physica* B **263-264** 146

Simple model of Mg_2Al_3: β and β' phases

J. Wolny, M. Duda, B. Kozakowski

Faculty of Physics and Computer Science, AGH University of Science and Technology, al. Mickiewicza 30, 30-059 Kraków, Poland

wolny@novell.ftj.agh.edu.pl

Abstract: Two crystalline phases: β and β' of the Mg_2Al_3 alloy are compared. There are three types of structural domains, whose structures are based on sets of hexagonal layers. The main chain contains 11 layers. It does not undergo any transformation during a phase transition from the β'-phase to the β-phase. Conversely, during this process, shorter chains vanish and transform into clusters, whose atomic positions are occupied partially. Any two adjoining domains merge in a coherent way. It does not apply to the third domain. At its border structural defects appear. They are the seeds of clusters emerging in the β-phase.

1. Introduction

The Samson structure, β-Mg_2Al_3, is one of the most complex intermetallic structures (Samson, 1965, Feuerbacher et al., 2007). Its cubic elementary cell (*Fd*-3m, no. 227 space group) contains 1168 atoms which are distributed over 1832 atomic positions. About 75% of atoms (879 to be exact) form the firm framework of the structure - 528 of them are Al and 351 are Mg. By "the framework" or "skeleton atoms" we understand a set of Samson's positions which are occupied by atoms with the probability of 100%. The remaining 289 (25%) atoms partially occupy 953 positions with the average occupation probability of 30%. They form clusters arranged in an elementary cell in a tetrahedral lattice. Their structure has been described in detail in (Feuerbacher et al., 2007) and the following references and in (Sikora et al., 2008). The lattice constant of the Samson structure is gigantic: a_c=2.8242(1) nm. When cooled down, at the temperature of 214°C (with the cooling rate of 5°C/min), the structure undergoes a phase transformation to the rhombohedral β'-Mg_2Al_3 (space group *R3m*, no. 160, which is a subgroup of the *Fd*-3*m* group - index 4) with a_r=1.9968(1) nm, c_r=4.89114(8) nm. It should be pointed out that the constant c_r of the rhombohedral structure is practically equal to the length of the main diagonal of the cubic structure $a_c\sqrt{3} = 4.89166 \text{ nm} \approx c_r$. Additionally, the ratio $a_c / a_r = \sqrt{2}$, which means that the frame of the unit cell is the same for both: cubic and rhombohedral structure.

Stacking of hexagonal layers in the structure of β-Mg_2Al_3 has been published recently (Wolny et al., 2008). In that paper we pinpointed the fact that there were three structural domains distributed along the main diagonal ([111] direction) of a cubic unit cell. The relative positions of the centers of these domains were shifted by 1/3 of the diagonal length. All length values were given in relation to the z axis unit of the hexagonal reference frame, which is equal to the length of the main cubic diagonal - for the β cubic phase, or its equivalent, lattice constant c - for the β' rhombohedral phase. It was proved, that the core of a domain consists of eleven parallel and evenly distant hexagonal planes occupied by skeleton atoms. 75% of the β phase atoms and 100% of the β' phase atoms are the skeleton ones. All skeleton atoms form a set of 22 parallel and evenly distant hexagonal layers. The set is clearly illustrated by the figures introduced in the further part of the paper.

For the purposes of the structural analysis presented in this paper we use atomic positions obtained from the gigantic elementary cell (Samson, 1965, Feuerbacher et al., 2007). These positions are widely accepted by different teams of crystallographers. Consequently, we adopted this model for our study of the structure, aiming at discovering some simple rules that would allow us to reconstruct such complex structures.

2. Hexagonal layers

Figures 1 a&b show the positions of all Al and Mg skeleton atoms of inner part of domain-I for β-Mg_2Al_3 projected onto the base layer. With great accuracy only the hexagonal sites marked by A, B and C are occupied by Al atoms (figure 1a). The Mg atoms behave similarly (figure 1b).

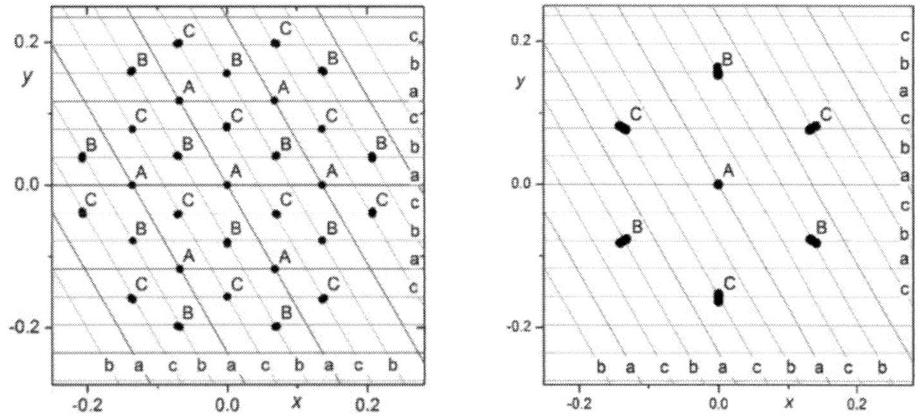

Figure 1. The skeleton Al (a) and Mg (b) atoms projected onto the plane (111) for the cubic structure of β-Mg_2Al_3. All the atoms fit to the nodes of hexagonal network.

Figures 2a&b outline an exact sequence of layers. The whole structure is divided into three domains. In figures 2a&b, only domain I is presented in details, because its structure is identical to the structure of domains II and III. These domains are represented here only by a set of lines: thick ones (Al atoms) and thin ones (Mg atoms).

Their relative shift along the z-axis is equal to 1/3. The whole set consists of 22 parallel and equally distant layers. Some of them are occupied by the skeleton atoms. The rest is filled by atoms forming clusters – the structures whose atomic positions are occupied with the probability less than one. The distance between two layers equals $1/22 \approx 0.0788$. The figures show only the skeleton atoms, which lie inside domains ($r \leq 0.22$): Al atoms – thick lines and open circles; Mg atoms – thin lines and full circles.

In case of the β-Mg_2Al_3 phase (figure 2a) the Al atoms are arranged into 11 layers forming the main chain. Only these layers are occupied by the skeleton atoms. The Mg atoms which are associated with the Al layers form atomic complexes denoted with α, β and γ symbols. The α complex consists of three hexagonal layers containing some A-type knots, which are occupied in a sequence: Mg/Al/Mg (see also figure 3a). Within the middle layer, among Al atoms, there are gaps occupied by Mg atoms. The outer Mg layers are shifted by around 0.06 with respect to the Al layers, because the atomic radius of Mg atoms is nearly 12% larger than the Al one. Consequently, Al gaps are too small for Mg atoms. There is not enough space for them within a common hexagonal layer. Mg atoms are situated either above or below a hexagonal Al layer. The β and γ complexes closely resemble the α complex, however in their case, these are positions B (β) or C (γ) that are taken. It is also worth emphasizing that for each complex (α, β and γ) the ratio Mg/Al is equal to 2/3, which is in accordance with the chemical compositions of the alloy Mg_2Al_3.

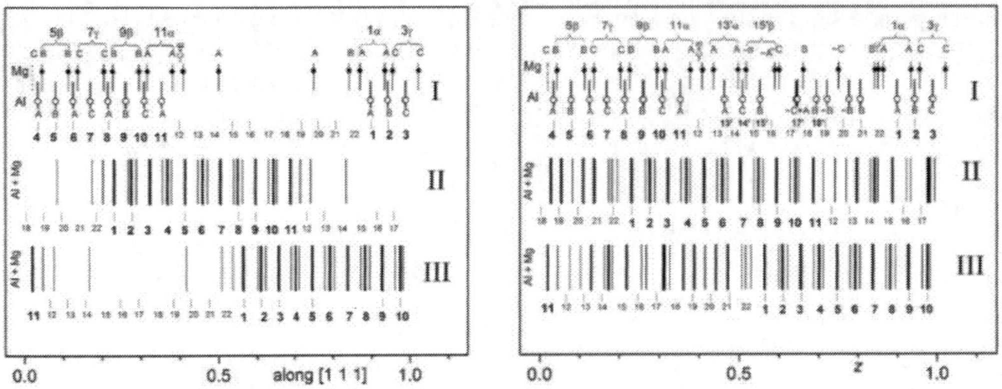

Figure 2. The sequence of 22 hexagonal layers and their arrangement within particular domains: I, II and III. Thick lines indicate Al layers, thin ones – Mg layers. All three domains are identical. They are shifted by 1/3 along the z axis with respect to each other. The structure of the inner area ($r \leq 0.22$) of the first domain is outlined in details in the figure. Open circles are associated with Al atoms and full circles with Mg atoms. In case of the β-phase (a) the skeleton atoms form 11 layers of the main chain. In case of the β'-phase (b), in addition to the main sequence, there are also shorter ones: layers 13'-15' and 17'-18'.

Among the α, β and γ complexes, there are hexagonal layers of type A, B or C partially filled by Al atoms, which are interspersed with a gap. The resulting sequence of 11 layers of the main chain is as follows: 1α2B3γ4A5β6A7γ8A9β10C11α. The middle part of this chain consists of a regular sequence: 3γ4A5β6A7γ8A9β within which Al atoms occupy A positions whereas Mg atoms are put in B or C positions in alternate layers. The periodic set of 11 hexagonal layers can also be described by Friauf polihedra, as it is shown in figure 3c. This is an energetically stable sequence. However, it is not consistent with the chemical composition of the alloy. If it filled the whole space, the ratio Mg/Al would achieve the value 1/2 and the Mg atoms would be in deficiency. The chemical balance is restored by Mg atoms scattered within outer parts of the structural domains. It should be emphasized that the factual chemical composition of the studied sample was not fully stoichiometric – it was equal to $Mg_{38.5}Al_{61.5}$ (Feuerbacher et al., 2007).

Half of the domain is occupied by two clusters containing most of the atoms with SOF<1. As it is shown in figure 3b, a Mg atom occupies the centers of hexagonal rings of Al atoms. The size of Mg atom is bigger than the size of Al atom making the lattice constant for the hexagonal lattice greater than for the hexagonal lattices of skeleton layers.

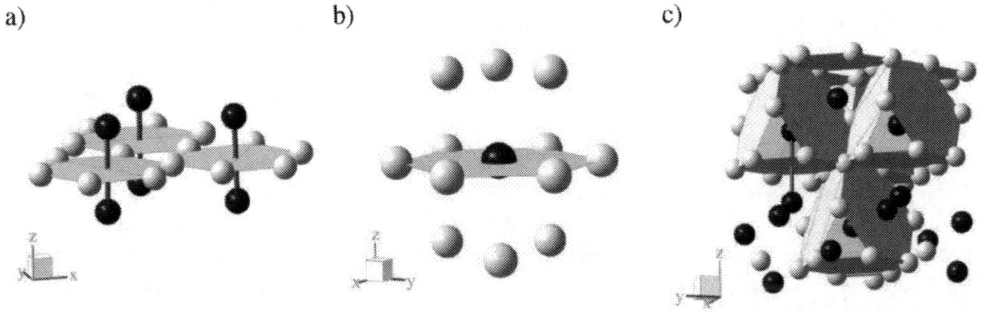

Figure 3. Local arrangement of atoms for a periodic set of 11 hexagonal layers (a) and for aperiodic clusters (b) for β-Mg_2Al_3. Decoration of periodic set of hexagonal layers by Friauf polihedra (c). Light balls – Al atoms, dark ones – Mg atoms.

The structure gets heavily defected above room temperature. The defects destroy the initial periodic sequence of layers. At higher temperatures Al atoms can easily swap positions between A and B (layer 2) or C (layer 10). It does not involve a considerable amount of energy. For a perfect set of close-packing of spheres such energy would be equal to zero. Therefore, Al swaps can easily occur at temperatures which are only slightly higher than room temperature. This kind of structural defect terminates the main chain of hexagonal layers and consequently layers 2B and 10C are the second to last layers in this chain.

The set of layers of the rhombohedral phase β'-Mg_2Al_3 is outlined in figure 2b. As in previous figures, only the inner atoms ($r \leq 0.22$) are highlighted there. Similarly to the cubic phase, this structure also has 11 hexagonal layers arranged in a sequence of the main chain. Additionally, the structure contains some shorter chains (three-layer long 13'-14'-15' and two-layer long 17'-18'). They are not observed in the high temperature phase (β). These short chains occupy positions situated between 22 layers of the main chain, which are described above. The transition between positions of the main chain's layers to the positions of the short chains' layers is realized by successive Al-Mg swaps in consecutive layers.

All the atoms that belong to the 11 layer-long main chain occupy A, B or C hexagonal positions (figure 2). Only layer 17 is occupied by those Al atoms that can be found in ~C positions. These positions do not belong to the set of hexagonal structure's knots. Nonetheless, they are quite close to them. A similar atomic arrangement can be observed in the case of layer ~20' (figure 2b) where Al atoms occupy positions ~B.

The single domains of two phases: β and β' of Mg_2Al_3 are compared in figure 4. The transformation from the high temperature phase (β) to the low temperature phase (β') results in reducing the symmetry of a single domain. A symmetrical domain becomes asymmetric and the clusters break apart into short chains of hexagonal layers of different sizes. Depending on the size of the chain, the observed process undergoes at different temperatures. Instead of classical phase transition one observs a diffuse-controlled transformation, which also depends on the annealing time of the sample.

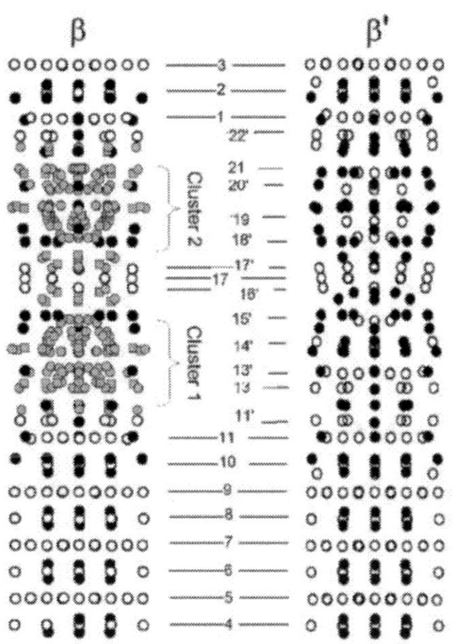

Figure 4. Comparison of single domains of β and β' phases in Mg_2Al_3. A set of 22 layers and two aperiodic clusters are marked in the figure.

3. Relative shift of the domains

The structure of Mg_2Al_3 consists of three domains shifted with respect to each other along the z axis by 1/3. Within the base plane these domains form a hexagonal set. It is shown in figure 5, where one can find the projected atomic positions of all domains forming layers 9B. As a reference frame for each domain a hexagonal lattice type A (thin lines) is marked in the figure.

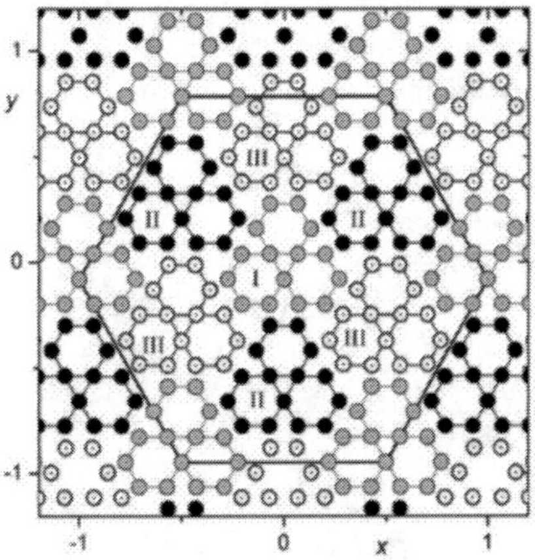

Figure 5. Layers 9B of neighboring domains are drawn within a common plane. The centers of the domains form a superstructure marked with thick lines. The base hexagonal lattice is marked with thin lines. The superstructure is $22/3 \approx 7.3$ times larger than the base hexagonal lattice.

An average lattice constant a_0 can be established by adjusting the hexagonal lattice to the atoms' coordinates. Having done that for 11 consecutive layers of the main chain, the value of this constant can be estimated as $a_0 \approx 0.1368$. This value is identical to the lattice length a of close-packed spheres arranged along c_h into 22 hexagonal layers.

Equal distances between atoms, regardless of their lying strictly within hexagonal layers or outside them, are proved by an assumption proposed also by Steurer (2007), which states that atoms are stacked into almost a close-packed structure. The value of nearest neighbors distance obtained for Al atoms projected into hexagonal plane is equal to 0.273 nm (i.e. equal to a_0) which is very close to the nearest neighbors distance observed for the FCC aluminum structure (0.286 nm). A slightly lower value of that quantity measured for Mg_2Al_3 (of about 4%) can be explained by low roughness of the basic plane.

The behavior similar to the Samson structure has already been observed in some other complex phases, e.g.: Cu_4Cd_3 (Anderson, 1980) or $NaCd_2$ (Yang, Anderson & Stenberg, 1987), where the regions corresponding to the hexagonal domains consisted of fragments of the Laves phase structures. The Laves structures are frequently described in terms of the stacking of hexagonal layers.

It also seems that the skeleton atoms, which are forming hexagonal layers, play an important role for other Laves phases. For example, there should be a certain simple explanation of the commensurate modulated vectors for Mg_2Al_3 and $Mg_{17}Al_{12}$ described by Donnadieu et al. (2002).

4. Conclusions

Mg_2Al_3 structure consists of three types of hexagonal rod-like domains. The core of each domain contains the main chain of 11 evenly distant hexagonal layers occupied by skeleton Al atoms arranged into a sequence: $1\alpha2B3\gamma4A5\beta6A7\gamma8A9\beta10C11\alpha$. The building element of the structure consists of hexagonal chains of aluminum atoms completed with dimmers of magnesium atoms, as it was shown in the figure 3. The connection between such complexes is realized by hexagonal Al layers, for which every second atomic position is empty. An arrangement of such building blocks can be also described by Friauf polihedra, which are very common for the Laves phases. The resulting structure of 11 layers is rather stable and does not change during the transition from the low temperature phase (β') to the higher temperature one (β). The most important problem is, however, the chemical composition of the core of the eleven layers. The corresponding value of the chemical composition is $MgAl_2$ rather than Mg_2Al_3 and the magnesium atoms are in deficiency. The Mg atoms are mostly located in the outer part of the domains.

The most common defect, which ends the well ordered sequence of hexagonal layers in a single domain, is associated with the process of atomic swap – an exchange of positions between an Al atom and a Mg one. This atomic disorder makes one chain of hexagonal layers end and another chain begin. Due to the Al-Mg swap, the shorter few-layer-long chains are formed. These short chains of layers are shifted with respect to the main chain and they occupy intermediate positions for β'-phase. At high temperature such short chains become unstable and transform into clusters, whose crystal positions are only fractionally occupied (β-phase).

A refinement process conducted upon all the layers of the main chain allowed us to ascertain the average value of the lattice constant for the hexagonal lattice equal to $a_0 = 0.1368$, when measured in relative units or 0.2732 nm. Assuming that a_0 is also the diameter of an Al atom of a close-packed structure, we obtain the distance between hexagonal layers equal to 0.2223 nm. Such value fits perfectly to the distance between 22 layers along the z direction (i.e. $c_r/22 = 0.2223$ nm). This means that size of the unit cells of both rhombohedral (β') and cubic (β) phases is fully determined by the close packing of Al skeleton atoms.

The three rod-like domains have identical structures. They are shifted along the z axis by 1/3 of the period. In case of symmetrical domains (β-phase) their sequence is: I, II and III, which leads to the FCC unit cell. In case of asymmetrical domains (β'-phase), even if the domains are in the same sequence as for cubic structure, the symmetry of the unit cell is reduced to rhombohedral.

In the base plane there is a hexagonal superstructure which periodicity is $22/3 \approx 7.3$ times bigger than the hexagonal lattice constant for Al atoms in a single domain. Such superstructure of domains together with the symmetry of individual domains are responsible for cubic or rhombohedral symmetry of an average unit cell.

Acknowledgements
This work has been partially supported by Ministry of Science and Higher Education of Poland for contract No. N N202 174135 and also by CMA-NoE activities in AGH-UST.

References

[1] Anderson S 1980 An alternative description of the structure of Cu_4Cd_3 *Acta Crystallographica* **B36** 2513-2516
[2] Donnadieu P et al 2002 Commensurately modulated states in the Mg-Al system *Philosophical Magazine* **A82** 297-316
[3] Feuerbacher M et al 2007 The Samson phase, β-Mg_2Al_3, revisited *Z. Kristallogr.* **222** 259-288
[4] Samson S 1965 The crystal structure of the phase β-Mg_2Al_3 *Acta Crystallographica* **19** 401–413
[5] Sikora W et al 2008 Symmetry analysis in the investigation of clusters in complex metallic alloys *Journal of Physics: Conference Series* **104** 012-023
[6] Steurer W 2007 A lecture on CMA-NoE Annual Meeting Santorini Greece
[7] Wolny J et al 2008 Stacking of hexagonal layers in the structure of β-Mg_2Al_3 *Philosophical Magazine Letters* **88** 501-507
[8] Wolny J et al 2001 Phason and phonon contribution to diffraction pattern of $\sqrt{3}\times\sqrt{3}$-type hexagonal layers *Philosophical Magazine* **A81** 301-310
[9] Yang Q-B et al 1987 An alternative description of the structure of $NaCd_2$ *Acta Crystallographica* **B43** 14-16

Configuration energy analysis of β-Mg_2Al_3 cluster structure

B. Łabno, L. Pytlik, J. Adamowski, J. Wolny and M. Duda

Faculty of Physics and Applied Computer Science, AGH University of Science and Technology, al. Mickiewicza 30, 30059 Krakow, Poland.

E-mail: `bartek.labno@gmail.com`

Abstract. Stability of (Mg-Al) clusters in the β-Mg_2Al_3 structure has been studied with the use of the effective pair potential of interatomic interactions taken from ab-initio calculations. The total potential energy has been calculated for various possible (Mg-Al) cluster configurations in order to find the optimal ones. The computation has been started from the cluster located at the 8(b) site consisting of 72 partially occupied, overlapping atomic sites, as determined by X-ray studies. The configurations of fully occupied, non-overlaping atomic positions have been selected using the criterion of minimal potential energy. The simulated annealing algorithm has been used for further optimization of the atomic positions in 3D space. The energy optimization leads to the selection of several configurations with the energy close to the minimum. For such structures the configuration entropy has been calculated as a function of the total potential energy. More complex structures, consisting of the optimized clusters and various interface configurations, have been also studied.

1. Introduction

The β-phase of Mg_2Al_3 has been originally studied by Samson [1] and more recently in Refs. [2], [3]. The structure is described by the cubic space group Fd-3m (ITA 227) and the final X-ray profile fit is obtained using 23 Wyckoff positions, generating the total of 1832 atomic sites in the unit cell. Part of the sites (840) are considered as fully occupied by Mg or Al atoms with site occupation factor (SOF) equal to one, while the rest are treated as partially occupied (SOF<1). These two subsets of atomic positions exhibit considerable differences in the geometry of local neighborhood. The fully occupied sites obey the mutual distance constraints, i.e. no neighbor is located nearer than the minimum distance evaluated from ionic radii, thus the atoms occupy only non-overlapping positions. The fully occupied fraction forms a set of hexagonal planes [4], perpendicular to [1,1,1] directions, which provide a stable crystalline matrix (fixed framework) for the partially occupied sites.

The partially occupied sites strongly violate the minimum distance rules, since there is an extensive overlapping between neighboring sites, thus they cannot be occupied at the same time (alternative occupation). Fig.1(a) shows the two sets of sites: green (SOF=1) and red/blue (SOF<1). The cross-section on the left shows that the red/blue fraction is located in spherically shaped caverns, centered at the 8(b) positions, and their interconnecting channels. The set of 8(b) positions in the Fd-3m structure forms a diamond-type lattice with sites occupied by the spherical clusters of partially occupied atomic positions and their interconnections (bridges) located along the diamond lattice bonds. The essential part of the partially occupied fraction, consisting of two spherical clusters and their interconnection is shown in Fig.1(b). The ionic radii has been reduced to exhibit the details of the structure and the overlapping of atomic positions. A detailed analysis shows that each cluster comprises the central Mg(23) ion (ion numbering after Ref. [1]) at 8(b) position and 72 surrounding sites, which are partially occupied and strongly overlapping. The interface between the clusters, centered around the 16(c) position, consists of twelve well defined pairs of alternatively occupied sites and one double split position (green atoms near the center), giving a total number of 26 positions. The occupation factors listed in Table 2 in Ref. [2] show that the interface part exhibits two alternative forms of the twelve ion configuration (probabilities for each pair sum up to one) and one alternatively occupied Al(21) atom near the 16(c) position.

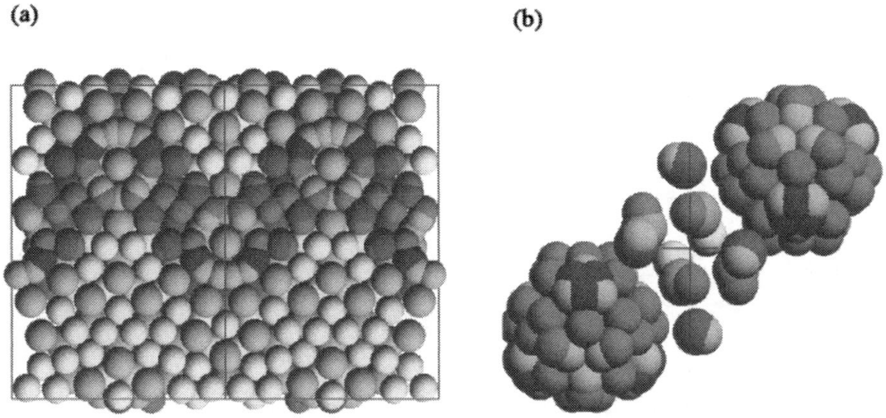

Figure 1. Model of the structure determined from X-ray diffraction. (a) [110] cross section of the unit cell, (b) Clusters with partially occupied sites. The different types of atoms are distinguished by their size (Al - 20% smaller), the colors denote occupation: green-fully occupied, red-partially occupied by Al, blue-partially occupied by Mg.

The ambiguous nature of the above-mentioned cluster structure can be resolved in two ways. First, it is possible that strongly distorted local clusters emerge with many atoms occupying "interstitial" positions. The second proposition is based on the presence of several forms of well-defined, stable clusters, which occupy at random the spherical caverns in the structure. A strong hint for the second proposition is given by the ordering phase transition to the β' phase, with symmetry lowered to the R3m space group. The symmetry analysis shows [5] that the lattice of 8(b) sites splits into two interpenetrating sublattices, occupied by two different types of clusters with possibly different number of ions. Such a suggestion is also expressed by the X-ray study of the β' phase [2] and NMR study [3].

The purpose of the present paper is to search for the most stable configurations of Mg-Al clusters, which could be used for a construction of alternative variants of the 8(b) position clusters. The problem of stability of β-Mg_2Al_3 clusters has been studied in [2], by testing the stability of the determined structure and its sensitivity to the Mg/Al substitution. The authors used both the full potential calculations and the effective inter-atomic potential obtained from ab-initio calculations [6]. In the present paper this effective pair potential is used to find the clusters with minimal energy and to construct more complex structures, consisting of both the spherical clusters and the interface parts arranged on the diamond lattice in the Fd-3m structure.

2. Computation Method

The total potential energy has been calculated for various cluster configurations in order to find the optimal ones. In the first stage of calculations, the two types of ions (Mg,Al) are located at atomic sites allowed by the symmetry group determined from the X-ray data, taking into account the minimum distance rules in order to avoid the overlapping. The potential energy of each configuration is computed from following pair interaction formula:

$$E = \sum_{\substack{i=1,\dots,N \\ j>i}} u_{ij}(r_{ij}) \tag{1}$$

where u_{ij} is the interaction energy between atoms i and j, taken from authors of Ref. [6]. The corresponding plots of the Al-Al, Mg-Mg, and Mg-Al pair potentials, which can be found in Ref. [2], are essentially different with respect to the minimum distance and potential oscillations with the atomic pair distance. The observed minimum distance values ensure that the energy calculations should be highly consistent with topological analysis based on ionic radii values.

2.1. Divide and conquer scheme for a spherical cluster

The present calculations have been started from the cluster located at the 8(b) site occupied by Mg(23) atom surrounded by 72 partially occupied, overlapping atomic sites. Therefore, in principle 2^{72} possible configurations should be generated and compared. Such calculations are characterized by a high computational complexity, therefore, we have carried out the search by scanning the configuration subsets with use of the divide and conquer computation scheme, described below. The computational procedure consists of the following steps:

(i) The set of 72 atomic positions is divided into four subsets (four quarters of the sphere), each including 18 atoms. The central Mg atom is added to each quarter, because its interaction with other atoms generate a considerable contribution to the total energy. The program scans all 2^{18} possible configurations of each quarter.

(ii) For each configuration the energy is calculated from formula (1). If the calculated energy is lower than zero, the configuration is considered as stable and saved, otherwise the configuration is discarded as unstable. If the interatomic separation is smaller then the minimum distance (equal to 0.25092 nm [3]), the energy is taken to be infinite and the configuration is also discarded.

(iii) Each stable configuration of the first spherical quarter is combined together with all stable configurations of the second quarter. This leads to $n_{q1} \times n_{q2}$ possible configurations, where n_{q1} and n_{q2} denote the number of stable configurations of the first and second quarter, respectively. For each combined configuration the total potential energy is calculated and only the configurations with the energy lower than zero are saved.

(iv) The configurations of the entire spherical cluster are created from the combinations of stable configurations of the two half-spheres. The structures with energies less than zero are saved, classified and next sorted to find those with the lowest energies. For the sets of stable structures we have calculated the configuration entropy S as a function of energy E, according to $S(E) = \ln g(E)$, where $g(E)$ is the density of states determined from the histogram of the number of configurations versus energy [7].

2.2. Creation of combined structures

The configurations obtained in Sec. 2.1 have been used to create the more complex structures by constructing the combinations of neighboring clusters and additional interconnecting atoms. First, the configuration of two spherical clusters is created, next the other partially occupied sites are added, which form the interfaces between the previously arranged clusters. Additionally, the surrounding fully occupied atomic sites are included. This form an external framework, which additionally stabilizes the entire cluster configuration. All the constructed complex structures are analyzed and the structures with the lowest energy are chosen for further optimization.

2.3. Simulated annealing

The energy optimization performed according to Subsection 2.1 selects the configurations of various classes (atomic compositions) with the energy close to the minimum. In the final stage, the configurations have been additionally optimized by varying the atomic positions in 3D space using the simulated annealing algorithm. Usually the annealing was started at the temperature of 700 K and the temperature was lowered by 1% every 50 steps. In each step, each atom was allowed to move within a 3D box with the box edge initially equal 0.1 nm. The box size was reduced by 0.5% every 150 algorithm steps, if less than 10% of results was accepted.

3. Results

3.1. Spherical clusters surrounding the 8(b) position

The comparison of stable configurations of the spherical cluster shows that it is not possible to create stable clusters consisting of more than 17 atoms (Mg(23) atom plus the surrounding 16 atoms). The limited volume of the spherical cavern, in which the cluster is located, does not allow for the creation of configurations with the number of atoms exceeding 17. This agrees with the estimated mean number of atoms, i.e., 17.32, derived from the X-ray data as a sum of SOF values over all atomic positions in the cluster.

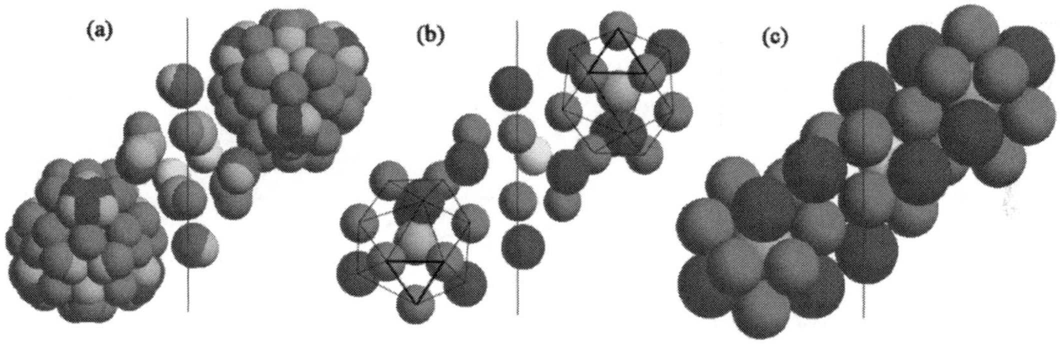

Figure 2. Illustration of three stages of 2xC17+I13 cluster construction. (a) Atomic positions as determined from X-ray studies with partial occupation and overlapping (color intensity grows with occupation probability). (b) Atomic cluster configuration as determined by energy selection from fully occupied, non-overlapping positions. (c) Cluster configuration after the random annealing procedure.

The most-stable C17 configuration consisting of 17 atoms and it is formed by the central Mg(23) atom surrounded by 4 Mg(19) atoms and 12 Al(7) atoms. There are six equivalent and symmetry related configurations with the lowest energy and some more configurations attributed to five higher energies levels (within 0.01 eV). The detailed inspection shows that all these configurations are close to the well-known Friauff polyhedron, with Mg(19) atoms shifted from their high-symmetry positions (see Fig.2b). Further adjustment of atomic positions has lead to further energy lowering (see Table 1). The Mg(19) atoms have been slightly shifted to nearby positions, located 0.3067 nm from the central Mg(23), while the Al(7) atoms have been moved to 0.3214 nm from center. The calculated Mg(19) - Mg(23) - Mg(19) angles clearly indicate that the Mg(19) atoms form a perfect tetrahedron. Together with the Al(7) atoms they form a Friauf polyhedron (see Fig.2b,c). The majority of clusters with energies close to the minimum are those with 16 or 17 atoms. The 16-atom configuration C16 with slightly higher energy comprises the defected C17 polyhedrons with one Al(7) atom missing. Other clusters worth mentioning are the B17 and B16 clusters, that emerge in the structure analysis of the β'-phase [2]. These clusters include a smaller number of Mg(19) atoms that are partially replaced by Al(20) atoms, (B17) or other vertex configurations (B16), with preservation of the triple symmetry axis along [1,1,1] direction.

3.2. *Combined structures*

The results for the combined clusters are presented in Table 1. The basic clusters are denoted by C17 and C16. The added interface atoms visible in the central part of Fig.2(b) are denoted by I13. As each C17 cluster in the diamond network is connected by four bonds, the other interconnecting atoms have been denoted as SH24. Finally, the framework of fully occupied atomic positions is denoted by FA. It can be seen that for the more complex structures energy per atom goes down since the contribution of the surface energy decreases. The addition of fully occupied positions leads to the further lowering of the energy per atom, what confirms the role of the external framework that stabilizes the entire cluster configuration. The energies of B17 and B16 configurations are initially higher than that of the most stable C17 cluster, which clearly suggests that their important role is reached by the adjustment to the neighboring atoms with SOF=1.

Table 1. Total potential energy E calculated for the analyzed cluster configurations (cluster labels are explained in text).

cluster	No. of Mg atoms	No. of Al atoms	E [eV]	E/atom [eV]	E/atom [eV] after an-nealing
C17	5	12	-1.6934	-0.099612	-0.146759
C16	5	11	-1.59278	-0.099549	-0.146903
B17	2	15	-1.57666	-0.092745	-0.115002
B16	4	12	-1.47049	-0.091906	-0.138194
B17 + B16 +IB	12	34	-4.80369	-0.104428	-0.153859
C17 + C16 + I13	16	30	-5.255129	-0.114242	-0.164605
2xC17 + I13	16	31	-5.39753	-0.114841	-0.165422
2xC17 + I13 + 2xSH24 + FA	127	92	-39.73249	-0.181427	-0.216556

Two examples of calculated configuration entropy $S = ln\ g(E)$ are displayed in Fig. 3. The right hand side plot refers to all possible configurations of isolated spherical clusters, which contain less than 18 atoms. The left hand side plot includes configurations of the entire 2xC17+I13 complex, which result from the interface variations with the spherical C17 clusters fixed (see Fig.2c). In both the cases, it is visible that the preferred lowest energy configurations are separated by an energy gap from the higher energy configurations, which can be still observed in samples quenched from the temperatures above the β - β' transition point.

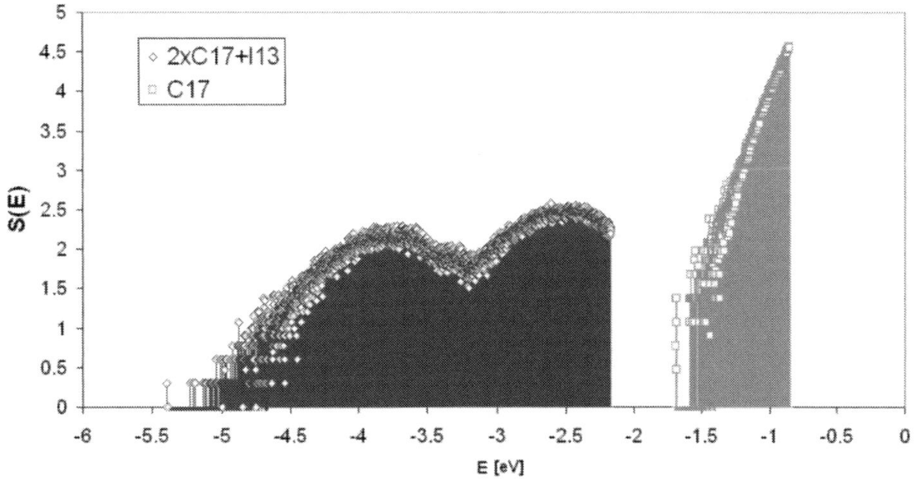

Figure 3. Configuration entropy S calculated as a function of total potential energy E for spherical clusters C17 and the 2xC17+I13 complex.

4. Summary

The clusters in the β-Mg_2Al_3 structure have been analyzed with use of the effective interatomic potential. The divide and conquer scheme has been used for the selection of low-energy spherical cluster configurations, which were optimized by the simulated annealing algorithm. We have found that the optimal C17 configurations consist of five Mg atoms and twelve accompanying Al atoms, forming a Friauf polyhedron. Alternative 16 and 17 atom configurations with the comparable energy have also been obtained. The results have been used for the creation of complex structures, containing two spherical clusters and the connecting atomic interface. The 2xC17+I13 structure exhibits the symmetry of high temperature β phase and contains only a subset of positions found in the X-ray studies. This indicates that the disordered phase includes also some other clusters, which order in the low temperature β'-phase by preferring the occupation of some Mg(19) sites by Mg and replacing the other by Al ions. Further studies are required for configurations including the entire unit cell with periodic boundary conditions.

Acknowledgements

Part of computations in this work was performed at the Academic Computer Center Cyfronet AGH. Grant number: MNiSW/SGI3700/AGH/084/2006. This works has been partially supported by EU Complex Metallic Alloys (CMA) Project and the respective Polish national fund.

References

[1] Samson S 1965 *Acta Crystallogr.* **19** 401-413
[2] Feuerbaecher M et al. 2007 *Z. Kristallogr.* **222** 259-288
[3] Jeglic P, Komelj M, Klanjsek M, Tkalec U, Vrtnik S, Feuerbacher M and Dolinsek J 2007 *Phys. Rev.* B **75** 014-202
[4] Wolny J, Kozakowski B, Duda M and Kusz J 2008 *Phil. Mag. Lett.* **88** 501-507
[5] Sikora W, Malinowski J, Kuna A and Pytlik L 2008 *J. Phys.: Conf. Ser.* **104** 012-023
[6] Mihalkovič M, Al-Lehyani I, Cockayne E, Henley C L, Moghadam N, Moriarty J A, Wang Y and Widom M 2002 *Phys. Rev.* B **65** 104205
[7] Reichert M and Gähler F 2003 *Phys. Rev.* B **68** 202-214

AUTHOR INDEX

Institute of Physics
Dirac House, Temple Back
Bristol BS1 6BE UK

CURRAN ASSOCIATES INC
proceedings.com